Intelligent Systems Reference Library

Volume 60

The aim of this series is to publish a Reference Library, including novel advances and developments in all aspects of Intelligent Systems in an easily accessible and well structured form. The series includes reference works, handbooks, compendia, textbooks, well-structured monographs, dictionaries, and encyclopedias. It contains well integrated knowledge and current information in the field of Intelligent Systems. The series covers the theory, applications, and design methods of Intelligent Systems. Virtually all disciplines such as engineering, computer science, avionics, business, e-commerce, environment, healthcare, physics and life science are included. The list of topics spans all the areas of modern intelligent systems such as: Ambient intelligence, Computational intelligence, Social intelligence, Computational neuroscience, Artificial life, Virtual society, Cognitive systems, DNA and immunity-based systems, e-Learning and teaching, Human-centred computing and Machine ethics, Intelligent control, Intelligent data analysis, Knowledge-based paradigms, Knowledge management, Intelligent agents, Intelligent decision making, Intelligent network security, Interactive entertainment, Learning paradigms, Recommender systems, Robotics and Mechatronics including human-machine teaming, Self-organizing and adaptive systems, Soft computing including Neural systems, Fuzzy systems, Evolutionary computing and the Fusion of these paradigms, Perception and Vision, Web intelligence and Multimedia.

Indexed by SCOPUS, DBLP, zbMATH, SCImago.

All books published in the series are submitted for consideration in Web of Science.

Janmenjoy Nayak · Bighnaraj Naik · Vimal S. ·
Margarita Favorskaya
Editors

Machine Learning for Cyber Physical System: Advances and Challenges

Springer

Editors
Janmenjoy Nayak
Department of Computer Science
MSCB University
Baripada, India

Bighnaraj Naik
School of Computer Sciences
Veer Surendra Sai University of Technology
Sambalpur, India

Vimal S.
Department of Artificial Intelligence
and Data Science
Sri Eshwar college of Engineering
and Technology
Coimbatore, India

Margarita Favorskaya
Institute of Informatics
and Telecommunications
Reshetnev Siberian State University
of Science and Technology
Krasnoyarsk, Russia

ISSN 1868-4394 ISSN 1868-4408 (electronic)
Intelligent Systems Reference Library
ISBN 978-3-031-54037-0 ISBN 978-3-031-54038-7 (eBook)
https://doi.org/10.1007/978-3-031-54038-7

This Springer imprint is published by the registered company Springer Nature Switzerland AG
The registered company address is: Gewerbestrasse 11, 6330 Cham, Switzerland

Paper in this product is recyclable.

Foreword

The hustle of new developments in the Cyber-Physical System poses new challenges for data scientists and business delegates to influence smarter perception; challenging a real-time dashboard of information extracted from data in movement. In CPS, network and system security is of supreme importance in the present data communication environment. Hackers and intruders can have many successful attempts to disrup the operation of networks and web services through unauthorized intrusion. CPS exists everywhere in different sizes, with different functionalities and capabilities. Moreover, IoT is responsible for the communication between connected devices while exchanging data that requires Internet, wireless connections, and other communication mediums. Mainly, CPS makes use of IoT devices to fetch data and efficiently process it for implementing it in a particular area. The sensors and connected devices in CPS collect data from various gateways installed in the network and then analyze it for better decision-making. CPS comprises a new cohort of sophisticated systems whose normal operation depends on vigorous communications between their physical and cyber components. As we increasingly rely on these systems, asserting their correct functionality has become essential. Inefficient planning and control strategies may lead to a harmful cause.

The last decade has seen enormous research in the field of deep learning and neural networks in most of the engineering domains. Nowadays, various aspects of our lives depend on complex cyber-physical systems, automated anomaly detection, and developing a general model for security and privacy applications are crucial. For accurate and efficient data analysis, ML-based approaches are the best suitable way to protect and secure the network from any uncertain threats. The real cyber-physical system should have both physical and digital parts inter-connected in each part and process, and the system itself should have the capacity to change its behavior to adapt to changing requirements. Machine learning occupies a major role in estimating the cyberattacks that target the cyber-physical system and such attacks are challenging throughout the world. Machine learning for anomaly detection in CPS includes techniques that provide a promising alternative for the detection and classification of anomalies based on an initially large set of features.

In recent years various applications and algorithms have been proposed to mitigate those attacks through inferential data analysis. Data Driven capabilities for securing the cyber- physical system are possible through emerging ML approaches. The need for security in integrated components occupies a major criterion for CPS, I can find some chapters contributing good approaches for security mitigations such as an In-depth Analysis of Cyber-Physical Systems: Deep Machine Intelligence-based Security Mitigations. Also, risk assessment in CPS and security using ML is a holistic approach to the challenges that can be overcome in recent years.

Due to the hastening progress in machine learning and its application in dealing various issues of CPS, this book's publication is much needed at the present time. This volume provides a literature review covering the complete history of CPS security, an overview of the state-of-art, and many valuable references. Also, all the fourteen peer-reviewed chapters report the state of the art in CPS and anomaly detection research as it relates to smart city and other areas such as IoT, covering many important aspects of security. In my opinion, this book is a valuable resource for graduate students, academicians, and researchers interested in understanding and investigating this important field of study.

Junxin CHEN

Dr. Junxin Chen
Professor
School of Software
Dalian University of Technology
Dalian, China

Preface

The integration of computers and physical systems, known as a cyber-physical system (CPS), orchestrates a synergy between embedded computers and physical processes. This collaboration, often facilitated by feedback loops, sees a reciprocal influence where physical processes impact computations and vice versa. CPS applications span a myriad of sectors, including automotive systems, smart city, manufacturing, healthcare instruments, military and safety control operations, traffic control, power system and control, water management, and more. Combining engineering models from diverse disciplines with computer science methods, CPS emerged in 2006 and addresses the fundamental challenge of harmonizing the cyber and physical worlds. In the ever-evolving landscape of technology, the term CPS stands out as foundational and enduring, contrasting with contemporaneous terms like IoT, Industry 4.0, and the Industrial Internet. While these terms focus on implementation approaches or specific applications, CPS delves into the intellectual problem of unifying engineering traditions across cyber and physical realms. Artificial Intelligence (AI) and Machine Learning (ML) have significantly impacted our society and economy. The rise in AI and ML decision-making capabilities has prompted discussions about potential harms and the imperative to enhance transparency in these processes. The evolving landscape includes the possibility of self-building technologies and cognitive architectures simulating truly intelligent human-like performance, raising concerns about the emergence of collective entities through wearable connected devices. One such technology provoking these concerns is the Industrial Internet of Things (IIoT), garnering substantial interest in academia, government, and industry. IIoT utilizes IoT technologies to enhance manufacturing and industrial processes, closely linked to the paradigm shift denoted by Industry 4.0 (I4.0). The term I4.0 signifies not only a shift in industrial production, but also strategic initiatives, emerging business assets, and a distinct historical and social period. The literature underscores the importance of comprehending the inevitable and autonomous evolution of artificial cognition in complex socio-technical systems. Examples of AI and ML working in tandem with IoT devices abound, such as Tesla cars utilizing predictive analytics for optimal driving conditions and smart buildings predicting optimal heating or cooling times, especially relevant in the context of Covid-19. Future applications envision AI and

ML in CPS contributing to health monitoring, robotics, intelligent edge devices, and disaster correction. With the proliferation of IoT-connected devices and the added dimension of IIoT enhancing productivity and efficiency, the conventional five levels of CPS architecture seem outdated. The need for a new CPS architecture is evident, and considerations must account for the changing roles of AI and ML in creating economic benefits. It encompasses the cyber-physical attributes of IIoT, intertwining with the social aspects of its deployment, reflecting the future cognitive landscape of IIoT/I4.0.

Machine learning is the hottest to provide multifaceted solutions for Information Security. With the continuous monitoring of data frameworks for anomalies and threats, machine learning techniques are efficient in detecting and frightening the threats. The capacity of machine learning for processing real-time data is extremely helpful for the detection of threats, harm breaches, and malware for the prevention of huge losses. Machine learning techniques are highly adaptive for training the endpoint security setups to identify various malicious activities, on which it has already proved its efficacy. Machine learning provides automated options for condition monitoring preparation, predictive maintenance, image processing and diagnosis digital twins, model predictive control, medical diagnosis, questionable analysis, and prognosis. Machine learning is purely a computer-based approach where its industrial application needs extensive domain knowledge with remote computational power. Industry 4.0 foresees a broad gamut of application of the digital world for intelligent computing and control, health monitoring and analytics, digital revolution to the real-world application, digital forensics, smart city, and optimum industrialization processes. Machine learning techniques have evolved as key enablers for infusing intelligent self-learning and analytical capabilities in industrial processes. The cyber-physical system is a recent hot topic of research with remote-based applications, especially with IoT and other Edge computing. The last two decades have witnessed an exceptional growth of CPS, which are foreseen to modernize our world through the creation of new services and applications in a variety of sectors such as cloud computing, environmental monitoring, big data, cybercrime, e-health systems, intelligent transportation systems, among others. CPSs interrelate various physical worlds with digital computers and networks to facilitate automatic production and distribution processes. For remote monitoring and control, most of the CPS is not working in seclusion, but their digital part is connected to the Internet. There is always a huge chance of getting unacceptably high residual risk in critical network infrastructures, though the impediment and monitoring measures may condense the risk of cyberattacks. In such scenarios, machine learning helps to enable the system to endure adverse events with the maintenance of an acceptable functionality. Moreover, the integration of CPS and Big data has emerged many new solutions for cyberattacks. The interconnection with the real world, in industrial and critical environments, requires reaction in real time. This book focuses on the latest approaches of machine learning to novel CPS in real-world industrial applications. Moreover, it will fulfill the requirements of the resilience of CPSs in the cross-discipline analysis along with real-life applications, challenges, and open issues involved with cybersecurity threats. The book offers a structured and highly accessible resource for diverse applications to readers, including researchers, academics, and industry practitioners, who are interested in evaluating and ensuring the resilience of CPSs in both the development and assessment stages using advanced machine learning techniques. This book addresses the architecture and methodology of machine learning methods, which can be used to advance cybersecurity objectives, including detection, modeling, and monitoring and analysis of defense against various threats to sensitive data and security systems. This Volume comprises 14 chapters and is organized as follows.

In Chap. 1, Suresh Kumar Pemmada et al., have developed the AdaBoost ensemble learning technique with SMOTE to detect network anomalies. The AdaBoost technique is primarily used in the classification process, while SMOTE solves the class imbalance issue. The proposed method was tested on the NSL-KDD dataset. The performance of the proposed AdaBoost approach, as well as other conventional and ensemble learning approaches, was then validated using performance measures such as precision, recall, F1-score, and accuracy, and the results show that the proposed AdaBoost ensemble learning approach outperformed other ML algorithms and ensemble learning approaches.

Chapter 2 delves into the diverse world of cyber-physical systems, covering key aspects and recent challenges. In this Chapter, B. K. Tripathy et al. looked into the critical function of wireless sensor networks and reviewed various MAC protocols. Threats and security concerns in cyber-physical systems were also discussed, with a focus on the use of machine intelligence approaches to alleviate these challenges. Then an extensive discussion on approaches to machine learning and deep learning was presented and supported by experimental data. Moreover, the research on CPS attack classifications and prediction found that using a two-level class structure is more beneficial when using machine intelligence algorithms. Furthermore, when compared to ML techniques, the authors have highlighted the adaptability and usefulness of the DL method in resolving the complexity associated with k-level classifications in the field of Cyber-Physical Systems.

In Chap. 3, a systematic study of various unsupervised clustering techniques such as Partition, Density, Grid, Hierarchical, Model-based, and other approaches such as nearest neighbor methods or statistical techniques for anomaly detection that can be applied to machining process monitoring data has been performed by Juan Ramón Bermejo Higuera et al. Then the authors have discussed the criterion for comparing clustering algorithms that is based on 3 aspects such as the way the groups are formed, the structure of the data, and the sensitivity to the parameters of the clustering algorithm used. Further, the authors have also discussed the methodology used in implementing various unsupervised clustering approaches.

In Chap. 4, a robust classification framework for IoT device profiling is developed. In order to identify anomalous behavior in Smart Home IoT devices with an unusually high accuracy rate, Sudhir Kumar Das et al. wanted to enhance current studies in this chapter using a variety of machine learning techniques. In order to determine the frequency of changes in a single data point out of the four possible data points that were gathered by a single sensor, the authors conducted investigations by comparing and utilizing various classifiers, such as the KNeighbors Classifier, Decision Tree Classifier, Support Vector Classifier, AdaBoost Classifier, Random Forest with Extreme Gradient Boost Classifier, Random Forest Classifier, Gradient Boosting Classifier, Gradient Boosting Machine Classifier, and XGB Classifier. The outcomes demonstrated that, when compared to alternative methods, the Gradient Boosting Classifier algorithm employing random search achieved improved detection accuracy, suggesting a considerably lesser vulnerability to such changes.

Chapter 5 develops a useful collection of machine learning models for easy offshore wind industry deployment, with the goal of addressing a major gap in the current literature. The decision-making process on safety precautions, such as when to schedule maintenance or repairs or alter work procedures to lower risk, will subsequently be guided by these models. Furthermore, the models with the best performance for the majority class in the imbalanced dataset and the minority class in the imbalanced dataset have been highlighted by Barouti and Kadry in this chapter. From the experimental results, the authors concluded that the classifiers outperformed neural networks and deep neural networks. Furthermore, the chapter also emphasizes the possible effects of these tools on the industry's long-term profitability and the significance of creating efficient machine learning models and enforcing stricter data records to increase safety in the offshore wind sector. The chapter also points out that the excellent performance of a few chosen models indicates the validity of the anticipated predictions and shows how machine learning models work well for safety-related decision-making in the offshore wind sector.

In Chap. 6, a Convolutional Neural Network (CNN) model for attack detection has been created by Ravi Kishore et al. The proposed method has been verified with the latest V2X dataset in order to investigate several attributes, such as the source and destination vehicle addresses, network service kinds, network connection status, message type, and connection duration. Initially, the authors have performed preprocessing of data in order to create the desired detection system. In summary, the simulation results show that the proposed CNN performs better than the state-of-the-art machine learning techniques, including Random Forest (RF), Adaptive Boosting (AdaBoost), Gradient Boosting (GBoost), Bagging, and Extreme Gradient Boosting (XGBoost), and reaches an exceptional degree of accuracy when applying anomaly detection.

In Chap. 7, the use of breakthroughs in autonomous systems to challenge the foundations of the human cognitive linguistic process is unpacked in order to stimulate the development of cyber-physical system models and algorithms. In order to accomplish this, Monte-Serrat and Cattani employed an argumentation technique to demonstrate that a particular structure, or pattern, frequently arises in the cognitive language processes of both intelligent systems and humans. The authors use this to demonstrate not only that the pattern ensures coherence in the decision-making process of cognitive computing, but also highlights the issues surrounding the biases of AI's black box and the intelligence of autonomous vehicles. Thus, it is feasible to control the interpretative activity of cyber-physical systems and the way they make decisions by elucidating the dynamics of the distinct cognitive linguistic process as a shared process for people and machines, resulting in the development of safe and sustainable autonomous cars.

In Chap. 8, Kumar et al., have introduced a potential approach to enhance the security of CPS in smart city environments. This was accomplished by using under-sampling ensemble approaches to overcome the class imbalance problem that machine learning algorithms faced. Class imbalance is resolved using the under-sampling-based ensemble technique, which lowers the majority class and creates a balanced training set. The suggested approach promotes minority performance while decreasing bias toward the majority class. Additionally, the proposed method resolves the issue of class imbalance and increases accuracy without the disadvantages associated with complex model development. The MSCA benchmark IDS dataset is used for the tests, and the results show that the under-sampling classifiers such as Self-Paced Ensemble Classifier, Bagging Classifier, and Balance Cascade Classifier are remarkably accurate in identifying network anomalies.

Chapter 9 is about the application of deep learning approaches in medical cyber-physical system due to the large dimensionality and noticeable dynamic nature of the data in these kinds of systems. Swapnarekha and Manchala have built an intelligent security architecture in this chapter that uses deep neural networks to detect cyberattacks in the healthcare industry. The WUSTL-EHMS 2020 dataset, which is made up of network traffic indicators gathered from the patient's biometric data, was then used to validate the proposed framework. Since the features in the dataset had fluctuating values, min-max normalization was first applied to the data. Further, authors have used Synthetic minority oversampling (SMOTE) because the dataset included in this study is unbalanced, with 2046 network attack samples and 14,272 normal samples. Finally, the effectiveness of the proposed framework in comparison to a number of conventional machine learning and ensemble learning approaches has been verified, and the results of the experiments show that the proposed DNN model outperforms the examined machine learning and ensemble learning approaches in terms of accuracy, precision, recall, F1-score, AUC-ROC, and accuracy.

Chapter 10 explains about safeguarding sensitive industrial data, and averting safety risks using advanced machine learning approaches. In this chapter, an ensemble learning-based model is designed by Geetanjali Bhoi et al., to detect anomalies in Industrial IoT network. The authors used gradient boosted decision tree with its optimized hyperparameters using a gravitational search algorithm. The suggested approach has been validated using the X-IIoTID dataset. Then the performance of the proposed model has been compared with various machine learning and ensemble approaches such as Linear Regression, Linear Discriminant Analysis, Naïve Bayes, Decision Tree, Stochastic Gradient Descent, Quadratic Discriminant Analysis, Multilayer Perceptron, Bagging, Random Forest, AdaBoost, Gradient Boosting, and XGBoost, and the experimental findings shows that the suggested approach attained superior performance in comparison with other approaches.

Chapter 11 presents a comprehensive review of the identification of patient-provided brain MR images and the categorization of patients' brain tumors through the use of AI and ML techniques. In order to create different AI and ML classifiers for this posture, brain pictures from the kaggle.com website were collected by Panda et al. The MR images are first subjected to preprocessing in order to improve their quality. Once preprocessing was finished, key characteristics were extracted by the authors to create the necessary feature vector, including technical, statistical, and transform domain features. Then, for training and validation, every feature vector is supplied to every one of the suggested models. Through simulation-based experiments conducted on the AI and ML classifiers, performance matrices have been obtained and compared. From the analysis of experimental findings, it is observed that Random Forest exhibits superior detection of brain tumor in comparison with other approaches.

Chapter 12 provides a thorough examination of many aspects of smart grid cybersecurity, allowing for a more in-depth understanding of the issues and potential solutions in this vital subject. Patnaik et al. investigated the complexity of smart grid infrastructure, noting challenges ranging from data availability and quality to the integration complexities of connected devices such as Advanced Metering Infrastructure (AMI), Information Technology (IT), Operational Technology (OT). In addition, the authors expanded their investigation into the wide spectrum of cyber threats, including the sorts of assaults and specific equipment vulnerable to these risks inside a smart grid architecture. Furthermore, the chapter investigated the use of artificial intelligence, including both machine learning and deep learning, as a transformative method for strengthening smart grid cybersecurity. Additionally, the authors have also identified potential avenues for improving the resilience of smart grids against cyber threats, such as federated learning, explainable AI, generative adversarial networks, multi-agent systems, and homomorphic encryption.

In Chap. 13, a CNN methodology that enhances the detection and reduction of cyberattacks in medical cyber-physical system devices has been developed by Dash et al. The suggested approach seeks to improve the security of IoT-enabled medical devices. With a focus on multi-class classification, the suggested approach has been designed to recognize DoS assaults, ARP Spoofing, Smurf attacks, and Nmap attacks. This contrasts with the current system, which uses binary class classification as its foundation and can identify many attack types. The suggested methodology has been assessed using the ECU-IoHT healthcare domain dataset. Initially, in order to resolve the inequality in class distribution, the authors adopted a random oversampling strategy for the dataset. Then the dataset has been standardized by applying the min-max scalar technique, which makes sure that all attribute values fall into the same scale. When compared to the current method, the experimental results show that the proposed CNN strategy produces a much higher accurate identification rate and a lower false detection rate.

In Chap. 14, a comprehensive survey regarding various datasets that have been made accessible for the purpose of addressing automatic vehicle classification problems, including automatic license plate recognition, vehicle category identification, and vehicle make and model recognition during the last decade has been presented by Maity et al. The authors have carried out the survey by categorizing datasets into two types such as still image-based, and video-based. The datasets based on still images are additionally divided into datasets based on front images and aerial imaging. An extensive comparison of the various dataset types, with particular attention to their properties, has been presented in this chapter. Additionally, the chapter lists difficulties and research gaps pertaining to automatic vehicle classification datasets. Along with offering a thorough examination of every dataset, this chapter also makes several important recommendations for future automatic vehicle classification research directions.

Baripada, India — Janmenjoy Nayak
Sambalpur, India — Bighnaraj Naik
Coimbatore, India — Vimal S.
Krasnoyarsk, Russia — Margarita Favorskaya

Contents

Chapter 1
SMOTE Integrated Adaptive Boosting Framework for Network Intrusion Detection

Suresh Kumar Pemmada, K. Sowjanya Naidu, and Dukka Karun Kumar Reddy

Abstract Network abnormalities may occur due to enormous reasons, such as user irregular behavioral patterns, network system failure, attacker malicious activities, botnets, or malicious software. The importance of information management and data processing systems has changed the enormous volume of data and its incremental increase. An IDS monitors and examines data to detect unauthorized entries into a system or network. In this article, the Ada-Boost ensemble learning technique is proposed with SMOTE to identify the anomalies in the network. The Ada-Boost algorithm is utilized mainly in the classification task, and SMOTE handles the class imbalance problem. The suggested approach outperformed various ML algorithms and ensemble learning approaches in relation of precision, recall, F1-score, and accuracy with 0.999 and 99.97% respectively when investigated with the NSL-KDD dataset.

Keywords Ada-Boost · Intrusion Detection System (IDS) · Machine Learning (ML) · Synthetic Minority Oversampling Technique (SMOTE)

S. K. Pemmada (✉)
Department of Computer Science and Engineering, GITAM (Deemed to be University), GST, Visakhapatnam, Andhra Pradesh 530045, India
e-mail: reshu.suri@gmail.com; spemmada@gitam.edu

K. S. Naidu
Department of CSE-IoT, Malla Reddy Engineering College, Medchal-Malkajgiri, Hyderabad, Telangana State 500100, India

D. K. K. Reddy
Department of Computer Science Engineering, Vignan's Institute of Engineering for Women(A), Visakhapatnam, Andhra Pradesh 530046, India

J. Nayak et al. (eds.), *Machine Learning for Cyber Physical System: Advances and Challenges*, Intelligent Systems Reference Library 60,
https://doi.org/10.1007/978-3-031-54038-7_1

1.1 Introduction

The growing concern about communication technology and security in information is the primary reason for the attacks or anomalies in the networks that affect the nation's private data storage, economic issues, and security. Because of the growing usage of information technology in our everyday lives, data security has become more important, posing risks to computerized systems. As a result, data protection has become a requirement, giving priority to threats in computerized systems. The extensive use of computer networks in internet security poses serious threats to computing facilities and networking environments. Hence, intrusion detection and preserving security for the networks play a vital role in various anomalies and attacks. Detection of anomalies is an essential activity in data analysis that is used to identify suspicious or anomalous activity separated by normal data. Anomalies are defined in different ways by different researchers, but Salehi and Rashidi [1] provides the most accepted description as "A deviation in an observation that is significantly different from the norm to the extent that it raises suspicions of being produced by a distinct mechanism is referred to as an anomaly". Anomalies are treated as a high priority as they identify the changes in the data patterns and can prompt immediate actions in the application domains. Anomaly-based systems have been developed using machine learning algorithms to magnify and not to depend on human interaction [2]. Anomaly detection is used extensively in many fields and, more importantly, in intrusion detection, image processing, fraud detection, medical and public health, sensor networks, etc. The network anomaly detection method considers the input data that allows data of different types to be processed. The techniques of data processing are based on various available anomaly detection techniques.

Because of the increased public awareness of the value of safeguarding online transactions and information, attackers have changed their tactics for network attacks. Over the past few years, technological advancements have allowed attackers to develop more inventive and stealthy ways of attacking computer networks. Various types of attacks prominently used by the attackers are as follows. A DoS attack targets systems or networks with the intent to disrupt their normal operations and deny access to legitimate users. A LAND attack is a DoS attack that involves sending a poisoned faked packet to a network, forcing it to lock up. A smurf attack disrupts network operations by taking advantage of vulnerabilities in the Internet Protocol and the Internet Control Message Protocol (ICMP), rendering the network non-functional and constituting a type of DoS assault. A teardrop attack makes a computer resource inaccessible by flooding requests and data from a network or server. In a UDP storm attack, many packets are sent to the targeted server by User Datagram Protocol (UDP) to overload the ability of that system to process and react. Worm attack spreads copies of itself from computer to computer [3]. Satan attack is a tool used to find loopholes or vulnerabilities in one's computers. IP sweep attack (also known as ICMP sweep attack) occurs when the attacker sends an echo request to multiple destination addresses for ICMP. The reply exposes the target IP address to the attacker if these requests are answered by the target host, who gets the echo request. Saint attack screens every system live on the network for TCP and UDP services. It launches a series of probes for any service it finds running to detect something that might allow an intruder to gain unlawful access. Ftp Write attack is an FTP protocol exploit where an attacker can use the PORT command to indirectly request access to ports by using the victim's computer, which acts as a request proxy. Warez Master (WM), and Warez Client (WC) attacks are two types of assaults that take advantage of flaws in "anonymous" FTP on both Windows and Linux. Rootkit attacks are stealthy programs designed to obtain a network device's administrative rights and access. In a Mailbomb attack, a mail bomb sends a massive amount of emails to a specific person or computer.

Over half of the global population resides in cities, and it is anticipated that this figure will increase as people continue to move to urban regions seeking improved employment prospects and educational resources. Smart city facilities can be extended to several fields, such as transportation, tourism, health, environment, safety, home energy management, and security [4]. Several components of a smart city include various sensors in applications such as structural health awareness, real-time nose mapping, smart parking, smart street lights, route optimization, etc. With the emergence of these, wireless technologies have reached the public eye and gradually incorporated into every corner. So, there is always a scope for several unauthorized access to such devices, which may lead to data inconsistency and the evolvement of suspicious activities. An IDS is designed to assist ongoing monitoring and detection of cyber-attacks over the smart city (especially IoT networks) to supplement the security protocol provision. IDS is a security approach used to discover suspicious device behavior and intercept the attacking source promptly to secure the network [5].

Three distinct categories can be used to classify ML algorithms employed in anomaly detection systems: those that utilize supervised learning, those that apply unsupervised learning techniques, and those that incorporate a combination of both in a hybrid approach [6]. The techniques of supervised detection of anomalies train the classifiers with the labeled information. For both anomalous and normal data, the testing and training of data utilized in these methods should be submitted to the necessary mark. ML algorithms that are unsupervised do not require labeled datasets. They focus on analyzing and discovering the structure of the data. Hybrid strategies are made up of two or more aspects, each of which performs a certain function. One component is used for classification, clustering, and preprocessing, and another component for optimization tasks. Hybrid approaches are used to make the best of each of the algorithms mentioned above and boost machine efficiency. A variety of ML techniques have been employed to identify different kinds of network threats [7]. ML algorithms in a particular, decision tree [8], k-nearest neighbor [9], random forest [10], support vector machine [8], multi-layer perceptron [11], etc. are the majorly applied method for intrusion detection. Different techniques, learning processes, and different features of input do not provide the same results concerning the various classes of attacks. However, such algorithms have various disadvantages such as *data acquisition* (collected data may be bogus and imbalance in nature), *error-prone* (data must be clean that needs data preprocessing), *algorithm selection* (difficult task to select a proper algorithm for a particular problem), *time-consuming* (requires an extensive amount of time to handle with substantial datasets). Ensemble learning models are the combination of two or more other ML models, which overcome the disadvantages of ML algorithms. The major advantage of an ensemble-based framework is to raise the prediction accuracy by contributing the model performance through successively learning of sub-models with additive, and sequential mode from its predecessor trees. Among other ensemble learning methods, Ada-boost is the best-suited algorithm for boosting, where the performance of the decision trees is suitably used in classification problems. The motivation behind this work is to introduce IDS and give a profound understanding of some sophisticated ML techniques for intrusion detection. IDS are essential for safeguarding infrastructure, and their susceptibility to attacks is heightened due to the growing complexity of modern network environments. As most conventional ML approaches resulting a high accuracy for unbalanced datasets due to learning algorithms focuses on majority classes. But instead of learning a single complex classifier with parameterized criterion, learning several simple classifiers and combine their individual output to produce the classification decision leads to a systematic, efficient and automated approach for IDS.

IDS often deal with extremely unbalanced data, in which occurrences of one class exceed cases of another. This imbalance might cause the model to be biased towards the one class (normal behavior), resulting in poor detection of the other class (intrusive behavior). This motivated me to use SMOTE to address the challenges of imbalanced data. SMOTE is a prominent oversampling approach that is used to address this problem by providing synthetic instances of the minority class and so balancing the dataset. AdaBoost, on the other hand, is an ensemble learning approach that combines several weak classifiers to generate a strong classifier. It works especially well in cases with a complex decision boundary. In the context of intrusion detection, AdaBoost can assist enhance detection accuracy by focusing on difficult-to classify instances and dynamically modifying the weights of the training examples depending on prior classification mistakes. Furthermore, when AdaBoost and SMOTE are combined, they can provide an optimal set of synthetic samples, modifying the updating weights and compensating for skewed distributions. This combination will reduce the error rate, while maintaining the better accuracy rate.

The significant contribution of the paper includes:

- Proposed Adaptive Boosting for efficient identification of intrusive activities in a network.
- SMOTE is used to balance the disparity in the data.
- A deep study of the NSL-KDD dataset and it's influencing characteristics for classification attacks has been done.
- ML approaches such as Logistic Regression (LR), Stochastic Gradient Descent (SGD), Multi-Layer Perceptron (MLP), Linear Discriminant Analysis (LDA), Quadratic Discriminant Analysis (QDA), K-Nearest Neighbor (KNN), Gaussian Naive Bayes (GNB); and ensemble methods such as Gradient Boosting, Bagging, and Stacking have been examined for a thorough performance study to demonstrate the proposed method's superiority over others.

The rest of this document is organized as follows: Sect. 1.2 delves into the existing literature on network anomaly detection. Section 1.3 details the projected methodology, while Sect. 1.4 outlines the experimental framework, including the dataset used, data preprocessing techniques, performance metrics for method validation, parameter configurations, and the simulation environment. Section 1.5 presents the results obtained using the projected method alongside comparisons with other models. Finally, Sect. 1.6 provides a conclusion to the chapter.

1.2 Literature Study

The capacity to detect network abnormalities is critical for ensuring network stability. The majority of intrusion detection research for predictive approaches is done using comparable training and test datasets.

Adhi Tama et al., [12] aimed to study and highlight the usefulness of stacking ensemble-based approach for anomaly-based IDS, where the base learner model uses Deep Neural Network (DNN). It shows the stacking-based DNN model for intrusion detection for two-class detection problems as normal and malicious data. They validated the proposed model on NSL-KDD, UNSW-NB15, and CICIDS 2017 datasets using different evaluation metrics. According to the results, the suggested model outperformed the underlying DNN model and other current ML algorithms in the literature.

Jain and Kaur [13] explored the disseminated AI-based group methods to identify float's presence in the organization traffic and recognize the organization-based assaults. The investigation has been directed in three stages. Initially, Random Forest (RF) and LR classifiers are utilized as initial phase learner, and Support Vector Machines (SVM) are used as the second level phase learner. Next, K-means clustering based on sliding window is used to handle the concept drift. Finally, techniques based on ensemble learning are used to identify the attacks in the network. Experimentation has been conducted on CIDDS-2017, generated testbed data, and NSL-KDD. The assessment has been directed at different machines by shifting numerous agent centers to realize the learning time dormancy in the distributed environment. The test results demonstrated that the SVM-based model had shown better exactness.

Several methods have been suggested to identify normal data with anomalies to detect network intrusions. Zhong et al. [14] discussed the integration framework of several ML techniques. They utilized a damped incremental statistics method to abstract features from network tragics and then used an autoencoder using labeled data to identify network traffic anomalies. The proposed algorithm combines the LSTM classifier and autoencoder classifier and finally displays the experimental results for the proposed algorithm.

Khammassi and Krichen [15] suggested a multi-objective Feature Selection method as a learning algorithm using a logistic regression wrapper approach and a non-dominated genetic algorithm as a search methodology. The proposed method is tested in two phases, and the results are compared for both binary-class and multi-class classifiers using Naive Bayes, RF, and C4.5 classifiers on UNSW-NB15, CIC-IDS2017, and NSL-KDD data sets. The binary class datasets display better accuracy compared to multi-class datasets. Table 1.1 outlines a variety of strategies and evaluative studies that have been suggested by different researchers.

From Table 1.1, it is seen that most of the research has been focused on the use of the KDD-CUP 99 and NSL-KDD datasets. However, many of them are involved with the issues like complexity and finding highly accurate solutions.

1.3 Proposed Method

Adaptive Boosting (AdaBoost) was proposed by Freund et al. [25]. The base learners build on a weighted distribution dataset, where the instance weights on the dataset depend on the prediction of previous base learner instances. If a particular instance be misclassified, the subsequent model will assign a higher weight to that instance; else, if the classification is right, the weight will be unaltered. The final decision-making is accomplished by the weighted vote of the base learners, where the weights are determined by the misclassification rates of the models. In AdaBoost, DTs serve as the foundational classifiers, and the models that achieve higher predictive accuracy are assigned greater weights, whereas those with lower accuracy are given lesser weights. Figure 1.1 depicts the proposed approach framework. An IDS has the capability to scrutinize both user behaviors and system activities, identify established patterns of attacks, and spot nefarious activities within the network. The primary objective of an IDS lies in overseeing the network and its individual components, recognizing a range of network breaches, and alerting the respective personnel upon the detection of such security incidents. Several smart city sensor data have been preprocessed using different steps, normalizing non-numerical labels and balancing class labels with the

Table 1.1 Literature study of the identification of network anomalies

Intelligent method	Compared method	Classification/ Regression	Dataset	Evaluation metrics	Ref
Stacking	K-Means clustering, GMM	Classification	NSL-KDD, UNSW-NB15	Accuracy	[16]
Sparse framework	RF, Decision trees (DT), Gaussian-based models	Regression	UNSW-NB15	Accuracy	[17]
Dimensionality reduction	CANN, GARUDA, UTTAMA	Classification	KDD, NSL-KDD	Accuracy, Precision, Recall	[18]
LSTM	Genetic algorithm	Classification	NSL-KDD, UNSW-NB15	Accuracy	[19]
Sparse auto encoder	–	Regression	UNSW-NB15 AND NSL-KDD	Accuracy	[7]
Generative adversarial Network architectures	Deep autoencoding Gaussian mixture model, Autoencoder	NA	UNSW-NB15	Precision, Recall	[20]
Sparse auto encoder	Signature-based intrusion detection methods	Regression	NSL-KDD	Accuracy, Precision, Recall	[21]
Auto encoder	–	Classification	NSL-KDD	Accuracy	[22]
Union and Quorum techniques	KNN, RF, Decision tree, GNB, and LR	Classification	UNSW-NB15	–	[23]
Hybrid feature selection with Naïve bayes	Feature selection	Classification	UNSW-NB15	Accuracy	[24]

target variables. The prepared data is then put into Ada-Boost, an intelligent ensemble framework. If the proposed method detects an attack, the network administrator will be notified, and the monitoring system will be alerted. In addition, intrusion prevention systems scan incoming network packets to detect malicious or anomalous activity and provide alerts.

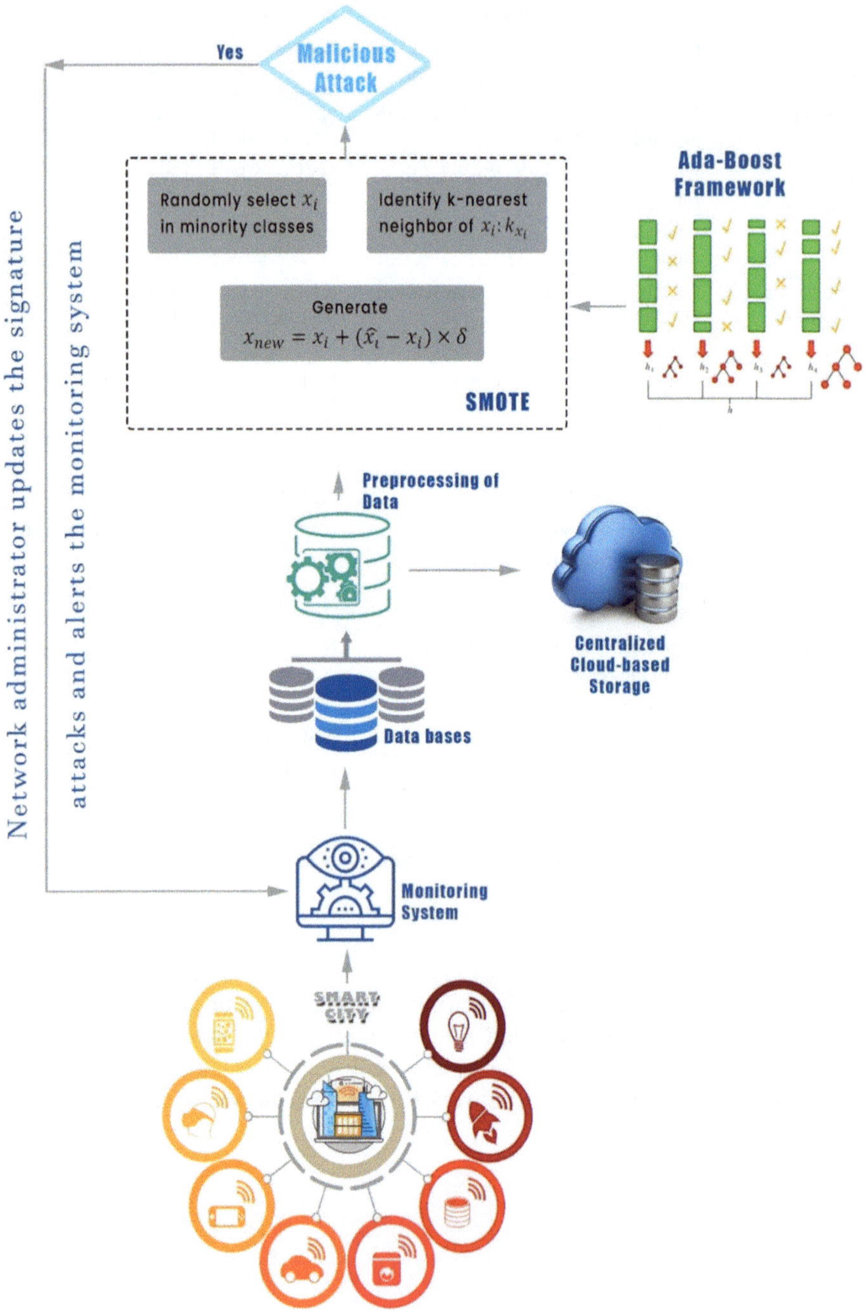

Fig. 1.1 The proposed method's framework

AdaBoost Algorithm

1: Initializing weights to $w_i = \frac{1}{N}$, where i ranging from $1\ to\ N$, representing the total number of instances in the dataset. Initialize an initial constant value for f, which serves as the ensemble function.

2: Untill $m < M$ do: where $M$is the number of models to be grown

2.1 A model is created and fitted with a weak classifier $H_m(x_i)$ to the training data using sample weights w_i for all data points, where x_i and y_i represents the training and training labels of the dataset.

2.2 Train the sub classifiers $H_m(x_i)$, where $m = 1,2..M$

2.3 Computing the error e of the training set summed over all data points x_n in training set with:

$$e_m = \frac{\sum_{i=1}^{N} w_i^{(m)} * C(y_i \neq H_m(x_i))}{\sum_{i=1}^{N} w_i^{(m)}}$$

if $C(condition)$ is false, increase weights

2.4 Computing the weight of the mth sub-classifier as $\psi_m = log\left(\frac{1-e_m}{e_m}\right)$, where ψ is a specific loss function.

2.5 The weights are updated for T training instances in the next $(m+1)$ model with:

$$w_i^{(m+1)} = w_i^{(m)} * exp(\psi_m * C(y_i \neq H_m(x_i))),$$

where $i = 1,2..N$

2.6 Renormalize the weights $\sum_{i=1}^{N} w_i = 1$

3: Following M iterations, compute the functional output with: $f(x) = sign(\sum_{m=1}^{M} \psi_m * H_m(x))$ for making predictions using the final model.

SMOTE employs an interpolation strategy to generate new samples, thereby augmenting the minority classes through an oversampling technique. The minority samples that are identified are grouped together prior to the employment in the formation of new minority class samples. SMOTE is generating synthetic samples rather than replicating minority samples. Detailed algorithm is presented below.

SMOTE

Dataset D
Minority Class Instances M
for every sample x in M
    Identify k-nearest neighbors that are minority samples of x_i in M
    Select one random sample from k_{x_i}, and it is x_{new}.
    $x_{new} = x_i + (\widehat{x}_i - x_i) \times \delta$
end for

1.4 Experimental Setup

This section discourses the dataset, data preprocessing, simulation environment, parameter setting of various classifiers, and the evaluation metrics for validating the proposed method's performance using different ensemble and ML classifiers.

1.4.1 Experimental Data

Different statistical studies have exposed the inherent disadvantages of KDD cup 99, which affected many researchers' detection accuracy of intrusion detection models [26]. NSL-KDD represents an enhanced iteration of the original KDD, incorporating essential records from the predecessor dataset of the KDD Cup 99. This work is simulated on the NSL-KDD [27] using an ensemble learning algorithm called Adaboost and validated the method proposed by comparing different state-of-the-art ML algorithms, SGD, KNN, RF, LDA, QDA, DT, GNB, LR, and MLP. The dataset contains 41 features referring to 'basic features', 'features related to the content', 'traffic features related to time', and 'traffic features based on the host of each network connection vector'. The detailed feature and its type are presented in Table 1.2.

This dataset is having 148,517 instances and in these instances, various attack types such as 'normal'–77,054; 'back'–1315; 'land'–25; 'Neptune'–45,871; 'pod'–242; 'smurf'–3311; 'worm'–2; 'teardrop'–904; 'processtable'–685; 'apache2'–737; 'udpstorm'–2; 'satan'–4368; 'ipsweep'–3740; 'nmap'–1566; 'portsweep'–3088; 'mscan'–996; 'saint'–319; 'guess–passwd'–1284; 'ftp–write'–11; 'imap'–965; 'phf'–6; 'multihop'–25; 'warezmaster'–964; 'warezclient'–890; 'spy'–2; 'xlock'–9; 'xsnoop'–4; 'snmpguess'–331; 'snmpgetattack'–178; 'httptunnel'–133; 'sendmail'–14; 'named'–17; 'buffer-overflow'–50; 'loadmodule'–11; 'rootkit'–23; 'perl'–5; 'sqlattack'–2; 'xterm'–13; 'ps'–15; 'mailbomb'–293 are mentioned as the dependent variable. Except normal class labels remaining attack types are converted as 4 class labels such as **'DoS'**—'pod', 'smurf', 'back', 'land', 'udpstorm', 'processtable', 'Neptune', 'teardrop', 'apache2', 'worm', 'mailbomb'. **'U2R'**— 'xterm', 'ps', 'buffer-overflow', 'perl', 'sqlattack', 'loadmodule', 'rootkit'. **'Probe'**— 'nmap', 'satan', 'mscan', 'ipsweep', 'portsweep', 'saint'. **'R2L'**— 'xsnoop', 'named','snmpguess', 'imap', 'multihop', 'warezclient', 'spy', 'xlock', 'snmpgetattack', 'phf', 'guess-passwd', 'ftp-write', 'httptunnel', 'warezmaster', 'sendmail'. So, the dependent variable has 5 classes which are normal, R2L, U2R, DoS, Probe.

Table 1.2 NSL-KDD dataset

Attribute	Type
'Protocol-type'	Nominal
'Service'	
'Flag'	
'Land'	Binary
'Logged-in'	
'Root-shell'	
'Is-Host-Login'	
'Is-Guest-Login'	
'Su-Attempted'	
'Duration'	Numeric
'Num-Root'	
'Num-File-Creations'	
'Num-Shells'	
'Num-Access-Files'	
'Num-Outbound-Cmds'	
'Num-Compromised'	
'Count'	
'Srv-Count'	
'Serror-Rate'	
'Srv-Serror-Rate'	
Attribute	*Type*
'Rerror-Rate'	Numeric
'Srv-Rerror-Rate'	
'Same-Srv-Rate'	
'Diff-Srv-Rate'	
'Srv-Diff-Host-Rate'	
'Dst-Host-Count'	
'Dst-Host-Srv-Count'	
'Dst-Host-Same-Srv-Rate'	
'Dst-Host-Diff-Srv-Rate'	
'Dst-Host-Same-Src-Port-Rate'	
'Dsthostsrvdiffhostrate'	
'Dsthostserrorrate'	
'Dsthostsrvserrorrate'	
'Dsthostrerrorrate'	
'Dsthostsrvrerrorrate'	
'Wrong-Fragment'	
'Urgent'	

(continued)

Table 1.2 (continued)

Attribute	Type
'Hot'	
'Num-Failed-Logins'	
'Src-Bytes'	
'Dst-Bytes'	

1.4.2 Data Preprocessing

Preprocessing of data is essential because it allows the quality of raw experimental data to be enhanced. Data preprocessing can also have a major impact on the competence of the algorithm. Verified for null values in the dataset, and there are no NAN (Not a Number) values. 'land', 'numfailedlogins', 'urgent', 'numoutboundcmds' features have mostly zeros, so excluded these features from the data. 'Protocoltype', 'service', 'flag' are categorical values in the dataset. Converting categorical variables into numerical form is a crucial step in data preprocessing. One common method for achieving this transformation is known as Label Encoding, where each unique category is assigned, a distinct integer based on the order of the alphabet. The 'attack' is the target variable with class labels such as 'normal', 'DoS', 'Probe', 'R2L', and 'U2R'; the detailed explanation is presented in Sect. 4.1. The considered dataset dependent variable class labels are highly imbalanced with risk level distributions—'U2R': 119, 'R2L': 3880, 'Probe': 14,077, 'DoS': 53,387, and 'Normal': 77,054 instances as shown in Fig. 1.2. Distribution of Classes Prior to SMOTE Technique. One method to tackle the challenge of imbalanced classes involves augmenting the less represented classes in the dataset. SMOTE is employed to mitigate this imbalance. This technique generates synthetic examples in the dataset based on existing minority class instances prior to model training. The assessment of risk levels after the application of SMOTE is depicted in Fig. 1.3, along with the altered class distribution post-SMOTE application.

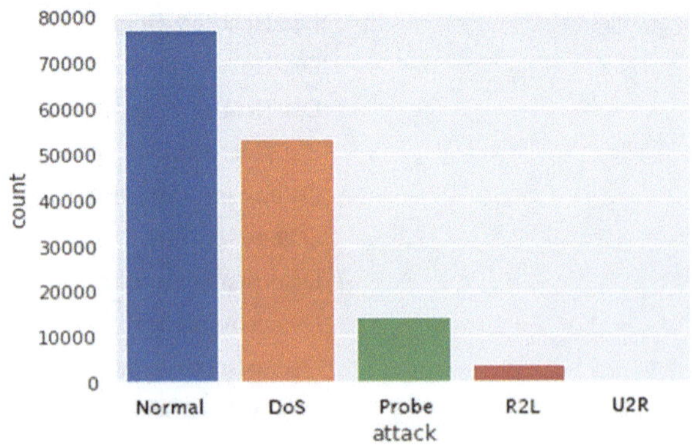

Fig. 1.2 Distribution of classes prior to SMOTE technique

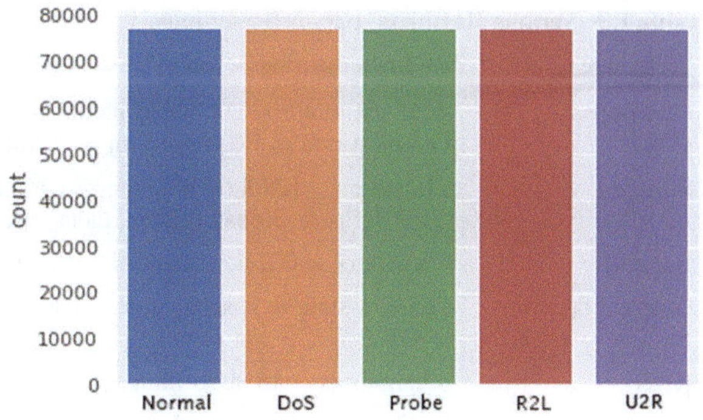

Fig. 1.3 Distribution of classes by SMOTE technique

1.4.3 Simulation Environment and Parameter Setting

The research was carried out on a computer equipped with Windows 10 Pro (64-bit) and powered by an Intel(R) Core (TM) by 8 GB of RAM. Simulations of the suggested models, alongside those for comparison, were executed within an environment based on Python. This setup encompassed the Numpy and Pandas libraries (utilized for data manipulation and analysis); the sklearn library (employed for the implementation of machine learning classifiers and data preprocessing tasks); pycm (used for evaluating multiclass classification metrics); Matplotlib and Seaborn (for graphical representation of data); and imblearn (applied for addressing class imbalance through random oversampling). Additionally, the classification-metrics library was used for assessing performance and analysis. The techniques under consideration, including the novel approach and those for comparison, underwent evaluation on a dataset partitioned in an 80% training to 20% testing split. The parameters for both the novel technique and the benchmark ensemble and ML methods are detailed in Table 1.3.

1.4.4 Performance Measures

Experimentation is carried out on the proposed approach and with several ML techniques. The projected method and comparative approaches are validated on various evaluation metrics such as true negative (TN), true positive (TP), false negative (FN), false positive (FP), false-positive rate (FPR), recall (TPR), f1-score, precision, accuracy, micro and macro average roc curve concerning every class, and overall accuracy [28].

Table 1.3 Various classifiers' parameter settings

Technique	Parameter setting
Ada-Boost	Base_estimator = DT, n-estimators = 50, max-depth = 15, Learning-rate = 1.0, algorithm = 'SAMME.R'
Stacking	Classifiers = KNN (n_neighbors = 15, algorithm = 'kd_tree'), GNB, RF, use-probas = True, meta_classifier = LR, use-clones = False
Bagging	Classifiers = DT, n-estimators = 500, random-state = 1
Gradient boosting	Random-state = 1, subsample = 0.8, solver = 'newton-cg', n-estimators = 120
KNN	n-neighbors = 15, algorithm = 'kd-tree'
MLP	Batch-size = 10, activation = 'logistic', and random-state = 2, solver = 'adam'
LDA	Tol = 0.0001
QDA	Tol = 0.0002
LR	Random-state = 1, solver = 'lbfgs'
GNB	Priors = None, var-smoothing = 1e–09
SGD	Random-state = 1, penalty = 'l1'

1.5 Result Analysis

The study demonstrates the performance of the AdaBoost classifier relating to various ML and ensemble learning techniques presented in Tables 1.4, 1.5, 1.6, 1.7, 1.8, 1.9, 1.10, 1.11, 1.12, 1.13 and 1.14. The SGD, GNB, and LR classifiers show a large miss classification rate performance for all the classes. The complete in-depth results of these classifiers are shown in Tables 1.4, 1.5 and 1.6, where the accuracy of SGD, GNB, and LR are 26.26, 30.58, and 33.15, respectively. This shows the inability to interpret and classify such large data with conventional ML methods.

Table 1.4 Evaluation factors of SGD

SGD	Normal	DoS	Probe	R2L	U2R
TN	47,063	58,156	50,799	56,829	38,552
TP	1960	10,879	634	1345	5419
FN	13,408	4547	14,732	14,166	9964
FP	14,623	3472	10,889	4714	23,119
FPR	0.237	0.056	0.177	0.077	0.375
Recall	0.13	0.71	0.04	0.09	0.35
F1-score	0.12	0.73	0.05	0.12	0.25
Precision	0.12	0.76	0.06	0.22	0.19
Accuracy	0.64	0.9	0.67	0.75	0.57
Overall accuracy	26.2634				

Table 1.5 Evaluation factors of GNB

GNB	Normal	DoS	Probe	R2L	U2R
TN	59,900	14,114	60,983	61,396	58,334
TP	154	14,755	932	32	7692
FN	15,214	671	14,434	15,479	7691
FP	1786	47,514	705	147	3337
FPR	0.029	0.771	0.011	0.002	0.054
Recall	0.01	0.957	0.061	0.002	0.5
F1-score	0.018	0.38	0.11	0.004	0.582
Precision	0.079	0.237	0.569	0.179	0.697
Accuracy	0.779	0.375	0.804	0.797	0.857
Overall accuracy	30.5825				

Table 1.6 Evaluation factors of LR

LR	Normal	DoS	Probe	R2L	U2R
TN	48,504	52,845	42,139	51,072	61,671
TP	6547	13,234	1326	3938	24
FN	8821	2192	14,040	11,573	15,359
FP	13,182	8783	19,549	10,471	0
FPR	0.214	0.143	0.317	0.17	0
Recall	0.426	0.858	0.086	0.254	0.002
F1-Score	0.373	0.707	0.073	0.263	0.003
Precision	0.332	0.601	0.064	0.273	1
Accuracy	0.714	0.858	0.564	0.714	0.801
Overall accuracy	33.1547				

Table 1.7 Evaluation factors of QDA

QDA	Normal	DoS	Probe	R2L	U2R
TN	61,615	55,484	46,773	61,320	56,544
TP	941	15,326	14,982	4166	15,159
FN	14,427	100	384	11,345	224
FP	71	6144	14,915	223	5127
FPR	0.001	0.1	0.242	0.004	0.083
Recall	0.061	0.994	0.975	0.269	0.985
F1-score	0.115	0.831	0.662	0.419	0.85
Precision	0.93	0.714	0.501	0.949	0.747
Accuracy	0.812	0.919	0.801	0.85	0.931
Overall accuracy	65.6345				

Table 1.8 Evaluation factors of LDA

LDA	Normal	DoS	Probe	R2L	U2R
TN	60,343	61,212	60,590	57,627	60,292
TP	14,152	14,304	14,162	13,246	13,038
FN	1216	1122	1204	2265	2345
FP	1343	416	1098	3916	1379
FPR	0.022	0.007	0.018	0.064	0.022
Recall	0.921	0.927	0.922	0.854	0.848
F1-score	0.917	0.949	0.925	0.811	0.875
Precision	0.913	0.972	0.928	0.772	0.904
Accuracy	0.967	0.98	0.97	0.92	0.952
Overall accuracy	89.4204				

Table 1.9 Evaluation factors of MLP

MLP	Normal	DoS	Probe	R2L	U2R
TN	61,038	61,390	61,420	60,053	61,189
TP	14,314	14,736	15,266	15,089	14,523
FN	1054	690	100	422	860
FP	648	238	268	1490	482
FPR	0.011	0.004	0.004	0.024	0.008
Recall	0.931	0.955	0.993	0.973	0.944
F1-score	0.944	0.969	0.988	0.94	0.956
Precision	0.957	0.984	0.983	0.91	0.968
Accuracy	0.978	0.988	0.995	0.975	0.983
Overall accuracy	95.9431				

Table 1.10 Evaluation factors of k-NN

k-NN	Normal	DoS	Probe	R2L	U2R
TN	61,488	61,470	61,475	61,429	61,368
TP	14,930	15,309	15,215	15,330	15,284
FN	438	117	151	181	99
FP	198	158	213	114	303
FPR	0.003	0.003	0.003	0.002	0.005
Recall	0.971	0.992	0.99	0.988	0.994
F1-score	0.979	0.991	0.988	0.99	0.987
Precision	0.987	0.99	0.986	0.993	0.981
Accuracy	0.992	0.996	0.995	0.996	0.995
Overall accuracy	98.697				

Table 1.11 Evaluation factors of GB

Gradient boosting	Normal	DoS	Probe	R2L	U2R
TN	61,678	61,624	61,616	61,432	61,612
TP	15,232	15,407	15,338	15,441	15,382
FN	136	19	28	70	1
FP	8	4	72	111	59
FPR	0.00013	0.00006	0.00117	0.0018	0.00096
Recall	0.99115	0.99877	0.998178	0.99549	0.99994
F1-score	0.995295	0.99925	0.996751	0.99417	0.99805
Precision	0.999475	0.99974	0.995328	0.99286	0.99618
Accuracy	0.998131	0.9997	0.998702	0.99765	0.99922
Overall accuracy	99.6704				

Table 1.12 Evaluation factors of bagging

Bagging	Normal	DoS	Probe	R2L	U2R
TN	61,672	61,617	61,678	61,513	61,659
TP	15,336	15,415	15,350	15,497	15,379
FN	32	11	16	14	4
FP	14	11	10	30	12
FPR	0.000227	0.00018	0.000162	0.00049	0.0002
Recall	0.997918	0.99929	0.998959	0.9991	0.99974
F1 score	0.998503	0.99929	0.999154	0.99858	0.99948
Precision	0.999088	0.99929	0.999349	0.99807	0.99922
Accuracy	0.999403	0.99971	0.999663	0.99943	0.99979
Overall accuracy	99.9001				

Table 1.13 Evaluation factors of stacking classifier

Stacking	Normal	DoS	Probe	R2L	U2R
TN	61,680	61,621	61,679	61,533	61,666
TP	15,344	15,419	15,365	15,507	15,382
FN	24	7	1	4	1
FP	6	7	9	10	5
FPR	0.0001	0.00011	0.00015	0.00016	0.00008
Recall	0.998438	0.99955	0.999935	0.99974	0.99994
F1 score	0.999023	0.99955	0.999675	0.99955	0.99981
Precision	0.999609	0.99955	0.999415	0.99936	0.99968
Accuracy	0.999611	0.99982	0.99987	0.99982	0.99992
Overall accuracy	99.952				

Table 1.14 Evaluation factors of AdaBoost

AdaBoost	Normal	DoS	Probe	R2L	U2R
TN	61,683	61,627	61,678	61,537	61,668
TP	15,356	15,418	15,365	15,509	15,383
FN	12	8	1	2	0
FP	3	1	10	6	3
FPR	0	0	0.0002	0.0001	0
Recall	0.9992	0.9995	0.9999	0.9999	1
F1 score	0.9995	0.9997	0.9996	0.9997	0.9999
Precision	0.9998	0.9999	0.9993	0.9996	0.9998
Accuracy	0.9998	0.9999	0.9999	0.9999	1
Overall accuracy	99.9702				

The QDA classified overall model performance with each and individual class are shown in Table 1.7. The DoS, U2R, and Probe produce a TPR of 99.4, 98.5, and 97.5. The classes Normal, R2L, and U2R produced the least FPR with 0.001, 0.004, and 0.083. The overall accuracy is 65.63, whereas individual accuracy of 93.1 and 91.9% is achieved for U2R and DoS classes.

The LDA classifies the DoS, Probe, and Normal classes precisely to some extent, i.e., these classes obtain TPR and individual accuracy with 98, 97, and 96.7%, as shown in Table 1.8. The DoS class shows an FPR of 0.007, an F1-score of 94.9, and 97.2 precision. The individual accuracy of U2R and R2L classes produces an accuracy of 95.2 and 92. The LDA gives an overall accuracy of 89.4.

Table 1.9 shows the MLP classifier's result analysis in the Probe, DoS, and U2R classes, with accuracy of 99.5, 98.8, and 98.3, respectively. 15,266 instances are correctly classified, and 268 are wrongly classified with FPs for the Probe class. The DoS and U2R class show that 14,736 and 14,523 are correctly classified, and 238 and 482 are misclassified and given false positives. Each class shows a TPR greater than 93% and an FPR of less than 0.025. The F1-score and precision values of individual classes derive precise values, which leads the MLP classifier with an overall accuracy of 95.94.

Table 1.10 shows the result analysis of the k-NN, where 'DoS' and 'R2L' classes are classified precisely and with an FPR of 0.003 and 0.002. The 'Probe' and 'U2R' classes are properly predicted with 15,309 and 15,284 instances, respectively, whereas 158 and 303 instances are misclassified as false positives. The k-NN classifier categorizes each class almost and achieves an accuracy rate of 98.69, whereas the class 'Normal' achieves a distinct accuracy of 99.2, 'DoS', and 'URL' with an accuracy of 99.6. The classes' Probe' and 'U2R' with a distinct accuracy of 99.5.

The GB classifier predicted almost all the classes precisely with recall, precision, F1-score, and the accuracy of these classes is greater than 99%. Table 1.11 shows the results of each class performance metric for the GB classifier. The classes' DoS' and 'U2R' achieved an individual accuracy of 99%, 'Normal' and 'Probe' achieved an individual accuracy of 99.8, and 'R2L' with an individual accuracy of 99.7. The FPR is less than 0.01 for all the classes and achieved an overall accuracy of 99.67.

Table 1.12 shows the result analysis of the Bagging classifier. The classes show high true positive instances; it can be concluded that very few instances of individual classes are misclassified. Thirty-two instances of the Normal class are predicted as attacks, and 14 instances of attacks have been classified as Normal. 11 instances of the DoS class are classified with Normal and with other attack classes, and 11 instances of other classes have been predicted as DoS attacks. The U2R attack is classified well compared to other classes, as it has 12 FP and 4 FN instances. The F1-score, TPR, precision, and individual accuracy show greater than 99% for each class, where the classifier achieves an overall accuracy of 99.9.

The stacking classifier result analysis are illustrated through Table 1.13, which illustrates that the classes are classified precisely except with a few misclassifications, with an overall accuracy of 99.95. The FP of the Normal class shows that 6 instances of attacks are predicted as Normal, and the false-negative instances with 24 are predicted as attacks. 7 instances of the DoS class are classified with Normal and with other attack classes, and 7 instances of other classes have been predicted as DoS attacks. The false-positive of the Probe class shows that 9 instances of other classes are predicted as Probe and the false-negative instances with 1 predicted as DoS attack. The U2R attack shows 5 false positives and 1 false negative instance i.e., 5 instances of other classes are predicted as U2R, and 1 instance of U2R is predicted as R2L.

Table 1.14 shows the analysis of the Ada-Boost classifier where the classes are classified precisely. The 'U2R' attack shows 3 FP and 0 FN instances i.e., 2 instances of 'Normal' and 1 instance of 'R2L' is predicted as 'U2R'. The false-positive of the R2L class shows that 6 instances of the 'Normal' class are predicted as 'R2L' attack, and the FN with 2 instances is predicted with one 'Normal' and one 'DoS attack'. The FP of the 'Probe' class shows that 10 instances of other classes are predicted as 'Probe', and the FN with 1 instance is predicted as 'DoS attack'. The FP and FN of the 'DoS' class are derived with 1 and 8 instances. The classes Normal, DoS, Probe, R2L, and U2R are wrongly classified with very few TP and TN instances. The overall accuracy of the AdaBoost classifier is 99.97.

Illustrated in Fig. 1.4 is the AUC-ROC curve for each model under consideration. A macro-average is determined by evaluating the metric separately for each class before taking the mean, whereas a micro-average compile the contributions from all classes to calculate the overall average metric used in the ROC curve. The micro-average ROC curve and macro-average for SGD, GNB, LR, QDA, LDA, MLP, and KNN are 0.52, 0.57, 0.57, 0.79, 0.93, 0.98 and 0.99, and the values for Bagging, stacking, and the proposed AdaBoost method has the values 1.0, respectively.

Figure 1.5 represents the classification measures of all the models. In all cases, proposed approach performed well compared to various EL and ML models. Figure 1.6 presents the respective overall accuracies of different ML and EL methods, such as SGD, GNB, LR, and QDA with 26.26, 30.58, 33.15, and 65.63% accuracy. The other methods, such as LDA, MLP, KNN, GB, Bagging, and Stacking, have an

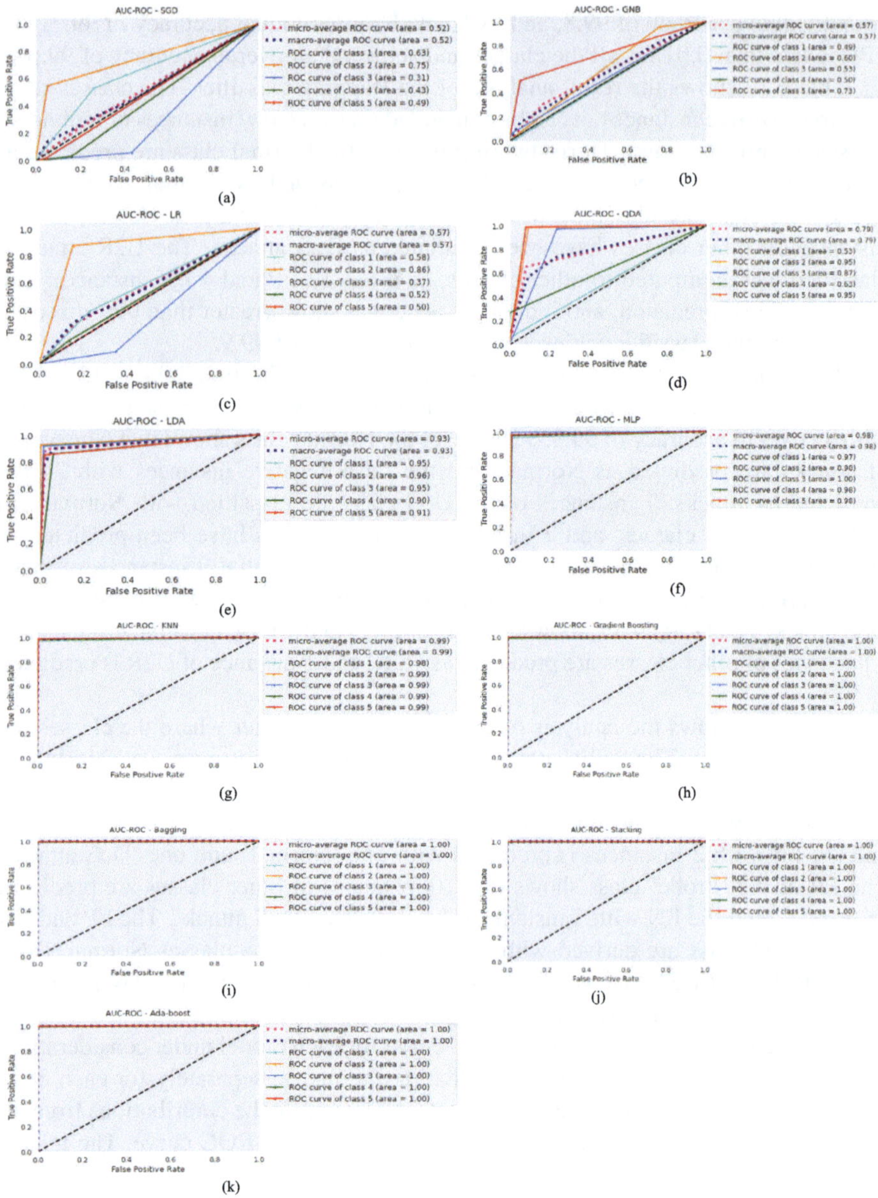

Fig. 1.4 AUC-ROC curves of **a** SGD, **b** GNB, **c** LR, **d** QDA, **e** LDA, **f** MLP, **g** K-NN, **h** GB, **i** Bagging, **j** Stacking, **k** Ada-Boost

accuracy range of 89.42–99.95%. The proposed Ada-Boost algorithm has obtained an accuracy of 99.97%, which is comparatively better than the existing methods.

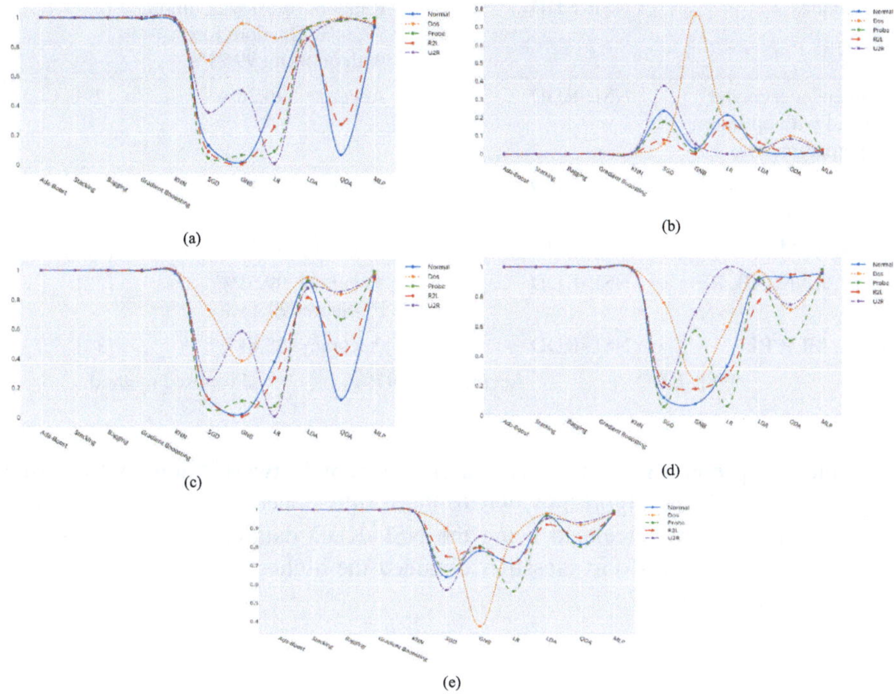

Fig. 1.5 **a** TPR against different models, **b** FPR against different models, **c** F1-score against different models, **d** Precision against different models, **e** Class accuracy against different models

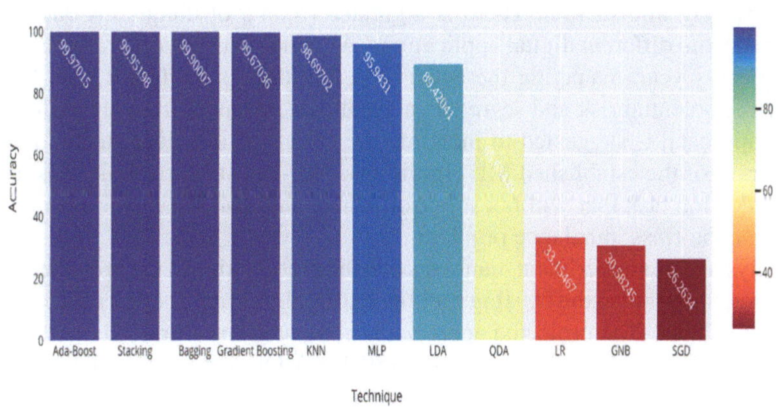

Fig. 1.6 Accuracy comparison of all the models

Table 1.15 Comparison of performance of the proposed method with previous articles

Intelligent method	Datasets	Evaluation factors	Ref
Union and quorum techniques	UNSW-NB15 and NSL-KDD	Accuracy: 99%, Random forest with union: 99.34%, Random forest with quorum: 99.21%	[23]
Autoencoder model trained with optimum hyperparameters	NSL-KDD	Accuracy: 96.36%	[29]
Hybrid supervised learning algorithm	NSL-KDD	Accuracy: 98.9%	[30]
KNN, SVM	NSL-KDD	Accuracy: 84.25%	[31]
SVM, KNN, NB, RF	NSL-KDD	Accuracy: 99.51%, F1-Score: 99.43%	[32]
KNN, MLP, RF	NSL-KDD	Accuracy: 85.81%	[33]
	NSL-KDD	Accuracy: 99.97%	Proposed method

Table 1.15 presents the previous research results on Network Anomaly Detection using various existing algorithms, where the results concerning the accuracy and other parameters are calculated using the NSL-KDD datasets are tabulated. The proposed method Ada-Boost classifier obtained the highest accuracy compared to various previous studies.

1.6 Conclusion

Data mining and ML approaches are actively trying to speed up the mechanism of discovering information. There is a greater volume of ubiquitous data streams produced from different digital applications. As computer network traffic is growing rapidly every year, managing the network in real time is a difficult task. Hence to reduce the potential risk and segregate normal data instances from anomalous ones, an EL approach is suggested to integrate the effects of individual techniques with the support of the established ML algorithms. Ada-Boost ML technique is used for a classification task that will boost performance, and SMOTE method is applied to overcome the class imbalance problem.

Analysis of this experiment manifests that the presentation of the projected method is relatively better than the existing traditional ML algorithms for precision, accuracy, and recall values. The projected approach is compared to several ML methods such as KNN (98.69%), MLP (95.94%), LDA (89.42%), QDA (65.63%), GNB (30.58%), and SGD (26.24%). When compared to the accuracy of the various techniques, the proposed Ada-Boost algorithm achieved an accuracy of 99.97%. The results are evidence that the projected approach for anomaly detection is better compared to other proposed methods.

Most conventional ML algorithms perform poorly because they prefer the majority class samples, resulting in low prediction accuracy for the minority class in the case of the unbalanced dataset. As a result, learning the critical instances becomes difficult. In fact, in order to reduce the overall error rate, assume equal misclassification costs for all samples, and oversampling increases the number of training instances, which increases computing time. Because the identical misclassification cost involved with each of the classes in the unbalanced datasets is not true and pushes the extreme computing limitations in identifying different assaults. AdaBoost combined with SMOTE produces an ideal set of synthetic samples by fixing for skewed distributions and modifying the updating weights. Although this approach can address class imbalance issues well, it may use a significant amount of system resources. Furthermore, studies may concentrate on improving the efficacy and efficiency of IDS by taking into account the many difficulties that ML-based IDS encounters and utilizing the newest ensemble learning algorithms, such XGBoost, LightGBM, etc. It's crucial to remember that these approaches would have to take the possibility of higher system resource usage into account.

References

1. Salehi, M., Rashidi, L.: A survey on anomaly detection in evolving data. ACM SIGKDD Explor. Newsl. **20**(1), 13–23 (2018). https://doi.org/10.1145/3229329.3229332
2. Reddy, D.K.K., Behera, H.S., Nayak, J., Routray, A.R., Kumar, P.S., Ghosh, U.: A fog-based intelligent secured IoMT framework for early diabetes prediction. In: Ghosh, U., Chakraborty, C., Garg, L., Srivastava, G. (eds.) Internet of Things, pp. 199–218. Springer, Cham (2022)
3. Nayak, J., Kumar, P.S., Reddy, D.K.K., Naik, B., Pelusi, D.: Machine learning and big data in cyber-physical system: methods, applications and challenges. In: Cognitive engineering for next generation computing, Wiley, pp. 49–91 (2021)
4. Baig, Z.A., et al.: Future challenges for smart cities: Cyber-security and digital forensics. Digit. Investig., 22 (September 2019), 3–13 (2017). https://doi.org/10.1016/j.diin.2017.06.015
5. Elsaeidy, A., Munasinghe, K.S., Sharma, D., Jamalipour, A.: Intrusion detection in smart cities using Restricted Boltzmann Machines. J. Netw. Comput. Appl., 135(September 2018), 76–83 (2019). https://doi.org/10.1016/j.jnca.2019.02.026
6. Chkirbene, Z., Erbad, A., Hamila, R.: A combined decision for secure cloud computing based on machine learning and past information. In: 2019 IEEE Wireless Communications and Networking Conference (WCNC), vol. 2019-April, pp. 1–6 (2019). https://doi.org/10.1109/WCNC.2019.8885566
7. Tun, M.T., Nyaung, D.E., Phyu, M.P.: Network anomaly detection using threshold-based sparse. In: Proceedings of the 11th International conference on advances in information technology, pp. 1–8 (2020). https://doi.org/10.1145/3406601.3406626
8. Peddabachigari, S., Abraham, A., Thomas, J.: Intrusion detection systems using decision trees and support vector machines. Int. J. Appl. Sci. Comput. **11**(3), 118–134 (2004)
9. Liao, Y., Vemuri, V.R.: Use of K-nearest neighbor classifier for intrusion detection. Comput. Secur. **21**(5), 439–448 (2002). https://doi.org/10.1016/S0167-4048(02)00514-X
10. Negandhi, P., Trivedi, Y., Mangrulkar, R.: Intrusion detection system using random forest on the NSL-KDD dataset, pp. 519–531 (2019)
11. Guezzaz, A., Asimi, A., Asimi, Y., Tbatous, Z., Sadqi, Y.: A global intrusion detection system using PcapSockS sniffer and multilayer perceptron classifier. Int. J. Netw. Secur. **21**(3), 438–450 (2019). https://doi.org/10.6633/IJNS.201905

12. Adhi Tama, B., Nkenyereye, L., Lim, S.: A Stacking-based deep neural network approach for effective network anomaly detection. Comput. Mater. Contin., **66**(2), 2217–2227 (2021). https://doi.org/10.32604/cmc.2020.012432
13. Jain, M., Kaur, G.: Distributed anomaly detection using concept drift detection based hybrid ensemble techniques in streamed network data. Cluster Comput., 1–16 (2021). https://doi.org/10.1007/s10586-021-03249-9
14. Zhong, Y., et al.: HELAD: A novel network anomaly detection model based on heterogeneous ensemble learning. Comput. Networks **169**, 107049 (2020). https://doi.org/10.1016/j.comnet.2019.107049
15. Khammassi, C., Krichen, S.: A NSGA2-LR wrapper approach for feature selection in network intrusion detection. Comput. Networks, **172**(February), 107183(2020). https://doi.org/10.1016/j.comnet.2020.107183
16. Kaur, G.: A comparison of two hybrid ensemble techniques for network anomaly detection in spark distributed environment. J. Inf. Secur. Appl., **55**(September), 102601(2020). https://doi.org/10.1016/j.jisa.2020.102601
17. Othman, D.M.S., Hicham, R., Zoulikha, M.M.: An efficient spark-based network anomaly detection. Int. J. Comput. Digit. Syst. **9**(6), 1175–1185 (2020). https://doi.org/10.12785/ijcds/0906015
18. Nagaraja, A., Boregowda, U., Khatatneh, K., Vangipuram, R., Nuvvusetty, R., Sravan Kiran, V.: Similarity based feature transformation for network anomaly detection. IEEE Access, 8, 39184–39196 (2020). https://doi.org/10.1109/ACCESS.2020.2975716
19. Thaseen, I.S., Chitturi, A.K., Al-Turjman, F., Shankar, A., Ghalib, M.R., Abhishek, K.: An intelligent ensemble of <scp>long</scp>-short <scp>-term</scp> memory with genetic algorithm for network anomaly identification. Trans. Emerg. Telecommun. Technol., (September), 1–21(2020). https://doi.org/10.1002/ett.4149
20. Truong-Huu, T., et al.: An empirical study on unsupervised network anomaly detection using generative adversarial networks. In: Proceedings of the 1st ACM workshop on security and privacy on artificial intelligence, pp. 20–29 (2020). https://doi.org/10.1145/3385003.3410924
21. Gurung, S., Kanti Ghose, M., Subedi, A.: Deep learning approach on network intrusion detection system using NSL-KDD dataset. Int. J. Comput. Netw. Inf. Secur., **11**(3), 8–14 (2019). https://doi.org/10.5815/ijcnis.2019.03.02
22. Zhang, C., Ruan, F., Yin, L., Chen, X., Zhai, L., Liu, F.: A deep learning approach for network intrusion detection based on NSL-KDD dataset. In: 2019 IEEE 13th International Conference on Anti-counterfeiting, Security, and Identification (ASID), vol. 2019-Octob, pp. 41–45. https://doi.org/10.1109/ICASID.2019.8925239
23. Doreswamy, Hooshmand, M.K., Gad, I.: Feature selection approach using ensemble learning for network anomaly detection. CAAI Trans. Intell. Technol., **5**(4), 283–293. https://doi.org/10.1049/trit.2020.0073
24. Bagui, S., Kalaimannan, E., Bagui, S., Nandi, D., Pinto, A.: Using machine learning techniques to identify rare cyber-attacks on the UNSW-NB15 dataset. Secur. Priv. **2**(6), 1–13 (2019). https://doi.org/10.1002/spy2.91
25. Freund, Y., Schapire, R.E., Hill, M.: Experiments with a new boosting algorithm. (1996)
26. Dhanabal, L., Shantharajah, S.P.: A study on NSL-KDD dataset for intrusion detection system based on classification algorithms. Int. J. Adv. Res. Comput. Commun. Eng. **4**(6), 446–452 (2015). https://doi.org/10.17148/IJARCCE.2015.4696
27. University of New Brunswick.: Canadian Institute for Cybersecurity. Research|Datasets|UNB. unb.ca, (2018)
28. Nayak, J., Kumar, P.S., Reddy, D.K., Naik, B.: Identification and classification of hepatitis C virus: an advance machine-learning-based approach. In: Blockchain and machine learning for e-Healthcare systems, Institution of Engineering and Technology, pp. 393–415
29. Kasim, Ö.: An efficient and robust deep learning based network anomaly detection against distributed denial of service attacks. Comput. Networks **180**, 107390 (2020). https://doi.org/10.1016/j.comnet.2020.107390

30. Hosseini, S., Azizi, M.: The hybrid technique for DDoS detection with supervised learning algorithms. Comput. Networks **158**, 35–45 (2019). https://doi.org/10.1016/j.comnet.2019.04.027
31. Su, T., Sun, H., Zhu, J., Wang, S., Li, Y.: BAT: Deep learning methods on network intrusion detection using NSL-KDD dataset. IEEE Access **8**, 29575–29585 (2020). https://doi.org/10.1109/ACCESS.2020.2972627
32. Kasongo, S.M., Sun, Y.: A deep long short-term memory based classifier for wireless intrusion detection system. ICT Express **6**(2), 98–103 (2020). https://doi.org/10.1016/j.icte.2019.08.004
33. Illy, P., Kaddoum, G., Moreira, C.M., Kaur, K., Garg, S.: Securing fog-to-things environment using intrusion detection system based on ensemble learning. *arXiv*, no. April, pp. 15–18

Chapter 2
An In-Depth Analysis of Cyber-Physical Systems: Deep Machine Intelligence Based Security Mitigations

B. K. Tripathy, G. K. Panda, and Ashok Sahu

Abstract Cyber Physical Systems (CPS) is a complex system whose components are both physical and software being intertwined has emerged as a crucial domain and is capable of exhibiting multiple and distinct behavioral modalities handling real-world applications. Over the past two decades, they have evolved into a cornerstone for research and industrial applications, embodying a convergence of physical, biological, and engineered components governed by a computational core. It is a network of interacting elements with physical input and output devices. These systems heavily rely on advanced sensor nodes, communication technologies and control units. Addressing the challenge of deploying sensors in spatially-distributed processes, wireless sensor networks (WSNs) have taken center stage in CPS. WSNs offer a cost-effective solution for monitoring a diverse range of applications, from battlefield surveillance to environmental oversight. The integration of sensor devices within CPS is pivotal in ensuring precision in control and enhancing reliability. Several transdisciplinary approaches like merging theory of cybernetics, mechatronics, design and process science are involved in a CPS and the process is called as embedded system. Concurrently, technological progress has given rise to sophisticated cyber threats, necessitating ongoing vigilance from researchers to safeguard both physical and virtual systems. This necessitates security mitigation to take measures for reducing these harmful effects or hazards. Deep machine intelligence means machine intelligence based upon deep learning techniques and is most recent AI techniques. This chapter delves into the challenges faced from the security aspects of CPS and their solutions based upon deep machine intelligence, presenting experimental findings through an intrusion detection dataset.

B. K. Tripathy (✉)
School of Computer Science Engineering and Information Systems, VIT, Vellore, Tamil Nadu 632014, India
e-mail: tripathybk@vit.ac.in

G. K. Panda
MITS School of Biotechnology, Bhubaneswar, Odisha 751024, India

A. Sahu
Research Scholar, Utkal University, Bhubaneswar, Odisha 751004, India

J. Nayak et al. (eds.), *Machine Learning for Cyber Physical System: Advances and Challenges*, Intelligent Systems Reference Library 60,
https://doi.org/10.1007/978-3-031-54038-7_2

Keywords Physical and computational resources · Wireless sensor network · Machine and deep learning intelligence · Cyber attacks · Threats and security concerns

2.1 Introduction

In late 2008, the National Science Foundation (NSF, US) acknowledged the importance of Cyber Physical Systems (CPS) as a significant domain for exploration and welcomed collaborative research proposals in 2009 [1]. Initially, this system involved with the amalgamation of computational and physical resources. Over time, it became more prominent in the research community and evolved into a rising technology for integrating into research and industrial applications. It can be described as systems involving tangible, biological and engineered elements, where their functions are seamlessly merged, observed and regulated through a computing device. Components are interconnected at all levels and computing is deeply ingrained in each physical element, potentially even within substances. These integrated computational units operate within a distributed environment, providing real-time responses. The behavior of it represents a fully integrated fusion of computational algorithms and physical actions.

In CPS, an array of advanced physical devices plays a pivotal role in enabling seamless interaction with and control of the physical world. Beyond the familiar devices like light-proximity sensors, microphones, GPS chips, GSM chips, cameras, touch screens, WiFi, Bluetooth, EDGE, and 4G/5G connectivity, there is a diverse spectrum of specialized hardware. This includes ultraviolet (UV) sensors for monitoring UV radiation, piezoelectric sensors for precise stress and strain measurements, Geiger-Muller counters for radiation detection, and colorimeters and spectrometers for in-depth color and spectral analysis. Additionally, devices like strain gauges, gas chromatographs, and mass spectrometers find applications in stress analysis, chemical analysis, and composition assessment, respectively. Further, sensors such as sonar sensors, seismic sensors, and turbidity sensors are deployed for underwater distance measurement, earthquake monitoring, and water quality assessment. Capacitive touch sensors, thermal imaging cameras, and Global Navigation Satellite System (GNSS) receivers enhance human–machine interaction, thermal analysis, and precise positioning. The list extends to hygrometers for humidity measurement, time-of-flight (ToF) sensors for 3D imaging, and accelerated stress testing (AST) chambers for extreme component testing, collectively forming the robust arsenal of technical components underpinning the functionality of CPS.

When dealing with spatially-distributed physical processes, the task of sensing can present significant challenges, particularly when deploying sensor devices across expansive areas. Design difficulties in CPS are discussed in [2]. To seamlessly integrate these devices into the system, a substantial number of sensors, actuators, or analogous physical components must be distributed over vast geographical regions. The use of wired sensors, while effective, can incur massive deployment costs and in certain circumstances, it may pose physical or legal constraints. Wireless sensor networks (WSNs), which offer a more simplified and cost-efficient deployment solution, emerge as a pivotal enabling technology for spatially-distributed cyber-physical systems [3, 4]. A WSN constitutes an integration of sensory units configured within a wireless network. WSNs have been utilized as an economical method for monitoring processes and occurrences spread across space. WSN-based current mission-critical applications include essential tasks such as military operations involving battlefield surveillance and the detection of chemical threats. Furthermore, these applications (Table 2.1) extend to environmental uses, such as monitoring forest fires and implementing precision agriculture. Additionally, healthcare applications involve monitoring human physiological data.

In the realm of CPS, it intuits a tightly integrated system comprising computational components and physical operations, with the computational elements exercising control over WSN and IoT in larger aspect under the physical processes. To enable effective control, it's imperative that the computational elements possess accurate and up-to-date information about the dynamic state of these physical processes. This requirement underscores the essential inclusion of sensor devices in every functional CPS. The computed information serves diverse purposes, including state estimation and fault detection, empowering the system to operate effectively and reliably.

As machine learning (ML) evolves from a theoretical concept into an essential tool for industrial automation, the need for a standardized method to export and define ML models independently of their original development framework has become increasingly evident. Today, numerous comprehensive frameworks, including those that didn't exist just a few years ago, offer both model description formats and runtime environments for the execution of these models. Fundamentally, a model is represented as a computational graph, outlining a sequence of operations that transform input data into an output, while the training process refines the model's parameters to minimize a defined loss function. It's worth highlighting that in the advancing field of ML, the entity responsible for model training need not be the same as the provider of the runtime system used in cyber applications for model execution.

With the advancement of technologies concerning physical, communication, and computational processes, varieties of cyber attacks are also getting advanced for unethical and intentional advantages. Cyber attacks pull the attention of researchers to address these issues, but still, these issues are needed to be addressed. It is also apparent that both these public systems, whether in the physical or virtual realm, are susceptible to threats and malicious attacks from adversaries that can result in severe consequences. In [5], the emphasis is placed on elucidating the predictions made by classifiers. In [6], a comprehensive examination of security and privacy concerns in CPS has been provided. Security issues and optimal practices in the IoT are addressed in [7]. Comparable strategies for security administration within the public domain of social networks (SN) are examined using encompassing methods such as protection against neighborhood attacks [8], the application of l-diversity in anonymized SN [9], and an efficient l-diversity algorithm based on rough sets [10].

The rest of this chapter can be encapsulated in the following manner: In Sect. 2.2, we bring forth an extensive analysis of Cyber-Physical Systems, focusing on their

Table 2.1 Cyber-physical systems in various domains

Domain	Applications	Characteristics
Agriculture	Production, food supply [11, 11]	Key features include precision farming, climate-smart agriculture, monitoring of soil and crops, detection and control of micronutrients, employment of mechatronics, drones and hyperspectral imaging,
Automotive-industry	V-cloud [13], automotive CPS [14] Industrial Automation [15] Chemical production [16], manufacturing and productions [17, 18]	Notable traits encompass heterogeneity, non-linearity, non-equilibrium, a wide range of scales, resource management, energy preservation, emission control and the spectrum of self-driving cars
Aviation	Air Transport [19], Commercial aviation [20, 21]	Prominent aspects involve precise control, high security and high-power computing
Defense	Surveillance-battlefield [22]	Focus areas include the control of unmanned aerial vehicles or drones, real-time data analysis, emergency augmentation, security and safety measures
Education	Security, Academic Measures, Human behavior [23–27]	Significant elements encompass real-time surveillance systems for attendance, administration and security, as well as behavior monitoring (peer-to-peer and student–teacher interactions)
Environmental monitoring	Situation aware [28], emergency handling [29]	Key attributes include minimal energy consumption, accuracy and timely response to environmental situations and emergencies
Infrastructure	Civil infrastructure [30], Road traffic congestion [31], Transportations [32, 33], Smart home [34]	Noteworthy features encompass en-route decision-making, real-time route prediction, vehicle movement network optimization, vehicular communication, traffic-route optimization and the analysis of voice commands and gestures for smart home appliances
Healthcare	CPeSC3 [35], eHealth [36], structural health monitoring [37–40]	Prominent characteristics include interoperable algorithms, technology integration for medical equipments, Electronic Health Record management and proactive false alarm detection

essential features and recent challenges. Section 2.3 delves into the integral aspects of WSN in conjunction with CPS and the corresponding intricacies of MAC protocols in this context. Section 2.4 is dedicated to the discussion of threats and security concerns in CPS and the utilization of machine intelligence and deep learning techniques to address these issues. In Sect. 2.5, we present the results of experiments through ML-based and DL-based models, substantiating these outcomes through experimental analysis and subsequent discussions.

2.2 In-Depth Insights into Cyber-Physical Systems

In this section, we dive into the foundational elements of CPS. This entails an in-depth look at the key components, operational structure, technological progress, domain applications, and the harmonious integration of hardware, software, and real-world processes. We emphasize the creation of intelligent systems and delve into the essential characteristics and challenges, particularly in ensuring secure computational and control processes within CPS.

Cyber refers to elements such as computation, communication, and control, which are discrete, based on logic, and operate in a switched manner. *Physical* pertains to systems, whether natural or human-made, that adhere to the laws of physics and function continuously over time. CPS represent systems in which the *cyber* and *physical* aspects are closely intertwined across all scales and levels. This marks a shift from merely applying cyber to the physical realm, moving away from the mindset of treating computing as off-the-shelf commodity "parts," and transitioning from ad-hoc approach to one that is grounded and assured in its development. Figure 2..1 represents an overview of these three terminologies.

In the context of a general overview, a CPS typically consists of a monitoring system, which usually includes one or more microcontrollers responsible for controlling and transmitting data obtained from sensors and actuators that interact with the physical environment. These embedded systems also require a communication interface to exchange information with other embedded systems or cloud-based platforms. The central and most crucial aspect of a CPS is the exchange of information, as data

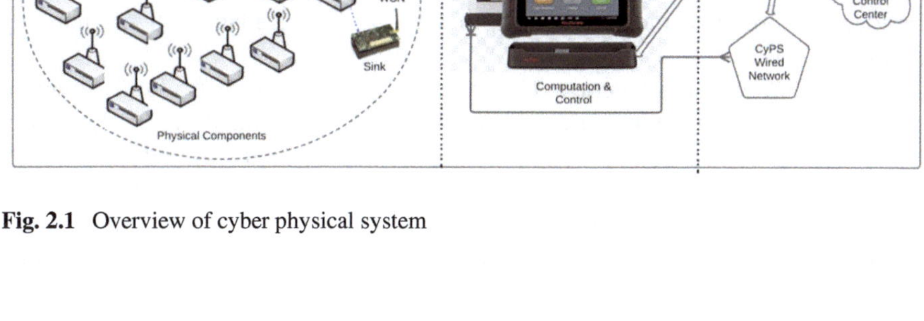

Fig. 2.1 Overview of cyber physical system

can be linked and analyzed in a centralized manner. In essence, a CPS is an embedded system capable of networking and communicating with other devices. This concept is in line with connecting either by ad-hoc network or the internet. Due to remarkable technological advancements such as sensor networks, IoT, wireless technology and cloud computing, wireless networks made a significant impact in many CPS. In Table 2.1, we analyze eight prominent verticals of applications concerning CPS.

2.2.1 Key Characteristics of CPS

Systems concerning are a defining feature of our contemporary world, blending the physical and digital realms in a seamless synergy. While they bring numerous advantages in terms of efficiency and automation, they also introduce a host of vulnerabilities and threats that necessitate a robust approach to cyber security and risk mitigation. Section 2.4 delves more on such issues. Understanding the definition, characteristics, importance, vulnerabilities, and threats of CPS is essential for effectively navigating the challenges and harnessing the potential of these systems in our modern society. Significant instances or case studies may be unveiled in [34].

Real-time Monitoring and Control: CPS excel at real-time data acquisition and decision-making, making them invaluable in applications where timely responses are critical [13]. This capability extends to various domains, such as manufacturing, healthcare, and transportation, where instantaneous adjustments can enhance efficiency and safety.

Interconnectedness of Devices and Systems: In CPS, the different components are interconnected, fostering communication between them. These systems work as a network, enabling collaborative decision-making and providing a holistic view of the environment. For instance, in a smart city, traffic signals, vehicles, and infrastructure communicate to optimize traffic flow [30, 31].

Integration of Sensors and Actuators: Sensors gather information about the physical world, while actuators enable CPS to act upon this data. These components are integral to the feedback loop that defines CPS, where data drives actions. In an agricultural context, CPS can monitor soil conditions (sensors) and autonomously control irrigation systems (actuators) based on this data [11, 12].

High Reliance on Software and Algorithms: Software forms the backbone of CPS, providing the intelligence that processes data, makes decisions, and controls physical elements. Advanced algorithms and machine learning are frequently employed to optimize the functioning of CPS, enabling adaptation to changing conditions[25].

Role in Critical Infrastructure: CPS plays a pivotal role in critical infrastructure sectors, such as energy, transportation, healthcare, and manufacturing. In the energy sector, smart grids leverage CPS to efficiently distribute electricity, reduce wastage, and accommodate renewable energy sources [29]. Transportation systems benefit from CPS through autonomous vehicles, traffic management, and real-time transit updates [32, 32]. In healthcare, CPS contributes to telemedicine, patient monitoring, and drug delivery systems [35, 35, 37]. Moreover, the manufacturing industry has been revolutionized by Industry 4.0, incorporating CPS to enhance efficiency and automation in production processes [15].

Advancements in Industry 4.0: The integration of CPS into the fourth industrial revolution, also known as Industry 4.0, signifies a significant shift in manufacturing [17, 18]. It emphasizes the use of smart technology, data analytics, and automation to create 'smart factories' where machines, products, and systems communicate with each other [14]. This leap in technology enhances productivity, quality control, and cost-efficiency. It's transforming the manufacturing sector and is poised to become a fundamental aspect of modern industrial production.

2.2.2 Critical Challenges in CPS Landscape

In this part, we address the challenges and emerging trends in CPS, such as security and privacy concerns, as well as the increasing role of machine intelligence in shaping the future of these systems. It's crucial to understand that the design of CPS encompasses three primary facets. The first aspect focuses on the hardware, embedded in the system, with the goal of expanding available computational resources (such as processing power, memory, sensors, and actuators) while keeping costs, size, and energy consumption in check [41]. In [2], key considerations and hurdles faced in the development of CPS are explored, providing insights into the complexities associated with integrating computational and physical elements. The second aspect deals with communication, whether wired or wireless, aiming to efficiently transmit messages between distributed devices, quickly and with minimal energy usage. In [3], efforts have been made to trace the evolutionary path from WSN to the broader domain of CPS. Their discussion encompasses on the transition, implications and advancements as sensor networks become integral to the broader concept of CPS. Researchers in [41], focus on energy consumption and optimization analysis, within the context of energy efficiency aspects of wireless CPS.

The third aspect centers on the design of a distributed system, enabling the implementation of CPS functions like remote monitoring and control of distributed processes. However, achieving perfect communication, such as a 100% packet reception rate, isn't the sole objective. Instead, it necessitates real-time guarantees of secure communication and distribution.

In these scenarios, distributed applications often provide transportation mechanisms for collected sensor data. The primary challenge lies in reliably aggregating or disseminating messages across the network. Single-hop communication occurs when a source node is within the communication range of its destination, which is a straightforward case. However, deployed networks often cover large areas, and low-power radios typically have a limited communication range of just tens of meters (Table 2.2). Hence, multi-hop (MHp) communication becomes necessary, where a source node relies on other network nodes to forward its messages, hop by hop, until they reach the destination.

Table 2.2 Variability of Wireless Communication Parameters in WSN Technologies

Range	Text|Graphics|Internet| Hi-Fi|Audio|Video (Stream/Digital/Multi-channel)					
Long (KMs)	1.6–4.8				LMDS	
	30–50			GPRS/4G (LTE)		
	Varies		GPRS (2.5G/3G)			
	35	GSM/CDMA (2G)				
Data rate		9.6–236 Kbps	56–144 Kbps	100 Mbps-1Gbps	1 Mbps-1 Gbps	
Short (Mts)	35–100					802.11a/HL2 (Wi-Fi)
	35–100				802.11b (Wi-Fi)	
	10			Bluetooth 2.0		
	10		Bluetooth 1.0			
	10–100	Zigbee				
Data rate		20–250 kbps	1 Mbps	2.1 Mbps	11 Mbps	54 Mbps

MHp communication is a collaborative task that requires coordination among sensor nodes. If one node transmits a message while another wireless communication is ongoing, their transmissions may interfere, causing both to fail. Additionally, the radio frequency bands used in WSN cannot be isolated, and other networks may use the same frequencies, leading to external interference and packet losses. MHps are carefully planned in a sequence of unicasts (i.e., one-hop transmissions), usually along one or a few of the shortest paths between a message source and its destination. Various routing algorithms have been developed for WSN, including Dijkstra's and Floyd-Warshall's methods, Distance Vector, Ant Colony-based routing, Dolphin Swarm optimization, PSTACK algorithm, Bellman-Ford process, and LEACH routing protocol.

These approaches are efficient because it involves only the necessary nodes in relaying a message. However, the complexity increases when privacy and security must be provided to protect against external threats, especially since these sensor and communication mechanisms are accessible to the public. CPS are not without their vulnerabilities, and one of the most critical concerns is cyber security. These systems are exposed to various cyber threats, including hacking, malware, and data breaches. Malicious actors can exploit these vulnerabilities to compromise the integrity, availability, or confidentiality of data and control systems. For instance, a breach in an industrial CPS could lead to equipment malfunctions, safety hazards, and production disruptions. Breaches in CPS can have far-reaching consequences. Disruption of critical services is a significant concern, especially in communication and controls, while attacks on infrastructure CPS can lead to service outages or even physical damage and system collapse. In [42], efforts have been undertaken to investigate security concerns within the dynamic realm of the social internet of things. In [43], the assessment of bias detection and social trust measures has been conducted through the utilization of methodologies grounded in explainable AI.

In practice, multi-hop wireless networks are sensitive to changes in network topology, external interference, and traffic congestion. These factors limit the reliability of communication and have been a significant hurdle in the adoption of wireless technology in CPS. Synchronous transmissions in low-power communication have introduced a paradigm shift in WSNs, enabling efficient network-wide broadcast in a multi-hop network.

2.3 WSNs in the Context of CPS

Wireless sensor networks (WSN) diverge from conventional networks in several ways, which leads to the need for protocols and tools tailored to address their specific difficulties and constraints. Consequently, innovative solutions are imperative for addressing issues related to energy efficiency, real-time routing, security, scheduling, localization, node clustering, data aggregation, fault detection and data integrity in WSNs.

Machine learning (ML) offers a range of techniques to improve the network's capacity to adapt to the dynamic characteristics of its environment. Table 2.2 illustrates a summary of the coverage, data transfer rates, and accompanying characteristics for WSN-related technologies. The range can vary depending on several factors, including power, environment, and interference associated with wireless communication technologies.

As previously mentioned, the effectiveness of applications based on WSNs relies on the deployment of affordable, diminutive sensor nodes characterized by inherent limitations. In the context of our discussion, we focus on constrained energy reserves, data transfer speed and communication range. These constraints necessitate a strong emphasis on critical aspects like energy preservation to extend network longevity, efficient channel access coordination, collision prevention, priority management, quality of service provision, network-wide synchronization, scalability and energy-saving sleep–wake cycling.

2.3.1 MAC and Diverse Protocol Adaptations

The above challenges have been addressed by numerous researchers, resulting in various solutions documented in the literature. Among these solutions, the Media Access Control protocol (MAC) and its variants have been at the forefront. The

MAC sublayer within the data link layer is responsible for controlling access to the physical network medium, addressing the diverse needs of the sensor network and minimizing or preventing packet collisions in the medium. Numerous advancements in MAC protocols have been specifically tailored for WSNs and we pick few related aspects and detailed in Table 2.3.

Table 2.3 MAC protocols and descriptive approaches

MAC protocols	Description	Approach
Sensor MAC (S-MAC) [44]	Protocols based on content, Coordination of node transmissions, Utilize a shared timetable, Employ a structure organized in frames	Conventional
Timeout MAC (T-MAC) [45]	An energy-efficient MAC protocol demonstrates strong performance in predefined stationary networks, even when dealing with mobile elements	Conventional
RL-MAC [46] RMAC, HEMAC	An adaptive Medium Access Control protocol in WSN using reinforcement learning	Reinforcement learning
Fuzzy hopfield neural network (FHNN) [47]	Nodes are allocated time slots in a way that optimizes the cycle duration, avoiding any overlap in transmissions and minimizing processing time	Neural network-based MAC
Bayesian statistical model for MAC [48]	A MAC protocol based on contention for the management of active and idle periods in WSN	Bayesian statistical model
ALOHA and Q–Learning based learning based MAC with Infromed Receiving [49]	Adopts the attributes of minimal resource demands and reduced collision likelihood from ALOHA and Q-Learning	Conventional
Self-Adapting MAC Layer (SAML) [50]	The MAC engine facilitates the acquisition of MAC protocol based on current network condition, while a Reconfigurable MAC architecture (RMA) allows the switching between various MAC protocols	Simulation
Multi-token based MAC-Cum-routing protocol [51]	A message-passing technique for distributing tokens to active sensor nodes in a distributed approach, ensuring collision-free data transmission and consistent packet delivery while conserving energy	Simulation

2.4 CPS Based Security and Risk Mitigation

What we've come to understand about CPS is that a central objective is to seamlessly merge physical components equipped with sensors and communication capabilities, both in the physical and virtual realms, in order to create automated and intelligent systems. Setting aside the various other aspects and challenges associated with CPS, when we focus on its physical components, many developers aspire to incorporate sensory devices like light sensors, proximity sensors, microphones, GPS chips, GSM chips, cameras, and touch screens. In addition to these sensory components, communication units such as WiFi, Bluetooth, EDGE, and 4G/5G are integral parts of the system.

It's important to note that most of these physical units are readily available to the public, though some may have proprietary features. Furthermore, the communication infrastructures that the integrators heavily rely on are predominantly public, such as the internet and cloud services, with the exception of defense or highly secure solutions.

As a result, the integrated CPS system effectively exposes its identity to the public, becoming a potentially attractive target for unauthorized access. This openness gives rise to a broad spectrum of concerns, including security vulnerabilities, privacy compliance issues, and the risk of data breaches. This begs the question: *How can we ensure the security and privacy of these interconnected systems in an environment where so much is publicly accessible?*

2.4.1 Attack Types

Some well-known real-world incidents, such as Stuxnet (in 2010), BlackEnergy (in 2015), Industroyer (in 2016), Triton (in 2017), WannaCry (in 2017), NotPetya (in 2017), Colonial Pipeline Ransomware Attack (in 2021) serve as reminders of the vulnerabilities in our interconnected systems. In the following section, we delve into comprehensive hypothetical scenarios related to security breaches and employ machine learning and deep learning techniques to address these challenges.

These digital threats can be classified based on the intruder's objectives. In the first category, their goal is to completely disable the target device. In the second category, they seek admin or unauthorized access privileges to the target devices. Broadly speaking, these vulnerabilities can be classified into an exhaustive list. We bring out most eight types of attacks as physical, network-based, software-driven, data breaches, side-channel, cryptographic analysis, access-level, and strategic attacks. Table 2.4 outlines the current cyber-world attacks specifically associated with software and network-based incidents only [52–54].

Table 2.4 Software and network-based threats in CPS

Sn	Attack name	Targeted medium	Network layer impact
1	Backdoor attacks	Software-based	Data processing layer
2	Brute force search attacks	Software-based	Transport layer Network-layer
3	Control Hijacking attack	Software-based	Transport layer
4	Cryptanalysis attacks	Software-based	Application layer
5	DDoS attacks	Software-based	Application layer Network layer
6	Eavesdropping attack	Software-based	Physical layer
7	Malicious code injection	Software-based	Application layer
8	Malware attack	Software-based	Application layer Data processing layer
9	Path-based DoS attacks	Software-based	Application layer
10	Phising attacks	Software-based	Application-layer
11	Reprogram attack	Software-based	Application layer
12	Reverse engineering attack	Software-based	Application layer
13	SL injection attack	Software-based	Application layer
14	Spyware attack	Software-based	Application layer
15	Trojan horse	Software-based	Application-layer
16	Viruses	Software-based	Application-layer
17	Worms	Software-based	Application-layer
18	Blackhole	Network-based	Physical layer
19	DoS/DDoS attack	Network-based	Physical layer
20	Grayhole attack/ Selective forwarding	Network-based	Physical layer
21	Hello flood	Network-based	Physical layer
22	Man in the middle attack	Network-based	Physical layer
23	Replay attack	Network-based	Physical layer
24	RFID unauthorized access	Network-based	Physical layer
25	Routing information attack	Network-based	Physical layer
26	Sinkhole attack	Network-based	Physical layer
27	Spoofing-RFID	Network-based	Physical layer
28	Sybil attack	Network-based	Physical layer
29	Traffic analysis attack	Network-based	Physical layer
30	Wormhole attack	Network-based	Physical layer

2.4.2 Swift CPS Forecasting with Machine Intelligence

The fundamental application of machine intelligence (MI) through a process of learning system involves three main phases. In the initial phase, historical data is inputted to the MI system to facilitate the learning process of the algorithms. Subsequently, the system constructs a representation, and its precision is assessed. If the result is unsatisfactory, additional refinement is required. This repetitive process persists until the precision of the model reaches a stable state. The trained MI algorithm is then subjected to validation using new data to ensure it still delivers high accuracy. This serves as a crucial performance metric to avoid the algorithm becoming overly tailored to the dataset used for learning/training.

The MI algorithm can undergo training using a labeled dataset, where it receives accurate answers; this process is recognized as "*Supervised learning*". In supervised learning, the system is aware of both the input and the desired output. This method is commonly employed when there is an adequate amount of historical data available. In the case of unlabeled datasets, the model resorts to unsupervised learning, seeking associated or clustering patterns without having access to correct answers in the dataset. When dealing with unlabeled datasets, the model turns to unsupervised learning, aiming to identify related or clustering patterns without relying on access to correct answers within the dataset. The third type of MI is based on Reinforcement Learning (RL), which enables an agent to take actions and interact with its environment to determine the best strategies for a given task. This method doesn't derive knowledge from a given dataset. Instead, an RL agent learns by assessing the outcomes of its actions and determines its future actions through a combination of past experiences and novel decisions.

Conventional or shallow MI methods surpass statistical and knowledge-based approaches in terms of flexibility, achieving high detection rates, generalizing models from limited data and learning from examples. Nevertheless, these techniques come with their constraints, including the necessity for manual feature extraction, diminished detection accuracy when dealing with extensive and imbalanced datasets, inefficiency with multi-dimensional data, the prerequisite of background knowledge to determine cluster numbers, and a notable false positive rate. To address these limitations, Deep Learning (DL) techniques have been put forward. DL techniques offer several benefits, such as automated feature extraction, the capacity to effectively manage both labeled and unlabeled data, and robust processing capabilities, particularly when leveraging Graphics Processing Units (GPUs), in contrast to shallow machine learning methods. Figure 2.2 illustrates the concepts of AI, ML and DL with accompanying legends providing concise explanations of the methods.

Our primary emphasis lies in the efficient utilization of machine intelligence for cyber security systems in an experimental approach of intrusion identification over a WSN. This approach involves addressing the specific challenges and restrictions that are often tied to traditional methods within this domain. To enhance the effectiveness of attack identification, we turn to a diverse set of methods. *Gaussian Naive Bayes, Decision Tree, Random Forest, Support Vector Machine and Logistic Regression*

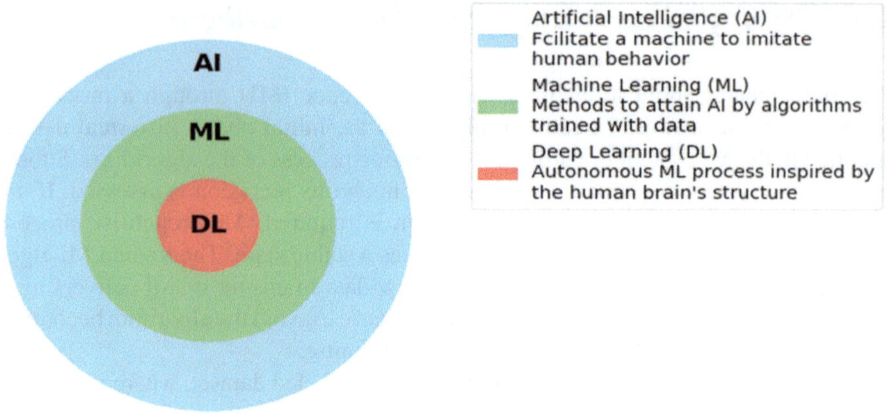

Fig. 2.2 Visualization of AI, ML and DL concepts

offer a more comprehensive toolkit for tackling intrusion detection challenges. Brief overviews of these methods are outlined.

The ***Gaussian-Naïve-Bayes*** learning procedure (*Colab-Python: GaussianNB*) centers around the assumption that the distribution of features follows a Gaussian (normal) distribution, which is a crucial statistical method used to compute the conditional probability of an event. The values of μ_b and σ_b are determined through maximum likelihood estimation as shown in Eq. 2.1.

$$P(a_i b) = \frac{1}{\sqrt{2\pi\sigma_b^2}} \exp\left(-\frac{(a_i - \mu_b)^2}{2\sigma_b^2}\right) \tag{2.1}$$

The ***Decision Tree*** learning process (*Colab-Python: DecisionTreeClassifier*) seeks to determine the attribute at the root node at each level, a step known as attribute selection. This selection can be carried out using one of two methods: information gain or the Gini index. Given n set of training samples, P_i is the probability of the occurrence of the *i-th* instance. Information gain involves computing the average information content, referred to as entropy (Eq. 2.2), while the Gini index quantifies the likelihood of incorrectly identifying a randomly chosen element (as depicted in Eq. 2.3).

$$Entropy = -\sum_{i=1}^{n} p_i * \log(p_i) \tag{2.2}$$

$$Gini_{Index} = 1 - \sum_{i=1}^{n} p_i^2 \tag{2.3}$$

The ***Support Vector Machine (SVM)*** algorithm (*Colab-Python: SVC*) relies on the primal optimization problem, which serves for both data regression and classification. It represents each data point as a point in a (X, Y) plane, n-dimensional. In Eq. 2.4, w signifies the gap between two support vector planes, x_i stands for the data point's value, y_i corresponds to the label assigned to each data point, b indicates the hyperplane's distance from the origin, and n is the total count of training examples.

$$Entropy = \frac{||w||}{2} s.t., y_i(w.x_i + b) - 1 \geq 0 \text{and} y_i(w.x_i + b) + 1 \geq 0 \tag{2.4}$$

The logistic regression model (*Colab-Python: LogisticRegression*) employs a linear model to handle classification tasks. It uses a logistic function (sigmoid curve) to model the probabilities associated with potential outcomes in a single trial, as shown in Eq. 2.5. In this context, a_0 represents the midpoint of the function, k indicates the logistic growth rate or the steepness of the curve and L signifies the maximum value attained by the function.

$$f(a) = \frac{L}{1 + e^{-k(a-a_0)}} \tag{2.5}$$

The Random forests process related to classification (*Colab-Python:* Random-ForestClassifier) every tree in the ensemble is constructed using a sample drawn from the training set with replacement, a technique known as a bootstrap sample. When making decisions at each node while building a tree, the optimal split is determined either from the complete set of input features or from a randomly selected subset of features, the size of which is defined by the *max_features* parameter.

The *Gradient Boosting* algorithm (*Colab-Python: GradientBoostingClassifier*) constructs an incremental model in a step-by-step manner, enabling the optimization of various differentiable loss functions. At each stage, a set of regression trees (equal to 'n_classes') is trained based on the negative gradient of the loss function. This process applies to various scenarios, such as binary or multiclass log loss. In the case of binary classification, a single regression tree is created as a special case.

Traditional approaches may face limitations in dealing with the evolving nature of cyber threats, and these advanced methods help overcome such constraints by providing enhanced accuracy, adaptability, and the ability to detect intricate patterns and anomalies in network traffic data. By employing this diverse set of techniques, we aim to fortify our threat detection capabilities and stay ahead in the ever evolving landscape of cyber security. In Algorithm-2 we provide a detailed approach for mitigating CPS based threats using methods from MI classification techniques.

Algorithm 2: CPS-Based Threat Mitigation with MI Classification

Input:
- CPS system components and Data sources (Sensors, controls, communication logs)
- CPS Raw data covering all anomalous scenarios
- MI model

Output:
- Trained MI models for threat detection
- Model performance evaluation
- Real-time monitoring and alerting

Begin

Step 1. Pre-Processing:
    Convert CPS raw data covering 2-level classification (normal and abnormal).
    Handling missing values with coverage.
Step 2. Feature Selection:
    Identify relevant features from processed data for threat detection.
Step 3. Select ML Model:
    Choose ML models suitable for CPS threat detection.
Step 4. Data Splitting:
    Split labelled data into training and testing sets for model evaluation.
Step 5. Model Training:
    Train specified ML model using labelled training data.
Step 6. Model Evaluation:
    Evaluate model specific performance measures.
Step 7. Real-time monitoring and alerting:
    Implement real-time monitoring using trained ML models
    Generate mitigation strategies (alerts or response mechanisms) for detected threats of attacks.

End

2.4.3 Swift CPS Forecasting with DL

Deep learning (DL) is a specialized field within machine learning that revolves around the development and training of artificial neural networks with multiple layers, commonly referred to as deep neural networks. The term "deep" reflects the incorporation of numerous interconnected layers in these networks. These deep architectures empower machines to autonomously learn and comprehend intricate patterns and features from input data, eliminating the need for explicit programming. Characterized by the use of neural networks with multiple hidden layers, DL models are adept at learning hierarchical representations of data. The core principle involves the automatic extraction of relevant features during the training process, a concept known as representation learning. Employing end-to-end learning, these models directly learn complex representations from raw input to produce predictions or decisions. The training process relies on back propagation, where the model iteratively adjusts its parameters based on the disparity between predicted and actual outcomes. This learning approach finds successful applications across diverse domains, including computer vision, natural language processing, speech recognition, and medical diagnosis. Notable architectures such as Convolutional Neural Networks (CNNs), Recurrent Neural Networks (RNNs), and Transformer models have propelled the field's advancements, showcasing the versatility and power of deep learning in tackling complex tasks. Table 2.5 provides a snapshot of popular deep learning models and their respective architectures and applications. For more detailed explanations can find comprehensive information in [55, 56].

Autoencoders (AEs) exhibit distinct strengths and characteristics within the realm of unsupervised learning. AE is a type of artificial neural network with at least an encoder and a decoder, considered a DL method. AEs are a class of unsupervised learning algorithms employed for efficiently learning representations of data, typically for dimensionality reduction or feature learning purposes. AEs are

Table 2.5 Chronological overview of some deep learning models

Sn	Model	Architecture/type	Application
1	Perceptron;1957	Single-layer neural network	Binary classification
2	MLP (Multi-Layer Perceptron;1965)	Feed forward neural network	General-purpose classification/regression
3	AE (Autoencoder;1980s)	Encoder-Decoder architecture	Dimensionality reduction, feature learning
4	RNN (Recurrent Neural Network;1986)	Recurrent connections	Sequential data, natural language processing
5	CNN (Convolutional Neural Network;1989)	Convolutional layers	Image recognition, computer vision
6	LSTM (Long Short-Term Memory;1997)	RNN variant	Sequential data, time-series analysis
7	Deep Belief Network (DBN;2006)	Stacked Restricted Boltzmann Machines	Feature learning, unsupervised learning
8	DQN (Deep Q Network;2013)	Reinforcement learning	Game playing, decision-making
9	GRU (Gated Recurrent Unit;2014)	RNN variant	Sequential data, machine translation
10	NTM (Neural Turing Machine;2014)	External memory access	Algorithmic tasks, reasoning
11	VAE (Variational Autoencoder;2013)	Probabilistic auto encoder	Generative modeling, data generation
12	ResNet (Residual Network;2015)	Skip connections	Image classification, very deep networks
13	GAN (Generative Adversarial Network;2014)	Generator-Discriminator setup	Image generation, style transfer
14	CapsNet (Capsule Network;2017)	Capsule-based architecture	Image recognition, handling hierarchical features

renowned for their simplicity and adaptability, with training occurring in an end-to-end manner, optimizing the reconstruction error by adjusting both the encoder and decoder weights simultaneously. In Algorithm-2 we explain the operations of this DL approach.

Algorithm 2: Autoencoder-Based Classification

Input:
- Unlabeled training data
- Autoencoder architecture parameters
- Labeled data for classification

Output:
- Trained autoencoder
- Trained classification model
- Predictions on new data

Begin

Step 1. Initialize Model:
Randomly initialize weights and biases for the autoencoder.
Step 2. Define Architecture:
Specify autoencoder architecture with encoder and decoder layers.
Step 3. Prepare Data:
Organize input data for training.
Step 4. Encoder-Decoder Structure:
Create functions for encoding and decoding.
Step 5. Loss Function:
Choose a loss function (e.g., mean squared error) for model training.
Step 6. Compile Model:
Compile the autoencoder model using an optimizer like Adam.
Step 7. Train Autoencoder:
Train the autoencoder to minimize the reconstruction loss.
Step 8. Extract Encoder Weights:
Extract learned encoder weights.
Step 9. Freeze Encoder Weights:
Freeze encoder weights for feature extraction.
Step10. Build Classification Model:
Construct a classification model on top of the frozen encoder.
Step 11. Classification Loss:
Select a classification loss function (e.g., categorical cross-entropy).
Step 12. Compile the classification model with an optimizer.
Step 13. Unfreeze Encoder weights for fine-tuning.
Step 14. Combine the encoder and classification model.
Step 15. Compile the end-to-end model with an optimizer.
Step 16. Train Combined Model:
Train the combined model using labeled data.
Step 17. Evaluate Model:
Assess model performance on a validation set.
Step 18. Hyper-parameter Tuning:
Adjust hyper-parameters for optimal results.
Step 19. Predictions:
Use the trained model for predictions on new data.
Step 20. Analysis and Deployment:
Analyze learned features and deploy the model for real-world applications.

End

2.5 Experiment and Results

The prediction process discussed in this section, involves continuously monitoring a network or system to identify any malicious activities and safeguard the computer network against unauthorized access by users. The predictive classifier needs to have the capability to differentiate between regular connections and suspicious or abnormal connections, which may indicate a potential attack or intrusion. MI and DL techniques discussed in previous sections are pivotal in integrating smart self-learning and automation capabilities into WSN based remote operations and industrial operations, particularly in environments with reduced or no human intervention. Although MI is commonly applied in cognitive fields, its influence on CPS is just beginning to be comprehended. An important challenge in implementing MI in industrial systems is that, unlike in computer science, its applications in industrial sectors necessitate a substantial amount of expertise in allied disciplines.

The experiments were carried out on a 64-bit Intel® Core™ i7-7500U CPU with 8 GB RAM operating in a Windows 10 environment. The implementation of the models took place in Google Colaboratory-Python using Scikit-learn, TensorFlow and Keras library.

2.5.1 *Benchmark Dataset*

To assess the undertaken models, we employed the Knowledge Discovery in Data Mining Cup 1999 dataset (KDDCup99) [57]. The dataset originated by the MIT Lincoln Laboratory, with support from the Defense Advanced Research Projects Agency (DARPA) and the Air Force Research Laboratory (AFRL). The undertaken dataset was compiled for the evaluation of computer network IDS. The purpose of the dataset is to assess the effectiveness of IDSs in a simulated Wireless Sensor Network (WSN). Each connection record in it consists of 41 features and is categorized as either a normal or an attack behavior. The original dataset includes approximately 3,11,029 records for testing and 4,94,020 records for training.

2.5.2 *MI Based Classification*

The undertaken dataset categorizes attacks into four primary types: (a) DoS: Denial of Service attacks, (b) R2L: Unauthorized access, especially from remote to local, (c) U2R: Unauthorized access aimed at obtaining local super-user privileges (referred to as User to Root) and (d) PROBE: Activities related to surveillance and probing, such as port scanning. These groups correspond to different types of attacks. Table 2.6 represents the mapping of dataset attributes to the type of attacks. In addition there are 97,278 normal instances.

Table 2.6 Mapping of attributes to attack types with associated instances

DoS		R2L		U2R		PROBE	
Attribute	Instances	Attribute	Instances	Attribute	Instances	Attribute	Instances
Back	2203	ftp_write	8	buffer_	30	Ipsweep	1247
Land	21	guess_	53	overflow	9	nmap	231
Neptune	107,201	passwd	12	loadmodule	3	portsweep	1040
Pod	264	imap	7	perl	10	satan	1589
smurf	280,790	multihop	4	rootkit			
teardrop	979	phf	2				
		spy	1020				
		warezclient	20				
		warezmaster					

Data Pre-Processing and Feature Mapping

The experimental dataset has undergone pre-processing, involving the elimination of redundant records and duplicates within the training data. This pre-processing step has aimed to achieve a balanced representation by proportionally equalizing the number of records in the training sets. The dataset categorizes source records based on user browsing patterns and initiation protocols such as ICMP, UDP, and TCP. As illustrated in Fig. 2.3, ICMP comprises the largest portion of packet features, making up 57.4% (283,602 packets) of the total, followed by TCP at 38.5% (190,064 packets) and UDP accounting for just 4.1% (20,354 packets). Furthermore, it is a widely acknowledged fact that a higher number of failed attempts increase susceptibility.

Unsuccessful login attempts, also known as brute force attacks or password guessing attacks, can pose various threats and security risks in the network. In our analysis, we've incorporated features related to service issues into the dataset, allowing us to differentiate between successful and unsuccessful user login attempts.

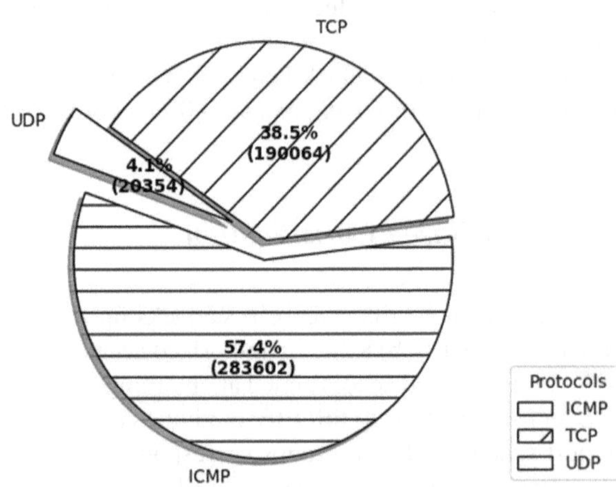

Fig. 2.3 Impact of Pre-processing on protocol distribution

Notably, a noteworthy finding emerges as we observe a substantial 4,20,784 instances of single or multiple unsuccessful login attempts, comprising a significant 85.2% of the overall dataset. This revelation underscores the prevalence of security risks associated with unsuccessful login activities and emphasizes the need for robust measures to mitigate potential threats and bolster network security.

A situation of traffic imbalance typically hinders classifiers from achieving robust detection rates, often leading to a bias toward the most prevalent or high-volume class within the dataset. As a consequence, minority class categories tend to exhibit lower prediction and detection rates. Nevertheless, from a security standpoint, it's essential to treat all potential threats or attacks as equally harmful to the network, regardless of their prevalence in the dataset.

Figure 2.4 presents a heat map providing a visual representation of the correlation among feature attributes within the dataset. The initial training and testing datasets contain overlapping and repeated entries. Notably, the test dataset introduces seventeen attack types not present in the training dataset. Upon employing a heat map for data correlation analysis, we identified 18,795 test cases and 125,971 train cases. These cases involve 38 continuous and 3 categorical attribute types, forming the basis for subsequent experimental processes. The categorical features underwent conversion into binary values, such as [1,0,0], [0,1,0] and [0,0,1]) by applying the one-hot-encoding method. Subsequently, when applying outlier analysis through the Median Absolute Deviation Estimator, along with min–max normalization and the one-hot encoding filter on the dataset, the train dataset was modified to 85,419 instances, while the test dataset was adjusted to 11,928 instances.

2-Level Classification

In this context, the attack labels are divided into two categories: normal and abnormal. The distribution of instances across four attack classes (DoS, Probe, R2L, U2R) reveals the following counts: 45,927, 11,656, 995, and 52, respectively. The collective instances of these attacks are classified as abnormal instances, constituting 46.54% of the total, while the normal class comprises 67,343 instances, accounting for 53.46%. The 2-level class-based attack dataset is further allowed to classify using seven ML based models. Figure 2.5 illustrates the performance analysis of ML models concerning Normal and Abnormal Attacks, depicting (a) Training-Time and (b) Testing-Time and accuracy of various ML based approaches is represented in Fig. 2.6.

K-Level Classification

In the event of a Denial of Service (DoS) attacks, the control server becomes inundated with a multitude of service requests, rendering it unable to cater to the needs of legitimate users. In the case of R2L (Remote to Local), attackers attempt to identify systems of interest within the network and then exploit recognized vulnerabilities. In U2R (User to Root), exploits are utilized to gain administrative control over a machine by a non-privileged system user. In the case of Probe intrusions, attackers aim to pinpoint target systems within the network and subsequently take advantage of recognized vulnerabilities. To enhance the detection rate, the various attack types

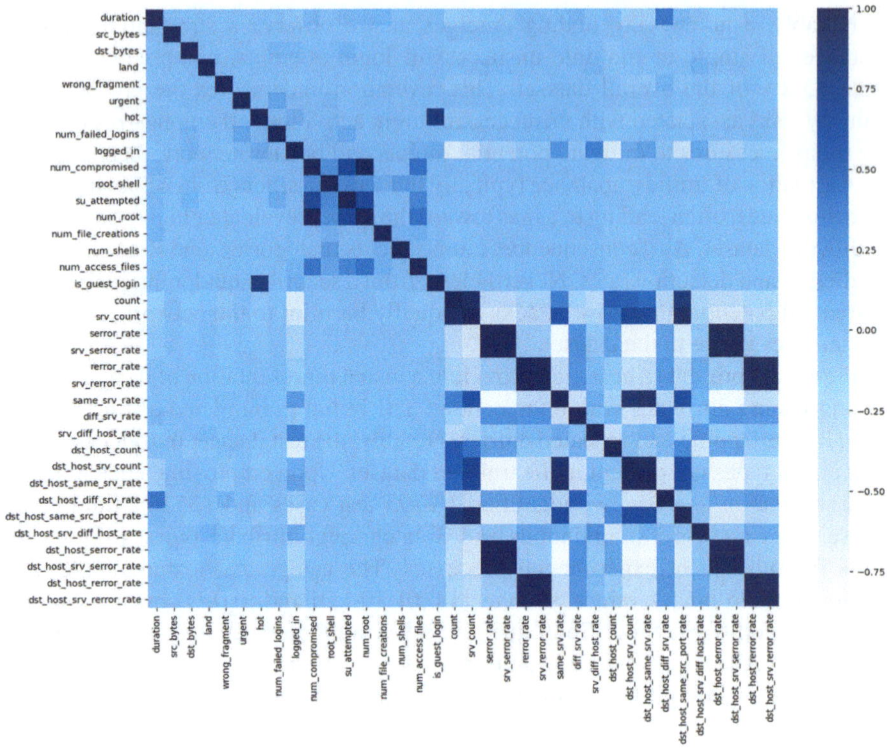

Fig. 2.4 Data correlation of undertaken dataset: heat map representation

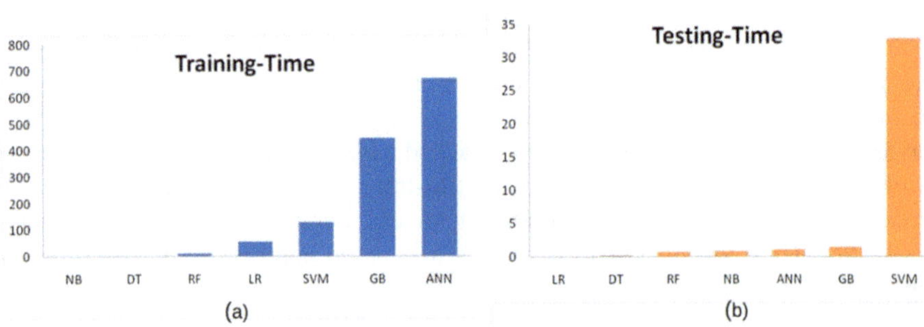

Fig. 2.5 Classification on normal and abnormal attacks: **a** Training-time **b** Testing-time

within these datasets have been organized into attack groups, grouping similar attack types together.

The goal is to identify highly correlated feature sets in this high-dimensional data and correlating the relevant attribute values with the least correlation feature sets. By comparing each variable with the highest correlation factor, considering differences up to 0.01343 from the highest value, we categorize into 5 class labels in accordance

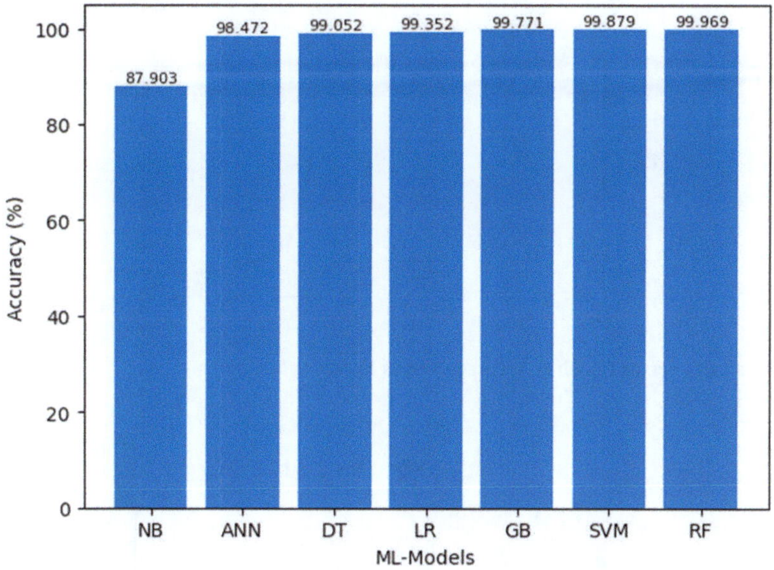

Fig. 2.6 ML based classification accuracy for normal and abnormal attacks

to the 4 dominant attack types (DoS, R2L, U2R and PROBE) and normal cases. This process helps highlight and classify critical attacks based on their correlation patterns within the dataset.

As the Auto Encoder model excels in unsupervised representation learning, autonomously capturing meaningful features, we leverage its working principles to enhance k-level classification tasks of the undertaken dataset. Accordingly, we split the dataset 75% for training and 25% for testing for AE deep learning experiments. This model comprises with input layer, encoding layer with 50 neurons, decoding and output layers and define with 'mean-squared-error' loss function with 'adam' optimizer. Figure. 2.7a and b are the outcome of loss versus epoch and accuracy vs epoch for 5-level class train and test datasets respectively.

In Fig. 2.8, we demonstrate a 3D scatter plot depicting the accuracy of k-level classification (four cases of attack and normal) by the AE-Model across various combinations of hidden layers. Comparative analysis reveals that the AE classifier with a configuration of 50, 20, 10 hidden layers exhibited superior performance in classifying attacks into five distinct levels. Further insight into the ROC analysis, specifically focusing on behavior in the network, normal and attack types is presented in Fig. 2.9a–e.

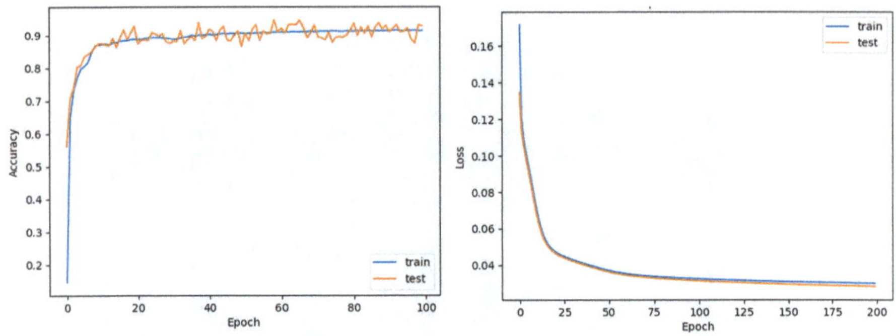

Fig. 2.7 **a** Loss versus Epoch **b** Accuracy versus Epoch

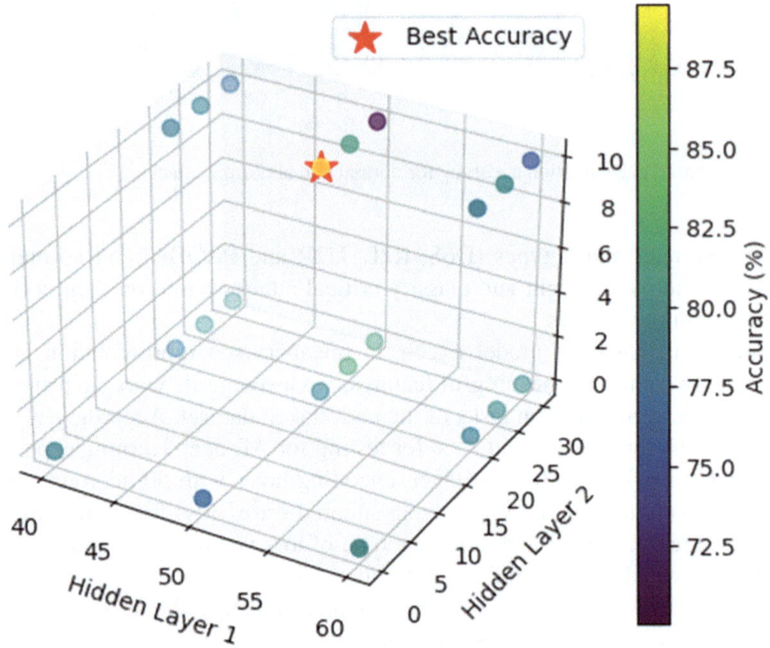

Fig. 2.8 AE-DL model Accuracy with varied hidden layers (3D Scatter Plot)

2.6 Conclusions and Future Scope

In this chapter, we explored the multifaceted realm of cyber-physical systems, highlighting their essential features and recent challenges. We delved into the integral role of wireless sensor networks within these systems and discussed various MAC protocols. Threats and security concerns in cyber-physical systems were also addressed, with a focus on the application of machine intelligence techniques to mitigate these

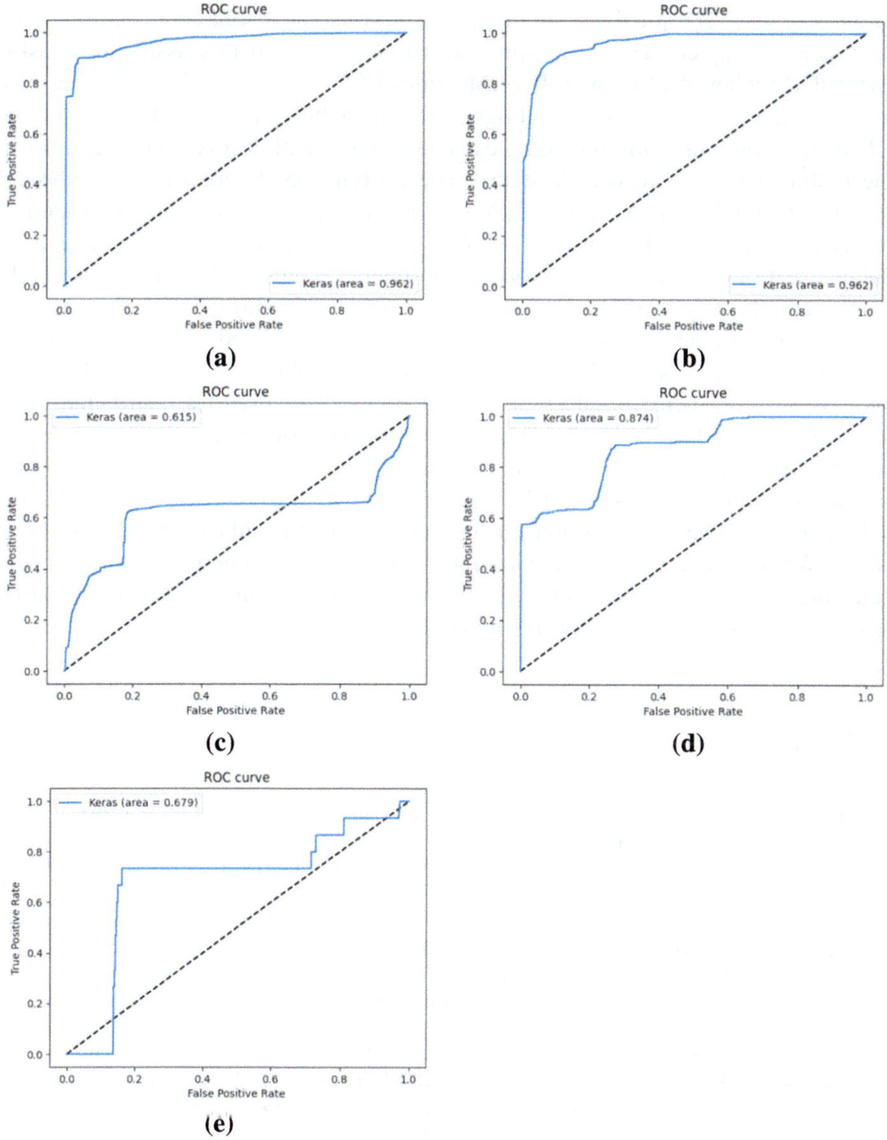

Fig. 2.9 DL based ROC in respect to **a** Normal and **b–e** Attacks in the network

issues. Machine learning and deep learning approaches were presented, substantiated with experimental analysis and comprehensive discussions. The examination of CPS attack classifications and prediction has revealed that employing a two-level class structure is most effective when utilizing machine intelligence processes. In this context, seven MI-based classification models exhibit commendable accuracy; however, their efficiency diminishes when tasked with handling more than two levels of classes. To overcome this limitation, an Auto Encoder-based deep learning (DL) approach is adapted. This DL approach proves successful in classification tasks, particularly when dealing with five categories of classes, achieving an impressive accuracy rate of 89.5%. This highlights the adaptability and effectiveness of the DL method in addressing the complexity associated with k-level classifications in the realm of Cyber-Physical Systems. The evolving landscape of cyber-physical systems, marked by the synergy of physical and computational elements, continues to inspire research and innovation to tackle the growing complexities and security demands of our interconnected world. Both methodologies (MI and DL) currently employed rely on a centralized computing environment. However, challenges may arise when implementing these methodologies in decentralized environments, especially in specific Wireless Sensor Network (WSN) applications. As a result, future endeavors could focus on developing methodologies tailored for distributed environments, where healthcare and sensor-based Internet of Things (IoT) devices are seamlessly integrated with other deep learning approaches. This approach can aim to address the constraints posed by decentralized settings, offering a more adaptable and comprehensive solution for the integration of healthcare and IoT devices within WSN applications. The incorporation of distributed computing environments can enhance the applicability and effectiveness of these methodologies in scenarios where centralized approaches may encounter limitations.

References

1. https://www.nsf.gov/pubs/2008/nsf08611/nsf08611.htm [Accessed on 12th Nov 2023]
2. Lee, E.A.: Cyber physical systems: Design challenges. In: Proceedings of the 11th IEEE Intl symposium on object oriented real-time distributed computing (ISORC), pp. 363–369. IEEE, (2008)
3. Wu, F.J., Kao, Y.F., Tseng, Y.C.: From wireless sensor networks towards cyber physical systems. Pervasive Mob. Comput. **7**(4), 397–413 (2011)
4. Jamwal, A., Agrawal, R., Manupati, V.K., Sharma, M., Varela, L., Machado, J.: Development of cyber physical system based manufacturing system design for process optimization. IOP Conf. Series: Mater. Sci. Eng. 997, 012048 (2020)
5. Ribeiro, M.T., Singh, S., Guestrin, C.: Why should i trust you? Explaining the predictions of any classifier. In: Proceedings of the 22nd ACM SIGKDD international conference on knowledge discovery and data mining, (2016). https://doi.org/10.1145/2939672.2939778
6. Nazarenko, A.A., Safdar, G.A.: Survey on security and privacy issues in cyber physical systems. AIMS Electron. Electr. Eng. **3**, 111–143 (2019)
7. Hukkeri, G.S., Goudar, R.H.: IoT:issues, challenges, tools, security, solutions and best practices. Intl J Pure Appl Math, **120**(6), 12099–12109 (2019)
8. Tripathy, B.K., Panda, G.K.: A new approach to manage security against neighborhood attacks in social networks. In: 2010 Intl Conf on advances in social networks analysis and mining, pp.264-269. IEEE, (2010)
9. Panda, G.K., Mitra, A., Singh, A., Gour, D., Prasad, A.: Applying *l*-Diversity in anonymizing collaborative social network. Int. J. IJCSIT **8**(2), 324–329 (2010)
10. Tripathy, B.K., Panda, G.K., Kumaran, K.: A rough set based efficient l-diversity algorithm. Intl. J. Adv Applied Sci, 302–313 (2011)

11. Rad, C.R., Hancu, O., Takacs, I.: Olteanu, G. Smart monitoring of potato crop: A cyber-physical system architecture model in the field of precision agriculture. Agric. Agric. Sci. Procedia **6**, 73–79 (2015)
12. Ahmad, I., Pothuganti, K.: Smart field monitoring using toxtrac: a cyber-physical system approach in agriculture. In: Proceedings of the 2020 Intl conf on smart electronics and communication (ICOSEC), Trichy, India, pp.10–12 (2020)
13. Abid, H., Phuong, L.T.T., Wang, J., Lee, S., Qaisar, S.: V-Cloud: vehicular cyber-physical systems and cloud com- putting. In: Proc of 4th Intl symposium on applied sciences in biomedical and communication technologies, Spain, (2011)
14. Work, D., Bayen, A., Jacobson, Q.: Automotive cyber phys- ical systems in the context of human mobility. In: Proceedings of the national workshop on high-confidence cyber-physical systems, Troy, Miss, USA, (2008)
15. Dafflon, B, Moalla, N, Ouzrout, Y.: The challenges, approaches, and used techniques of CPS for manufacturing in Industry 4.0: A literature review. Int. J. Adv. Manuf. Technol. **113**, 2395–2412 (2021)
16. He, G., Dang, Y., Zhou, L., Dai, Y., Que, Y., Ji, X.: Architecture model proposal of innovative intelligent manufacturing in the chemical industry based on multi-scale integration and key technologies. Comput. Chem. Eng. **141**, 106967 (2020)
17. Ren, S., Feng, D., Sun, Z., Zhang, R., Chen, L.: "A framework for shop floor material delivery based on real-time manufacturing big data. J. Ambient. Intell. Humaniz. Comput. **10**, 1093–1108 (2019)
18. Majeed, A., Lv, J., Peng, T.: A framework for big data driven process analysis and optimization for additive manufacturing. J. Rapid Prototyp. **24**, 735–747 (2018)
19. Sampigethaya, K., Poovendran, R.: Aviation cyber–physical systems: foundations for future aircraft and air transport. Proc. of IEEE **101**, 1823–1855 (2013)
20. Ying, D.S.X., Venema, D.S., Corman, D.D., Angus, D.I., Sampigethaya, D.R.: Aerospace cyber physical systems-challenges in commercial aviation, Cyber-Physical Systems Virtual Organization
21. Sampigethaya, K., Poovendran, R.: Aviation cyber–physical systems: Foundations for future aircraft and air transport. Proc. IEEE **101**, 1834–1855 (2013)
22. Huang, Y., Zhao, M., Xue, C.: Joint WCET and update activity minimization for cyber-physical systems. ACM Transactions, TECS **14**, 1–21 (2015)
23. Broo, D.G., Boman, U., Törngren, M.: Cyber-physical systems research and education in 2030: Scenarios and strategies. J. Ind. Inf. Integr. **21**, 100192 (2021)
24. Perry-Hazan, L., Birnhack, M.: Privacy CCTV and school surveillance in the shadow of imagined law. Law Soc. Rev. **50**, 415–449 (2016)
25. Singh, K., Sood, S.: Optical fog-assisted cyber-physical system for intelligent surveillance in the education system. Comput. Appl. Eng. Educ., 692–704 (2020)
26. Marwedel, P., Engel, M.: Flipped classroom teaching for a cyber-physical system course-an adequate presence-based learning approach in the internet age. In: Proc of the 10th European Workshop on Microelectronics Education (EWME), Tallinn, Estonia, pp.14–16 (2014)
27. Taha, W., Hedstrom, L., Xu, F., Duracz, A., Bartha, F.Y., David, J., Gunjan, G.: Flipping a first course on cyber-physical systems: An experience report. In: Proc of the 2016 workshop on embedded and cyber-physical systems education. Association for Computing Machinery, New York, NY, USA (2016)
28. Singh, V.K., Jain, R.: Situation based control for cyber- physical environments. In: Proc of the IEEE military communications conf (MILCOM '09), Boston, Mass, USA, (2009)
29. Meng, W., Liu., Xu, W., Zhou, Z.: A cyber-physical system for public environment perception and emergency handling. In: Proc of the IEEE Intl Conf on high performance computing and communications, (2011)
30. Hackmann, G., Guo, W., Yan, G., Sun, Z., Lu, C., Dyke, S.: Cyber-physical code sign of distributed structural health monitoring with wireless sensor networks. IEEE Trans. Parallel Distrib. Syst. **25**, 63–72 (2013)

31. Lin, J., Yu, W., Yang, X., Yang, Q., Fu, X., Zhao, W.: A real-time en-route route guidance decision scheme for transportation-based cyber physical systems. IEEE Trans. Veh. Technol. **66**, 2551–2566 (2016)
32. Kantarci, B.: Cyber-physical alternate route recommendation system for paramedics in an urban area. In: Proc of the 2015 IEEE Wireless Communications and Networking Conf (WCNC), USA, (2015)
33. Ko, W.H., Satchidanandan, B., Kumar, P.: Dynamic watermarking-based defense of transportation cyber-physical systems. ACM Trans. Cyber-Phys. Syst. **4**, 1–21 (2019)
34. Raisin, S.N., Jamaludin, J., Rahalim, F.M., Mohamad, F.A.J., Naeem, B.: Cyber-Physical System (CPS) application-a review. REKA ELKOMIKA J. Pengabdi. Kpd. Masy. **1**, 52–65 (2020)
35. Wang, J., Abid, H., Lee, S., Shu, L., Xia, F.: A secured health care application architecture for cyber-physical systems. Control Eng Appl Inform **13**(3), 101–108 (2011)
36. Lounis, A., Hadjidj, A., Bouabdallah, A., Challal, Y.: Secure and scalable cloud-based architecture for e-health Wireless sensor networks. In: Proc of the Intl Conf on Computer Communication Networks (ICCCN '12), Munich, Germany, (2012)
37. Bocca, M., Tojvola, J., Eriksson, L.M., Hollmen, J., Koivo, H.: Structural health monitoring in wireless sensor networks by the embedded goertzel algorithm. In: Proc of the IEEE/ACM 2nd Intl Conference on Cyber-Physical Systems (ICCPS '11), pp.206–214. Chicago, Ill, USA (2011)
38. Jindal, A., Liu, M.: Networked computing in wireless sensor networks for structural health monitoring. In: Proceeding of the IEEE/ACM transactions on networking (TON '12), vol. 20. pp.1203–1216 (2012)
39. Akter, F., Kashem, M.A., Islam, M.M., Chowdhury, M.A., Rokunojjaman, M., Uddin, J.: Cyber-Physical System (CPS) based heart disease's prediction model for community clinic using machine learning classifiers. J. Hunan Univ. Nat. Sci. **48**, 86–93 (2021)
40. Feng, J., Zhu, F., Li, P., Davari, H., Lee, J.: Development of an integrated framework for cyber physical system (CPS)-enabled rehabilitation system. Int. J. Progn. Health Manag **12**, 1–10 (2021)
41. Liu, J., Wang, P., Lin, J., Chu, C.H.: Model based energy consumption analysis of wireless cyber physical systems. In: Proc of 3rd IEEE Inl Conf on Big data security on cloud, IEEE Intl Conf on High Performance and Smart Computing (Hpsc), and IEEE Intl Conf on intelligent data and security, pp. 219–224. China (2017)
42. Panda, G.K., Tripathy, B.K., Padhi, M.K.: Evolution of social IoT world: security issues and research challenges, Internet of Things (IoT), pp.77–98. CRC Press, (2017)
43. Panda, G.K., Mishra, D., Nayak, S.: Comprehensive study on social trust with xAI: techniques, evaluation and future direction, (Accepted), explainable, interpretable and transparent AI system, pp.1–22 (Ch-10). CRC Press, (2023)
44. Ye, W., Heidemann, J., Estrin, D.: An energy-efficient MAC protocol for wireless sensor networks. In: 21st Annual joint Conf of the IEEE computer and communications societies, vol. 3. pp.1567–1576 (2002)
45. Van, T.D., Langendoen, K.: An adaptive energy-efficient MAC protocol for wireless sensor networks. In: Proc of the 1st Intl Conf on embedded networked sensor systems, pp. 171–180. ACM, New York, USA (2003)
46. Liu, Z., Elhanany, I.: RL-MAC: A reinforcement learning based MAC protocol for wireless sensor networks. Intl. J. Sensor Networks **1**(3), 117–124 (2006)
47. Shen, Y.J., Wang, M.S.: Broadcast scheduling in wireless sensor networks using fuzzy hopfield neural network. Expert Syst. Appl. **34**(2), 900–907 (2008)
48. Kim, M., Park, M.G.: Bayesian statistical modeling of system energy saving effectiveness for MAC protocols of wireless sensor networks. In: Software engineering, artificial intelligence, networking and parallel/distributed computing, studies in computational intelligence, vol. 209, pp. 233–245. Springer. (2009)
49. Chu, Y., Mitchell, P., Grace, D.: ALOHA and q-learning based medium access control for wireless sensor networks. In: Intl symposium on wireless communication systems, pp. 511–515 (2012)

50. Sha, M., Dor, R., Hackmann, G., Lu, C., Kim, T.S., Park, T.: Self adapting MAC layer for wireless sensor networks. Technical Report WUCSE-2013–75, Washington University in St. Louis. Tech Rep (2013)
51. Dash, S., Saras, K., Lenka, M.R., Swain, A.R.: Multi-token based MAC-Cum-routing protocol for WSN: A distributed approach. J. Commun. Softw Syst., 1–12 (2019)
52. Kumar, L.S., Panda, G.K., Tripathy, B.K.: Hyperspectral images: A succinct analytical deep learning study. In: Deep learning applications in image analysis. Studies in big data, vol. 129, pp.149–171. Springer, (2023)
53. Mpitziopoulos, A., Gavalas, D., Konstantopoulos, C., Pantziou, G.: A survey on jamming attacks and countermeasures in WSNs. IEEE. Commun. Surv & Tutor. **11**(4), 42–56 (2009)
54. Yin, D., Zhang, L., Yang, K.: A DDoS attack detection and mitigation with software-defined Internet of Things framework. IEEE Access **6**, 24694–24705 (2018)
55. Buduma, N., Locascio, N.: Fundamentals of deep learning: Designing next-generation machine intelligence algorithms. O'Reilly Media, Inc., O'Reilly (2017)
56. Sarker, I.H.: Deep learning: a comprehensive overview on techniques, taxonomy, applications and research directions. SN Comput. Sci. **2**, 420 (2021)
57. Irvine The UCI KDD Archive, University of California. KDD Cup 1999 Data, http://www.kdd.ics.uci.edu/databases/kddcup99/kddcup99/html/ [Accessed 20 April 2023]

Chapter 3
Unsupervised Approaches in Anomaly Detection

**Juan Ramón Bermejo Higuera, Javier Bermejo Higuera,
Juan Antonio Sicilia Montalvo, and Rubén González Crespo**

Abstract Industry 4.0 is a new industrial stage based on the revolution brought about by the integration of information and communication technologies (ICT) in conventional manufacturing systems, leading to the implementation of cyber-physical systems. With Industry 4.0 and cyber-physical systems, the number of sensors and thus the data from the monitoring of manufacturing machines is increasing. This implies an opportunity to leverage this data to improve production efficiency. One of these ways is by using it to detect unusual patterns, which can allow, among other things, the detection of machine malfunctions or cutting tool wear. In addition, this information can then be used to better schedule maintenance tasks and make the best possible use of resources. In this chapter, we will study unsupervised clustering techniques and others such as nearest neighbor methods or statistical techniques for anomaly detection that can be applied to machining process monitoring data.

Keywords Anomaly detection · Unsupervised methods · Clustering

3.1 Introduction

One of the applications of machine learning is anomaly detection. This task requires being able to identify anomalous behavior from non-anomalous behavior, which is not always trivial. The normal operating conditions of the industry 4.0 machines can

J. R. B. Higuera · J. B. Higuera · J. A. S. Montalvo · R. G. Crespo (✉)
Universidad Internacional de La Rioja, Avda. de La Paz 173La Rioja, Logroño, Spain
e-mail: ruben.gonzalez@unir.net

J. R. B. Higuera
e-mail: juanramon.bermejo@unir.net

J. B. Higuera
e-mail: javier.bermejo@unir.net

J. A. S. Montalvo
e-mail: juanantonio.sicilia@unir.net

J. Nayak et al. (eds.), *Machine Learning for Cyber Physical System: Advances and Challenges*, Intelligent Systems Reference Library 60,
https://doi.org/10.1007/978-3-031-54038-7_3

vary, machines allow working with a multitude of parts as well as with different materials and with different production sizes that will produce different monitoring data. This makes it impossible to know a priori which data are outside normal behavior. In addition, the data from this monitoring of machine operation are unbalanced, with much more data corresponding to normal operating behavior than to unusual operating behavior, which complicates their analysis. This situation means that in machine learning, anomaly detection is normally treated as a non-machine learning problem. This situation means that within machine learning, anomaly detection is normally treated as an unsupervised or semi-supervised learning problem (with only a few examples that are usually (having only a few examples that are usually of usual behavior).

Unsupervised learning is an important branch of machine learning with several applications. Techniques that fall under the umbrella of unsupervised learning do not assume that samples are labeled (for classification tasks) or have one or more associated values to predict (for regression tasks). Therefore, they cannot be used to design classifiers or regressors; they are used to find groupings of the data based on one or more criteria (e.g. Euclidean distance). That is why they can help us to divide data sets into two or more groups, as well as to detect outliers [1]. In anomaly detection tasks, unsupervised learning techniques help to identify patterns that are considered normal. For each regularly observed pattern associated with the normal operation of the system under observation, the unsupervised learning technique used may find several clusters. The idea is that unsupervised outlier detection approaches score data based solely on the inherent properties of the dataset. In all unsupervised learning tasks, we want to learn the natural structure of our data without using specially given features. Unsupervised learning is useful for exploratory analysis because it can automatically identify data structure. For example, if analysts are trying to segment consumers, unsupervised consumers, unsupervised clustering techniques would be a good starting point for their analysis. In situations where it is impossible or impractical for humans to suggest trends in data, unsupervised learning can provide initial insights that can then be used to test hypotheses. Unsupervised learning avoids the need to know which of the collected data are anomalous and requires less data to train. These algorithms allow unusual data to be defined dynamically and avoid the need for extensive knowledge of the application domain.

In addition, it is not only necessary to be able to identify atypical behavior. As mentioned above, industry 4.0 machines make different parts during their operation that will result in different measurements and the data does not always contain information in this regard. This makes it useful to be able to identify common, repeating patterns that correspond to specific part-processing signatures. One of the most used unsupervised machine learning techniques is clustering, where data is grouped according to a similarity measure.

Clustering techniques have been widely used for both static and dynamic data. Within the dynamic data, we find time series. This type of data is very common in a multitude of domains, including industry. The characteristics of time series can vary the way of tackling the problem. There is no one learning technique that is better than another for any given problem, which implies that it is required to test which technique is more effective for each problem. The final objective of this study is to detect anomalies in the data flow of a software-defined network. Initially, an unsupervised dataset is available, with different observations on the traffic flow of the software-defined network, which is examined and analysed. For this purpose, feature engineering is employed on the set, using certain technologies, applying a transformation on the data, and obtaining as a result a valid set for analysis in the following phases.

3.2 Methodology

Unsupervised learning does not know what class the data belongs to and its objective is to discover hidden patterns in the data. It has no direct feedback and one of the tasks of unsupervised learning is clustering.

Clustering is one of the most widely used techniques for pattern discovery of patterns. Clustering is the process of unsupervised partitioning of a dataset $D = \{\{F_1, F_2, \ldots, F_n\}$ within k groups $C = \{C_1, C_2, \ldots, C_k\}$ according to a similarity measure that maximizes the similarity between objects in the same group and minimizes the similarity with the data of the rest of the groups.

Objects within the same group must share characteristics or be related, have small differences between them, or at least be related to each other, have small differences between them, or at least be related to other objects in the group. The data set is considered to be groupable when there are continuous regions with a relatively high density, surrounded by a relatively high density and other continuous regions surrounded with a lower density.

In numerical data clustering, two types of groups can be distinguished.

- **Compact Groups**: all objects in the group are similar to each other and the group can be represented by its center.
- **Chained Groups**: each object in the group is more similar to another member of the group than to any other object in the other groups and can connect two objects of the group using a path.

In the modeling of the problem, the definition of the group as well as the separation criteria must be determined. Clustering methods are composed of several elements (see Fig. 3.1).

- **Data representation or pattern**: This is the set of characteristics of the data that is passed to the algorithm. It may require a prior dimensionality reduction by selection (choosing which data features are most effective for clustering) or feature extraction (transforming the data into new features that facilitate and improve the clustering process). The representation method affects the efficiency and accuracy of the algorithm.
- **Similarity measure**: A metric capable of measuring the similarity between pairs of data representations. This measure must be clear and have a practical meaning.

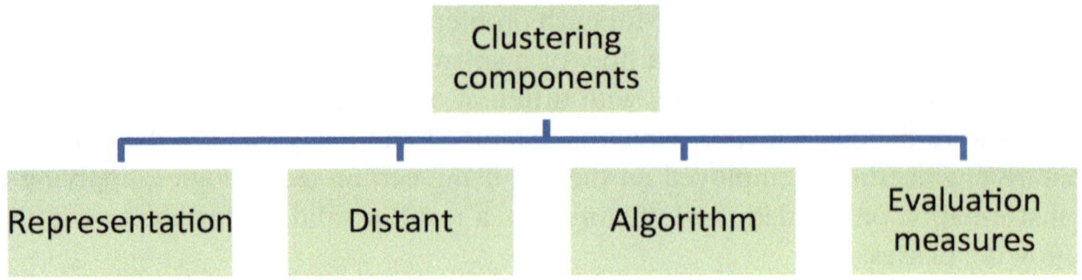

Fig. 3.1 Clustering components

- **Clustering algorithm**: This method allows us to divide the data representations into groups.
- **Evaluation measures**: Measures to analyze the validity of the clustering.

In addition to these components, some methods require a process of data abstraction that, once the grouping is done, allows us to obtain a compact and simpler representation of the data set.

From these components, a standard clustering process can be performed, which would consist of obtaining a representation of the data, the design and execution of the clustering algorithm using the appropriate similarity measure, the evaluation and validation of the results obtained, and a visualization and explanation of the results (see Fig. 3.2).

Depending on the problem, the clustering methods to be used should take into account several considerations, including the following considerations among which are:

- Scalability, since not all algorithms work well for a large volume of data.
- Ability to handle various types of data, whether categorical, numerical, or sequential, among others.
- Ability to discover clusters with arbitrary shapes, because many algorithms make assumptions about the shapes of the clusters, e.g., assuming that they are spherical, so that they do not work well for other types of shapes.
- Handling noise in the data.

In addition, many algorithms require information about the data domain to establish the parameters of the clustering algorithms, such as knowing the number of clusters. There is a great diversity of techniques to represent the data, to establish the

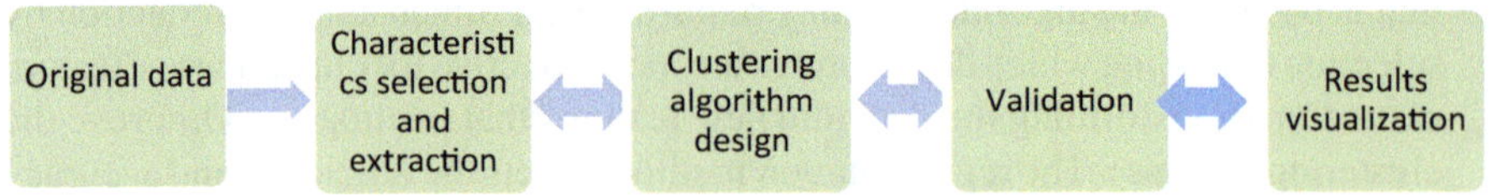

Fig. 3.2 Clustering process

similarity between pairs of data, and to form clusters of elements, as well as to evaluate the results. All these techniques are not always compatible with each other or work equally well. Some algorithms may present various configurations of clusters, depending on some other criterion such as the order in which the data are analyzed.

The criteria for grouping the data, the spacing between clusters, the similarity measure, or the space in which they work are often used to compare clustering methods. The choice of these components (representation method, algorithm, similarity measure, and evaluation measure) will depend on the problem. A method of clustering that works equally well for any situation does not exist.

3.2.1 Types of Algorithms

Clustering algorithms can be classified into different types, the best known of which are

- Partition
- Density
- Grid
- Hierarchical
- Model-based

In addition, several algorithms of different types can be combined to perform multi-step clustering. Each of these types of algorithms has several advantages and disadvantages, which make them more or less suitable depending on the problem. The choice of algorithm will depend on the data to be clustered. In Table 3.1 all types of algorithms considered in this chapter are summarized.

3.2.1.1 Clustering Based on Partitioning or Representative-Based Clustering

Partitioning methods divide the data into k groups where each group contains at least one element. The clusters are created all at once and there are no hierarchical relationships between the groups obtained.

To create these groups, a set of representative elements, also called prototypes of each of the groups, is used. These representatives can belong to the group or be created from the elements that compose it. However, the choice of these prototypes to achieve an optimal partitioning of the elements is unknown. Therefore, partition-based algorithms follow a two-step iterative approach. From the initially chosen prototypes, the elements are assigned to the cluster of the closest prototype (Assignment step) and after that, the prototypes are recalculated (Optimization step). These steps are repeated until a predefined requirement is met, such as an error or a limit on the number of iterations. The effectiveness of the method used depends not only on the prototype that is defined but also on the update method used to recalculate the

Table 3.1 Clustering algorithm types summarized

Clustering type	Techniques	Advantage	Limitations	References
Partition	K-means	They have low complexity, are fast, and usually give good efficiency	Not suitable for non-convex data and requires knowledge of the number of partitions	[2]
Density	DBSCAN, OPTICS, DENCLUE	They have high efficiency and are capable of clustering data with different shapes	Their results worsen when the density of the data space is not uniform and depends on the input parameters	[8]
Hierarchical	Birch Cure	They are deterministic algorithms that do not require knowledge of the number of clusters nor do they require the use of a prototype and have a high visualization capacity, allowing the representation of different clusters and their relationships using dendrograms	Once a cluster is merged or split, it is not possible to go backward, which negatively affects the quality of the clustering and makes it often used in hybrid clustering approaches	[3]
Grid	STING CLIQUE	They tend to have low complexity, are highly scalable and can take advantage of parallel processing	They are sensitive to the number of cells into which the space is merged. The smaller the number of cells, the higher the counting speed, but the lower the clustering accuracy	[4]

(continued)

Table 3.1 (continued)

Clustering type	Techniques	Advantage	Limitations	References
Probabilistic model	COB WEB EM	Probabilistic models allow the representation of subpopulations within a population	These types of algorithms require setting several parameters and are quite slow to process	[10]
Neural network model	SOM ART LVQ	They are trained, self-organizing, learn and forget They are robust and fault tolerant; the failure of one or several neurons does not imply a total failure in the neural network They are flexible, which allows them to easily adapt to new environments since they can be classified as independent systems They are used in data in which the pattern is obscure and imperceptible, exhibiting unpredictable or nonlinear behavior, such as in traditional time series models and chaotic data	These types of algorithms require setting several parameters and are quite slow to process	[9]

prototypes after each iteration of the algorithm. These algorithms are divided into hard clustering when each element belongs to one and only one group, and fuzzy clustering, when each element is assigned a percentage of probability of belonging to each of the clusters. They have low complexity, are fast, and usually give good efficiency, however, they are not suitable for non-convex data and require knowledge of the number of partitions. In addition, their efficiency is determined by the prototype used. The best-known algorithms in this category are k-means, where the group mean is used as the prototype, and k-medoids, where the group medoid and its fuzzy approaches such as fuzzy c-means are used. In Fig. 3.3, an example of clustering partition using the k-means algorithm can be examined.

3.2.1.2 Density-Based Clustering

These algorithms group data according to their connectivity and density; regions with high density belong to the same group. In other words, an element can continue to expand the group with its nearby elements when its neighborhood, which is the number of elements close to it, exceeds the threshold. They have high efficiency and

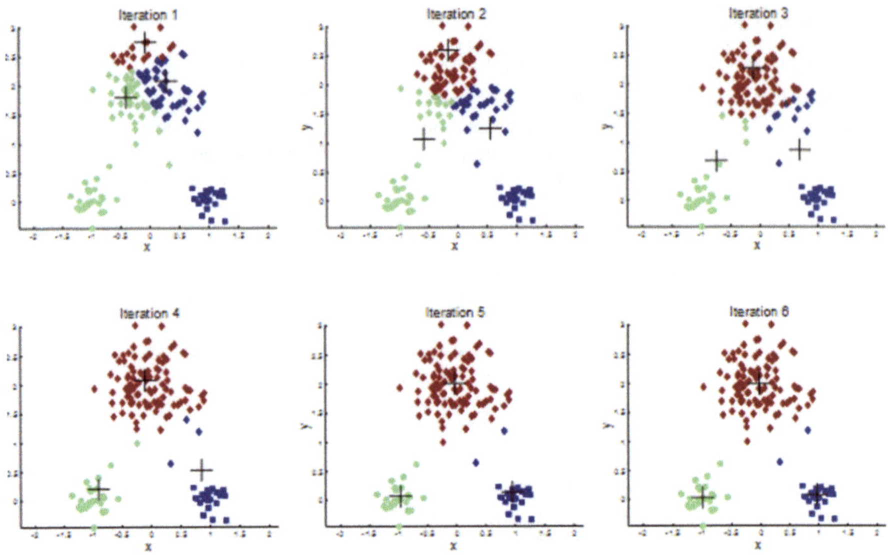

Fig. 3.3 Example of clustering partition based using k-means algorithm [2]

are capable of clustering data with different shapes, but their results worsen when the density of the data space is not uniform and depends on the input parameters. There are two approaches to density-based clustering; density-based connectivity using algorithms such as DBSCAN and OPTICS and a second type which is based on the density function, which is applied in algorithms such as DENCLUE, which also uses the influence function.

3.2.1.3 Hierarchical Clustering

These algorithms establish a hierarchical relationship of the set of elements, allowing to obtain several partitions of the groups (as opposed to partition-based clustering, where only one partition is obtained). These algorithms are divided into two types:

1. Agglomerative or bottom-up approach is initialized with an object in an independent group and in each iteration the closest groups are merged until a termination criterion is met.
2. Divisive or top-down approach is initialized with a single group containing all objects and at each iteration the groups are divided until a criterion is reached.

To perform the merging or splitting these methods can be based on several criteria such as distance or density. Most methods use distance. Depending on the distance there are several ways to determine the groups to be merged.

- **Simple Link**. When the distance between 2 groups is determined by calculating the distance between each object in the first group with all objects in the second group and selecting the minimum (this allows us to obtain clusters with a more elongated shape), it is defined as in Eq. (3.1).

$$DSL(C_i, C_j) = min\{x \epsilon C_i, y \epsilon C_j, dist(x, y)\} \tag{3.1}$$

- **Complete Link**. When the distance between 2 clusters is the longest distance between each pair of the elements that compose the groups (obtaining clusters with a more spherical shape). It is defined as shown in Eq. (3.2).

$$DSL(C_i, C_j) = max\{x \epsilon C_i, y \epsilon C_j, dist(x, y)\} \tag{3.2}$$

- **Average Link**. In this case, the distance between the groups is the average distance between each pair of objects in both groups.
- **Distance to centroid**. The center of each group (centroid) is determined and the distance between the two groups is calculated as the distance between their centroids.
- **Ward link**. Merges the two groups that account for a minimal increase in variance. This is calculated by comparing the variance of the groups before merging and after merging to find the pair of groups with the minimum increase in variance. To determine which groups to merge, the Lance-Williams formula can be used.

The Lance-Williams formula allows updating the dissimilarity matrix D after merging two groups at each iteration of the algorithm. Given the new group $C_{(i,j)}$ formed by the merger of groups i and j the dissimilarity with a group k is merging of groups i and j the dissimilarity with a group k, is calculated as in Eq. (3.3).

$$\begin{aligned} D\left(C_{(i,j))}, C_k\right) &= \alpha_i D(C_i, C_k) + \alpha_j D\left(C_j, C_k\right) \\ &+ \beta D\left(C_i, C_j\right) + \gamma \left|D(C_i, C_k) - D\left(C_j, C_k\right)\right| \end{aligned} \tag{3.3}$$

Depending on the type of bond used, the parameters of the formula are represented in Table 3.2.

Table 3.2 Parameters used according to the type of bond in the Lance-Williams formula

Methods	α_i	α_j	β	γ
Simple	1/2	1/2	0	–1/2
Complete	1/2	1/2	0	1/2
Average	$\frac{\|C_i\|}{\|C_i\|+\|C_i\|}$	$\frac{\|C_j\|}{\|C_i\|+\|C_i\|}$	0	0
Centroid	$\frac{\|C_i\|}{\|C_i\|+\|C_j\|}$	$\frac{\|C_j\|}{\|C_i\|+\|C_j\|}$	$\frac{\|C_i\|\cdot\|C_j\|}{(\|C_i\|+\|C_i\|)^2}$	0
Ward	$\frac{\|C_i\|+\|C_k\|}{\|C_i\|+\|C_i+\|C_k\|\|}$	$\frac{\|C_i\|+\|C_k\|}{\|C_i\|+\|C_i+\|C_k\|\|}$	$\frac{\|C_k\|}{\|C_i\|+\|C_i+\|C_k\|\|}$	0

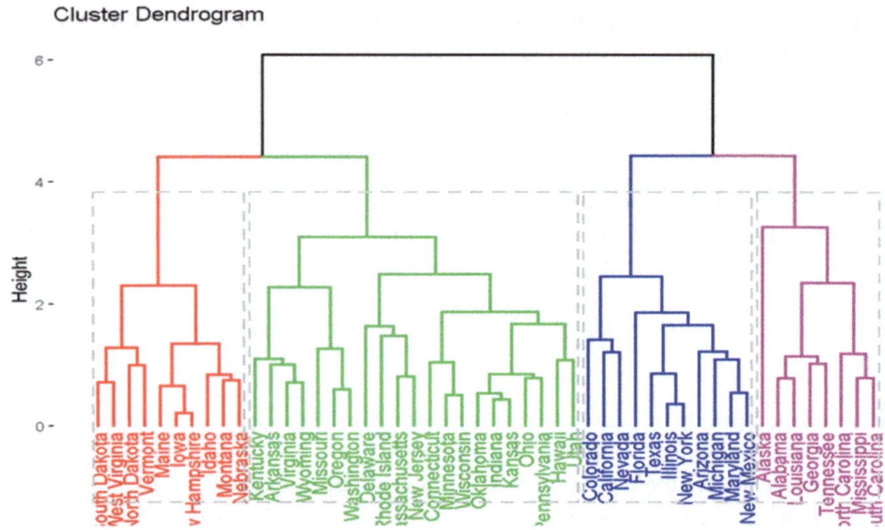

Fig. 3.4 Cluster dendrogram [3]

The main problem with these algorithms is that once a cluster is merged or split, it is not possible to go backward, which negatively affects the quality of the clustering and makes them often used in hybrid clustering approaches. Although the complexity of these algorithms is high, they are deterministic algorithms that do not require knowledge of the number of clusters nor do they require the use of a prototype, and have a high visualization capacity, allowing the representation of different clusters and their relationships using dendrograms (see Fig. 3.4). These dendrograms allow us to visualize the hierarchy of the clusters. However, they are not suitable from a moderate number of objects onwards, as the tree loses visualization capacity as the number of objects increases. Within this type, some of the best-known algorithms are Birch and Cure.

3.2.1.4 Grid-Based Clustering

These types of algorithms divide or quantize the space of the clustering elements into cells and perform a clustering of the cells (see Fig. 3.5). In other words, these algorithms focus on the data space instead of the data to perform the clustering. They tend to have low complexity and are highly scalable and can take advantage of parallel processing. However, they are sensitive to the number of cells into which the space is divided. The smaller the number of cells, the higher the counting speed, but the lower the clustering accuracy.

They consist of a series of basic steps:

1. Divide the space into a number of cells finite.

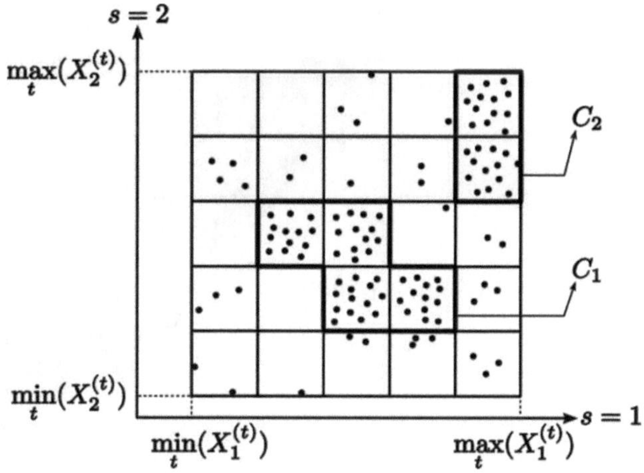

Fig. 3.5 Grid-based clustering in two dimensions [4]

2. Compute the density of each cell.
3. Classify cells by densities.
4. Specify grouping centers.
5. Traverse contiguous cells.

These algorithms do not require knowing the number of clusters in advance; however, they do require defining the number of grids and the density threshold. If the number of squares is too small, it may be the case that elements from different groups are placed in the same grid. However, the number not only increases the computational complexity but also the case of empty grids within clusters. In the case of the density threshold, the lower the threshold value, the fewer and larger clusters will occur and the fewer, larger clusters and less noise will be detected. However, if it is too high, there may be clusters or cluster elements that are identified as noise. The STING and CLIQUE algorithms fall into this category.

3.2.1.5 Model-Based Clustering

Model-based clustering algorithms attempt to recover an original model from the data, i.e. they assume a model for each of the groups and try to fit the elements to one of the models. They usually use two approaches, those based on probabilistic learning and those based on neural network learning [17, 18]. These types of algorithms require setting several parameters and are quite slow to process. Among the algorithms that are based on probabilistic learning are EM and COB WEB and those based on neural networks include ART and SOM.

In probabilistic clustering, it is assumed that the data are generated from a mixture distribution. A distribution is made up of k components, which are in turn distributions. The objective of probabilistic clustering is to obtain the mixture model to

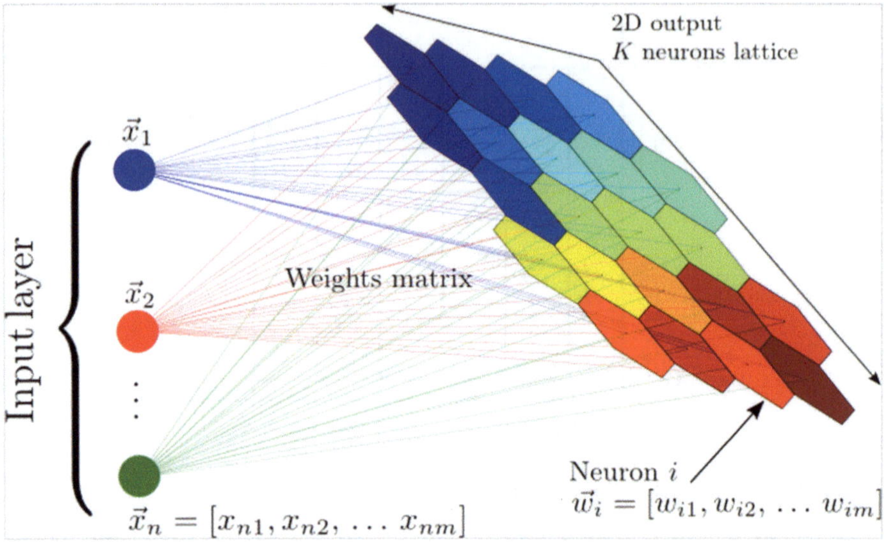

Fig. 3.6 SOM network architecture [5]

which the data belong. Probabilistic models allow the representation of subpopulations within a population. The most common component distribution for continuous data is the multi-variate Gaussian, giving rise to Gaussian mixture models (GMM).

The model based on neuronal networks are algorithms such as SOM (self-organizing maps) that consist of a single-layer neural network where the clusters are obtained by assigning the objects to be grouped to the output neurons (see Fig. 3.6). This is competitive unsupervised learning and requires as parameters the number of clusters and the grid of neurons. In a SOM network, the input layer and the output layer are fully connected. In SOMs, data are assigned to their nearest centroids and when the centroids are updated, the centroids are also updated. The objects that are close to the centroid are also updated when the centroids are updated. It thus presents a projection of the input space to a two-dimensional neuron map. It has the advantage that it is easy to visualize, one way to visualize it is through the Sammon projection.

In addition to SOM, neural network-based clustering has also been performed using learning Kohonen quantization (LVQ) vectors and with adaptive resonance models (ART).

3.2.1.6 Other Clustering Algorithms

More approaches have been used for clustering among which are the clustering based on graphs where the nodes represent the relationship between the points. Within this approach are algorithms such as CLINK, spectral clustering where a similarity graph is constructed, and then a spectral embedding is performed (applying eigenvectors of the Laplacian graph) and applying a traditional clustering algorithm or clustering based on swarm intelligence algorithms among others.

3.2.2 *Evaluation Metrics*

In general, in clustering problems the labels in the data are unknown. In this case, external indexes cannot be used, and instead internal indexes are used to measure the goodness of the clustering structure. A criterion for comparing clustering algorithms is based on 3 aspects; the way the groups are formed, the structure of the data, the sensitivity to the data, and the sensitivity to the parameters of the clustering algorithm used. The objective is to maximize the similarity within the group (cohesion) and minimize the similarity between the different groups (separation). Separation can be measured by calculating the distance between centers or the minimum distance between pairs of objects of different groups. Therefore, validation metrics are based on measuring cohesion, separation, or both.

For this, there are mainly two types of validation:

- **External Index** can be used when the truth is known (to which cluster the data belong), where the obtained solution is compared with the real one. Some external indexes are the purity of the group, the rand index, or the entropy, among others.
- **Internal indexes** do not use the ground truth to evaluate the result of the clustering process. These are based on evaluating high similarity between data of the same group and low similarity between different groups. These indexes include, among others, the silhouette index and Dunn's index.

In Table 3.3, the main validation metrics are summarized.

3.2.2.1 Internal Index Analysis

In general, in clustering problems the labels in the data are unknown. In this case, external indices cannot be used, and instead internal indices are used to compute the kindness of the clustering structure. There are eligibility criteria for comparing clustering algorithms based on 3 aspects; how groups are formed, the structure of the data, and the sensitivity to the parameters of the clustering algorithm used. The objective is to maximize the similarity within the group (cohesion) and minimize the similarity between the different groups (separation). Separation can be measured by calculating the distance between centers or the minimum distance between pairs of objects in different groups. Therefore, validation metrics are based on measuring cohesion, separation, or both.

In [11] the performance of the evaluation measures was compared in terms of various characteristics that the data may present to obtain the ideal number of the groups (for the comparison they used the K-Means algorithm, except for the

Table 3.3 Validation metrics

Internal index	They do not use the ground truth to evaluate the result of the clustering process. These are based on evaluating high similarity between data of the same group and low similarity between different groups		[11]
	Technique		
	Silhouette	It measures how well data is grouped. For this purpose, it calculates the average distance between groups. The values of this index can be represented by the silhouette plot, which shows how close each data is to the data of neighboring groups	
	Dunn	It is the relationship between the minimum separation between groups and group cohesion. The higher the Dunn index, the better the data are grouped	
	Davies-Bouldin	It should be minimized to make the groups more compact and farther apart, its main drawback is that it does not detect shapes well	
	S Dbw	It is calculated as the sum of the inter-cluster density, which is used to measure the separation between groups, and the average cluster dispersion, which is used to measure the dispersion of clusters This index checks that the density of at least one of the cluster centers is greater than the density at the midpoint of the cluster	
	Calinski-Harabasz	It is obtained as the ratio between the variance within the cluster and the variance outside the cluster. cluster. It consists of calculating the variance between the mean of each cluster concerning the mean of the whole set of data and dividing it by the sum of the variances of each cluster	
	Xie Beni	Divide the cohesion of the groups by the separation of the groups expressed as the ratio of the average distance within each group (sum of the distances of each data point, intra-cluster distance) by the minimum separation between cluster centers	
External index	They can be used when the truth is known (to which cluster the data long), where the obtained solution is compared with the real one		[12]
	Metric	Characteristics	Limitations
	Entropy (H)	It is similar to purity and is used to measure the homogeneity of the labels in the clusters obtained	It does not work well with unbalanced data. It also cannot be used to compensate cluster quality and cluster number, because the purity is high when there are more clusters

(continued)

Table 3.3 (continued)

	Purity	It is used to measure the homogeneity of the labels in the clusters obtained, that is, if the majority of objects in the group belong to the same class	Purity does not work well with unbalanced data. Also, purity cannot be used to balance cluster quality and number of clusters, because purity is high when there are more clusters
	F-score	Combines completeness and accuracy to assess clustering. It is appropriate for partitional clustering since it tends to split a large and pure cluster into many smaller disjoint partitions	F-score cannot be applied for cases where nested clustering is presented, and it cannot handle the problem of class size imbalance properly

skewed data where the experiment was performed with Chameleon) reaching several conclusions:

- **Monotonicity**: refers to how indices behave as the number of groups increases, indices that only compare one characteristic, separation, or cohesion increase or decrease steadily as the number of data increases while other indices reach a maximum or minimum when the correct number of groups is found.
- **Noise:** indexes that use minimum and maximum distances to calculate cohesion and separation are more sensitive to noise.
- **Density**: in general, most indexes work well for different data with different densities.
- **Impact of subgroups**: A subgroup is a group that is enclosed in another group where there is more than one subgroup. Indices that measure separation obtain maximums when subgroups are considered as a single group, when subgroups are considered as a single group which leads to incorrect results.
- **Skewed distributions**: When there are very large groups and very small groups, in general, most indices work well with skewed data, however, the Calinski-Harabasz index does not work well with this type of data.

The study revealed that of the indices it compared only S Dbw performed well for all of these characteristics. For arbitrary shapes, many of these measures do not perform well when measuring group separation and cohesion through the center of the group or pairs of points.

3.2.2.2 External Index Analysis

These measures can be calculated using the contingency matrix (see Table 3.4) where the columns of the matrix represent the clusters obtained and the rows are used for the labels. The cells of the matrix n_{ij} represent the number of clusters obtained and the rows are used for the class labels of the objects, thus the cells of the matrix n_{ij} represent the number of objects in the cluster j that belong to class i. of objects in cluster j that belong to class i:

- **Purity**. It is used to measure the homogeneity of the labels in the clusters obtained, that is, if the majority of objects in the group belong to the same class. To calculate it, the purity of each cluster is first calculated using the Eq. (3.4).

$$Pj = \frac{1}{n_j} x \max_i (n_j^i) \quad (3.4)$$

For example, the purity of a group j is the maximum number of objects in the cluster that belong to the same class i. Once the purity of each cluster has been calculated, we obtain the purity of the cluster purity is obtained by Eq (3.5).

$$Purify = \sum_{j=1}^{k} \frac{n_j}{n} \cdot P_j \quad (3.5)$$

where k is the number of clusters, n_j is the number of objects that have been grouped in cluster j, and n is the number of total objects.

- **Entropy (H)**. It is like purity and is used to measure the homogeneity of the labels in the clusters obtained. Both methods are frequently validated in K-Means. Similar to the purity to calculate the entropy, first the entropy associated with each cluster j is calculated with Eq. (3.6).

$$H = \sum_{i=1}^{c} \frac{n_j^i}{n_j} \times log \frac{n_j^i}{n_j} \quad (3.6)$$

and then calculate the global entropy with the Eq (3.7).

Table 3.4 Contingency matrix

	Cluster 1 … Cluster k	Σ
Clase 1 n_{11} …. n_{1k} n_{clase1}		
…….. ….. ….. ….		
Clase x n_{x1} …. n_{xk} n_{clasex}		
Σ	$n_{cluster1}$ $n_{clusterk}$	n

$$Entropy = \sum_{j=1}^{k} \frac{n_j}{n} x H_j \tag{3.7}$$

- **F-Score**. This measure combines completeness and accuracy to assess clustering. The completeness, defined as the ratio of objects of class i in cluster j to the total number of objects of class i,
- $\boldsymbol{Recall(i, j)} = \frac{n_j^i}{n^i}$, while precision is defined as the ratio between the objects of class i in cluster j and the total number of objects in cluster j
- $\boldsymbol{Precsision(i, j)} = \frac{n_j^i}{n_j}$, The higher the F-values of the clusters obtained, the better the clustering. The F-value is calculated as in Eq. (3.8).

$$Valor - F = \sum_{i=1}^{k} \frac{n_j}{n} \max \frac{2x\text{Recall}(i, j).\text{Precsision}(i, j)}{\text{recall}(i, j) + \text{Precsision}(i, j)} \tag{3.8}$$

3.3 ANN and CNN Models Integrated with SMOTE

Once the asymmetric dataset is obtained, the model can be connected to the SMOTE module. An unbalanced dataset can have many causes. Perhaps the target category has one set of data in the population, or the data is complicated to collect. You can ask SMOTE for analyzing an underrepresented category. The output of the module contains the original sample and additional samples. These new samples are composite minority samples. Before starting the technique, you must determine the number of these synthetic samples.

If the classification of information is not the same, we can talk about unbalanced information. This is a classification task and results in several problems with the model output. For example, a binary classification task has 100 instances. Class 1 contains 80 marked specimens. On the other side, the remaining labeled sample is class 2. This is a simple example of an unbalanced dataset. The ratio of 1st class to 2nd class would present 4:1.

If we talk about real test data or Kaggle competitions, the problem of class imbalance is very common. A real classification problem implies some classification imbalance. This usually happens when there is no matching profile for the category. Therefore, it is important to choose the right evaluation metric for your model. If the model has an asymmetric data set, its output is meaningless. But when this model is applied to real problems, the end result is waste. There is always class imbalance in different situations. A good example is looking at fraudulent and non-fraudulent transactions. You will find fraudulent transactions. This is the problem.

3.3.1 Oversampling Data: SMOTE

If we talk about real test data or Kaggle competitions, the problem of class imbalance is very common. A real classification problem implies some classification imbalance. This usually happens when there is no matching profile for the category. Therefore, it is important to choose the right evaluation metric for your model. If the model has an asymmetric data set, its output is meaningless. But when this model is applied to real problems, the end result is waste. There is always class imbalance in different situations. A good example is looking at fraudulent and non-fraudulent transactions. You will find fraudulent transactions. This is the problem.

Here are some of the benefits of SMOTE:

- Information is kept.
- This technique is simple and can be easily understandable and implemented in the model.
- This improves overfitting for synthetic examples. This helps to create new instances instead of copying.

Dup_size and K are two parameters of SMOTE. If you want to understand Dup_size and K, you need to learn how SMOTE works. SMOTE works from the perspective of existing cases and creates new ones randomly. The function creates a new instance at a specified distance from a neighboring instance. However, it is not yet clear how SMOTE will treat its neighbors in each established minority.

- The function considers the nearest neighbor at $K = 1$.
- The function considers nearest neighbor and nearest neighbor at $K = 2$.

Often, SMOTE initially experiences minority manifestations. Although loop iteration is an instance, the pattern creates a new instance between the original instance and its neighbors. The dup_size parameter specifies the number of times the SMOTE function will loop over the first instance. For example, if $dup\_size = 1$, the model will only synthesize four new data points, and so on. Finally. When building predictive models in ML, you may encounter non-balanced datasets. This data affects the output of the model. This problem can be solved by oversampling a small amount of data. So instead of generating duplicate data, use the SMOTE algorithm and generate synthetic data for oversampling. Here are some variations of SMOTE.

- Borderline-SMOTE
- SMOTE-NC.

3.3.2 Examples of Using SMOTE with ANN and CNN

A study by Naung et al. [13] is an example of the use of ANNs and SMOTE in simple ANN-based DDoS attack detection using SMOTE for IoT environments. In recent years, with the rapid development of the IoT era, attackers have mainly targeted the

Internet of Things environment. They optimize his Internet of Things devices as bots to attack target organizations, but due to limited resources to manage effective defense mechanisms on these devices, these devices are easily infected with IoT malware. Highly dangerous Internet of Things malware such as Mirai conducts DDoS attacks against targeted organizations using infected Internet of Things devices. Although many security mechanisms have been implemented in IoT devices, there is still a need for effective Internet of Things environment sensing systems. This detection system uses public datasets, machine learning techniques and a simple artificial neural network (ANN) architecture to detect such attacks. Bot Internet of Things, a modern botnet attack dataset, is used to detect DDoS attacks, but the dataset contains a small amount of benign data and a large amount of attack data, which makes it difficult to detect inaccurate data. Big issues like balance need to be addressed. In this work, Synthetic Minority Oversampling Technique (SMOTE) is used to solve the data imbalance problem to implement his machine learning based DDoS detection system. The results show that the proposed method can effectively detect DDoS attacks in Internet of Things environments.

In a study by Joloudari et al. [14], an efficient imbalance learning based on SMOTE and convolutional neural network classes, SMOTE is used to solve imbalanced data sets. Data imbalance (ID) is a problem that prevents machine learning (ML) models from achieving satisfactory results. ID is a situation where the number of samples belonging to one class significantly exceeds the number of samples belonging to another class. In this case, the learning of this state will be biased towards the majority class. In recent years, several solutions have been proposed to solve this problem, choosing to generate new synthetic configurations for minority classes or to reduce the number of majority classes to balance the profiles. Therefore, in this study, the effectiveness of methods based on a hybrid of deep neural networks (DNNs) and convolutional neural networks (CNNs) as well as several well-known solutions for unbalanced data involving oversampling and undersampling are investigated. Then, together with SMOTE, a CNN based model for efficient processing of unbalanced materials are presented. For evaluating the method the KEEL, Breast Cancer, and Z-Alizadeh San datasets are used. To obtain reliable results, 100 experiments using randomly shuffled data distributions are perfomed. The classification results show that Hybrid SMOTE outperforms various methods in normalized CNN and achieves 99.08% accuracy on 24 unbalanced datasets. Therefore, the proposed hybrid model can be applied to non-balanced binary classification problems in other real datasets.

3.4 Unsupervised Learning in Anomaly Detection in Practice

The purpose of this section is to detect anomalies in the data flow of a software-defined network. Initially, an unsupervised dataset is available, with different observations on the traffic flow of the software-defined network, which is examined and analyzed. For this purpose, feature engineering is employed on the set, using certain technologies, applying a transformation on the data, and obtaining as a result a valid set for analysis in the following phases. Once this phase of the data has been built, machine learning algorithms to be used are studied. Subsequently, the best combination of parameters to be applied to these algorithms is sought, comparing them with each other and generating the most optimal models possible, which can group the data samples with similar characteristics and obtain a valid set for analysis in the following phases and detect anomalies in the flow, thus meeting the established objectives. Through the models, we evaluate the results obtained with the scores of the different internal metrics selected. Finally, a comparison of the algorithms used, based on the results, execution times, and ease of understanding, highlights the most optimal and efficient one.

3.4.1 Method

The method to follow as shown in Fig. 3.7 consists of the following steps:

1. Data collection
   - Collection, description, and exploration of the data.
   - Verification of data quality.
2. Data preparation
   - Construction of the final data set encompassing all the necessary activities of data selection, cleaning, construction, integration, and formatting.
3. Modeling
   - Determination of evaluation metrics.

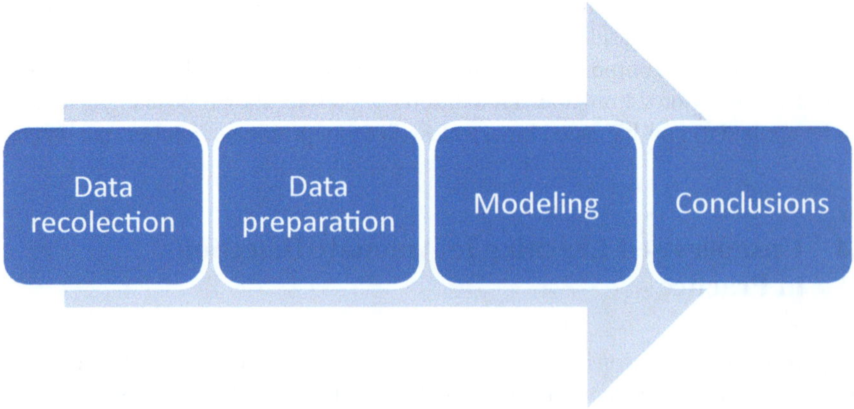

Fig. 3.7 Method

- Determination of hyperparameters.
- Creation of the different models.
- Evaluation of the results of each model.

4. Conclusions

- Consideration of the results obtained against the established objectives.
- Conclusions and lessons learned.

3.4.1.1 Data Collection

The dataset used for this project is a dataset already generated and downloaded from the following link [7]. It consists of three files with.csv extension, two of which contain attack data traffic (OVS.csv and metasploitable-2.csv) and the third of normal data traffic (Normal_data.csv). normal data traffic (Normal_data.csv). The first two are considered attacks on the OVS and attacks targeting the Metasploitable-2 server, respectively. These three files are put together to form a uniform data set. For this purpose, the Python library is used. This allows the data to be stored in an object called DataFrame, thus being able to form the data set. It allows working with large volumes of data, providing facilities when querying any column, row, or specific data. This set results in 343,889 annotations of data traffic flow, corresponding to the DataFrame flags, of which 138,722 belong to the OVS.csv file, 136,743 to the metasploitable-2.csv file and 68,424 to the Normal_data.csv file, 136,743 to metasploitable-2.csv and 68,424 to Normal_data.csv. In addition, it contains 84 features, corresponding to the columns of the DataFrame. The dataset used is public and attack-specific. It aims to provide its practice for anomaly detection systems applied in SDN networks, to verify the performance of intrusion detection systems. It contains categories of benign attacks, as well as different situations that can occur in the SDN platform scenario.

3.4.1.2 Data Preparation

After scanning the data, this section will check the quality of the data. To do this, we first check that the data set is complete by examining that it does not have any null values. In addition, it is also checked that it does not present variables with values such as "NaN". For this purpose, different techniques are applied, such as the function is null (), and isna() from the Pandas library or even with a heat map from the seaborn() library.

The purpose of this task is to generate, from the originally captured data, derived attributes, new records, or transformed values of existing attributes preparing the input to the modelling tools according to the requirements. The objective is to transform all variables into numerical. For example, this operation is performed for the source and destination port variables, 'Src Port' and 'Dst Port'. In this case, although the port values are divided into three ranges, two variables are generated for the source port and two for the destination port, applying the same pattern as for the IP addresses. as for the IP addresses. This optimizes the data set and avoids unnecessary correlations.

3.4.1.3 Modelling

When starting with the construction of the model, we start from a base in which the dataset used is completely unlabelled. For this purpose, different tests and runs of the of the different selected algorithms, exchanging the different parameters of each one of them. This phase is called hyperparametrization since the values used to configure the model are called hyperparameters. This term is defined as adjustable parameters that allow control of the training process of a model. They are values that are generally not obtained from the data,

Since the optimal value is unknown, it is necessary to use generic values, values that have worked correctly in similar problems, or to find the best or to find the best option based on trial and error. While, on the other hand, the parameters are the variables that are estimated during the training process with the data sets. Therefore, these values are obtained, and not provided manually.

Selected algorithms are:

1. K-means
2. DBSCAN
3. SOM

The validation techniques selected are:

- Silhouette
- Davies Bouldin
- Calinski and harabasz

K-means [6]. The function used to run this algorithm is provided by Scikit-Learn,

KMeans. The parameters to be taken into account, considering the most important ones of this algorithm are:

- n_clusters: represents the number of clusters to be formed, as well as the number of centroids to be generated. A range of values between 2 and 5 is provided, through a for loop, to perform several executions varying these values. The values provided are low since the objective is to obtain a low number of differentiated data sets.
- init: Represents the initialization method. It admits different values:

  k-means $++$: selects the initial cluster centers for clustering intelligently to speed up convergence.

  random: chooses random observations or flasks from the data for the initial centroids.

  Tests are performed considering both values.

- random_state: Determines the random number generation for centroid initialization. A seed is set to generate pseudo-random numbers, through numpy's random.seed() function. By providing the same seed, we obtain the same set of random numbers, allowing us not to vary the results in each identical run.

Table 3.5 shows k-mean results for the metrics selected.

Analyzing the results of the different runs, it can be deduced that the number of clusters provided is quite accurate since optimal scores are obtained in all metrics. That conclusion is taken due to the interpretation of the values of the different metrics, where the value of the scores for silhouete Calinski and Harabasz is quite high, in particular for the former, since that value is close to the most optimal limit. As for the results of the Davies Bouldin score, they follow the same line of this algorithm, since it is close to the most optimal limit, zero, presenting values very close to it. As for the inertia value, it is quite high, since the Euclidean distance tends to inflate in very high dimensional spaces, as is this case. Therefore, it is concluded that the init parameter in this case is irrelevant since similar results are obtained for the same number of clusters, making this parameter the most important one.

DBSCAN [15]. The function used to execute this algorithm is the one provided by Scikit-Learn, DBSCAN. The parameters to take into account considered the most important of this algorithm are:

- eps: Maximum distance between two samples for one to be considered in the neighborhood of the other.
- min_samples: Number of samples (or total weight), including the same point, in a neighborhood for a point to be considered as a central point.
- n_jobs: The number of jobs to run in parallel.

Table 3.6 shows DBSCAN results.

Table 3.5 K-Mean results

Init	Time (s)	Inertia	Silhouette	Davies bouldin	Calinski and harabasz	Clusters
Random	4.967	200,227,794,626,097,971,201	0.945	0.408	1,268,786.35	2
k-means ++	5.698	200,227,794,626,097,971,201	0.945	0.408	1,268,786.35	2
Random	6.801	89,389,001,485,593,477,122	0.958	0.521	1,634,208.40	3
k-means ++	7.622	89,389,001,485,593,477,123	0.957	0.521	1,634,208.40	3
Random	9.765	74,880,926,457,495,011,324	0.952	0.775	1,322,766.48	4
k-means ++	10.242	65,572,402,736,387,260,417	0.961	0.717	1,526,808.86	4
Random	16.282	69,190,860,588,814,958,593	0.951	0.699	1,080,726.95	5
k-means ++	9.731	53,282,084,568,454,586,367	0.952	0.720	1,429,067.62	5

Table 3.6 DBSCAN results

Time	Silhouette	Davies bouldin	Calinski and harabasz	Clusters
35.068 s	0.706	1.558	5334.08	148
22.024 s	0.636	1.547	4264.62.76	143
5.606 s	0.936	276	4305.37	28

To obtain the results of this algorithm, several difficulties have been encountered. The main one has been the size of the data set used since this implementation massively calculates all the neighborhood queries, and therefore, it increased the complexity of the memory, so much so that the execution process could not be carried out. Several solutions have been tried to obtain good results. The first one has been, an estimated adjustment of the hyperparameters mentioned above, even performing the famous "elbow technique", which technique, to provide a reasonable value for the EPS parameter. In addition, solutions proposed by Scikit have been tested, solutions proposed by Scikit-learn have been tested, such as pre-calculating the sparse neighbourhoods in fragments and thus using the metric with a 'precomputed' value. The way used to obtain good results, as shown in Table 3.5, has been to reduce the data set by 40, 30, and 10%, respectively which is not an optimal way, but to demonstrate that the algorithm is very efficient with smaller data sets. PCA algorithm has been used for reducing the dimensionality.

SOM. In this case, the function used to execute this algorithm is Minisom provided by the Minisom library [16]. The parameters to be considered as the most important of this algorithm are:

- x: Dimension x of the SOM.
- y: Dimension y of the SOM.
- input_len: Number of the elements of the input vectors. The number of features of the dataset used is provided.
- random_seed: Random seed to be used. Set in the same way as in the previous algorithms.

First, it is worth mentioning the dimensionality reduction performed by applying the PCA algorithm previously. The results obtained from this process are identical to the previous algorithm since it is applied directly to the initial data set. As for the performance obtained for this algorithm, it is shown in Table 3.7 through the results for the different metrics. The first column 'shape' indicates the size of the algorithm dimensions. This variable is the most important since at first glance it can be seen that as the size of the dimensions increases, the scores of the metrics shown in the following columns improve considerably. That is to say, for Silhouete, Calinski, and Harabasz the values increase according to reaching their most optimal values, while for Davies Bouldin the values decrease in seeking the same way to reach the most optimal value. In view of this reasoning, it is worth noting that as the size of the dimensions increased, the execution times intensified, being not very efficient and

Table 3.7 SOM results

Shape	Time	Silhouette	Davies bouldin	Calinski and harabasz	Clústeres
10 × 10	61.586 s	0.270	1.220	9128.90	95
100 × 100	2939.756 s	0.480	0.971	8361.95	990
300 × 300	81,631.615 s	0.572	0.991	33,747	2073

very costly to obtain results. On the other hand, the size of the clusters obtained varies proportionally to the established dimensions. Table 3.7 shows SOM results.

3.4.1.4 Conclusions

In this last phase of the methodology, an evaluation and critical analysis of the models created in the previous phase will be carried out. It compares the different algorithms used, basing the comparison on execution times, ease of understanding, and results obtained for the scoring of the different metrics used. With all the information provided from the previous phase, the k-means algorithm is taken as a reference. The k-means algorithm is taken as the one that provides the best results, as it is also the one that requires the shortest execution time. On the other hand, it is the easiest algorithm to understand due to its simplicity and low variable modification necessary to obtain optimal results.

Continuing with the SOM algorithm, the results are not entirely optimal. It should be noted that by increasing the dimensions of the algorithm, the results are optimized, but on the other hand, the execution times increase. These have been the worst of all the models used because the scikit-learn library does not have an implementation of this algorithm, as well as the dimensions of the data provided. Finally, it should be noted that it is quite simple to understand both the operation and the parameters to set in the function.

Finally, DBSCAN, as discussed in the previous phase, has yielded results out of the reach of the efficient and optimal, in terms of metric scores. The complexity of understanding the parameters to be set should be highlighted, since in min_samples it is advisable to have prior knowledge of the subject, as for eps, which would facilitate the choice of the values of the same, since, to adjust these values with large amounts of data, the algorithm can be heavy and quite inefficient.

After a final evaluation of the generated models, the K-Means model is the one that best fits the project objectives, grouping the network traffic of the data set into a compact and homogeneous number of clusters. In addition, it is best suited to large amounts of data, without increasing execution times too much, so it could be used in any field that requires data analysis and more specifically to detect anomalies in data traffic. It is also easy to understand and implement, which is always gratifying.

On the other hand, it is interesting to note that other algorithms such as DBSCAN can also be used in similar domains to the one developed in this work due to their high efficiency in terms of clustering observations into similar groups, although its efficiency improves in smaller amounts of data than those presented in this work.

3.5 Development Frameworks

To work with time series there are currently several libraries and resources that allow the preparation of data, the use of algorithms such as those indicated in this document, as well as the elaboration of mathematical models. These include Datetime, Pandas, Matplotlib, MatrixProfile, Numpy, Ruptures, Plotly, Stlearn and Sklearn.

- Datetime. It is a module that allows the manipulation and management of dates.
- Pandas. It is a Python package that allows you to work with structured data, creating fast, adaptable, and expressive data structures.
- Maplotlib. It is a library that contains a wide variety of graphics and allows the creation of two-dimensional graphics.
- MatrixProfile. Provides accurate and approximate algorithms for calculating the matrix profile of a time series, as well as for determining discords and motifs in the time series from it, and tools for visualizing the results.
- Numpy. It is a package that provides general-purpose array processing, i.e. a high-performance multidimensional array and methods for handling them that allow for easy computations.
- Ruptures. It is an offline change point detection library that provides approximate and accurate detection for parametric and non-parametric models.
- Plotly. A library for interactive visualisation with a wide variety of advanced graphics.
- Tslearn. It is a Python package for machine learning of time series. Among its many modules are time series metrics, including DTW and variants; a clustering module including K-means; a reprocessing module, including time series representations such as PAA and SAX; and a Shapelet-based algorithm package that requires Keras.
- Sklearn. Classification, regression, clustering, dimensionality reduction, and preprocessing algorithms (such as standardization and normalization) are included in this open-source library. It also includes techniques for comparing, validating, and choosing parameters for models. In addition to internal indices such as the silhouette coefficient, Calinski-Harabasz, and the Daves-Bouldin index, it includes clustering algorithms such as K-Means, affinity propagation, mean shift, spectral clustering, the Ward method, agglomerative clustering, and Gaussian and Birch mixtures.

References

1. Kibish, S.: A note about finding anomalies [Internet]. Medium. (2018). [Visited 23 May 2023]. Available on https://towardsdatascience.com/a-note-about-finding-anomalies-f9cedee38f0b
2. Berzal, F.: Partition based clustering. [Visited 23 May 2023]. Available on https://elvex.ugr.es/idbis/dm/slides/41%20Clustering%20-%20Partitional.pdf
3. Isaac, J.: Cluster jerarquico. (2021). [Visited 23 May 2023]. Available on https://rpubs.com/jaimeisaacp/760355
4. Bandaru, S., Kalyanmoy, D.: Towards automating the discovery of certain innovative design principles through a clustering-based optimization technique. Eng. Optim. **43**, 911–941 (2011). https://doi.org/10.1080/0305215X.2010.528410
5. Sancho, F.: Self Organizing Maps (SOM) in NetLogo. (2021). [Visited 23 June 2023]. Available on https://www.cs.us.es/~fsancho/?e=136
6. K-means.: [Visited 13 November 2023]. Available on https://scikit-learn.org/stable/modules/generated/sklearn.cluster.KMeans.html
7. DATASET.: [Visited 13 November 2023]. Available on https://aseados.ucd.ie/datasets/SDN/
8. DBSCAN.: [Visited 13 November 2023]. Available on https://www.kaggle.com/code/meetnagadia/dbsc [Visited 13 November 2023]. Available on: an-clustering
9. SOM.: [Visited 13 November 2023]. Available on https://www.kaggle.com/code/asparago/unsupervised-learning-with-som
10. Masich, I., Rezova, N., Shkaberina, G., Mironov, S., Bartosh, M., Kazakovtsev, L.: Subgroup discovery in machine learning problems with formal concepts analysis and test theory algorithms. Algorithms **16**, 246 (2023). https://doi.org/10.3390/a16050246
11. Liu, Y., Li, Z., Xiong, H., Gao, X., Wu, J.: Understanding of internal clustering validation measures. In: 2010 IEEE international conference on data mining, Sydney, NSW, Australia, pp. 911–916 (2010). https://doi.org/10.1109/ICDM.2010.35
12. Kashef, R.: Scattering-based quality measures. In: 2021 IEEE international IOT, electronics and mechatronics conference (IEMTRONICS), Toronto, ON, Canada, pp. 1–8 (2021). https://doi.org/10.1109/IEMTRONICS52119.2021.9422563
13. Soe, Y.N., Santosa, P.I., Hartanto, R.: DDoS attack detection based on simple ANN with SMOTE for IoT environment. Fourth International Conference on Informatics and Computing (ICIC) **2019**, 1–5 (2019)
14. Joloudari, J.H., Marefat, A., Nematollahi, M.A., Oyelere, S.S., Hussain, S.: Effective class-imbalance learning based on SMOTE and convolutional neural networks. Appl. Sci. **13**, 4006 (2023). https://doi.org/10.3390/app13064
15. DBSCAN.: [Visited 13 November 2023]. Available on https://scikit-learn.org/stable/modules/generated/sklearn.cluster.DBSCAN.html
16. MiniSOM.: [Visited 13 November 2023]. Available on https://pypi.org/project/MiniSom/
17. Jan, A., Muhammad Khan, G.: Real world anomalous scene detection and classification using multilayer deep neural networks. Int. J. Interact. Multimed. Artif. Intell. **8**(2), 158–167 (2023). https://doi.org/10.9781/ijimai.2021.10.010
18. Deore, M., Kulkarni, U.: MDFRCNN: Malware detection using faster region proposals convolution neural network. Int. J. Interact. Multimed. Artif. Intell. **7** (4), 146–162 (2022). https://doi.org/10.9781/ijimai.2021.09.005

Chapter 4
Profiling and Classification of IoT Devices for Smart Home Environments

Sudhir Kumar Das, Sujit Bebortta, Bibudhendu Pati, Chhabi Rani Panigrahi, and Dilip Senapati

Abstract The goal of this study is to create a strong categorization system specifically designed for Internet of Things (IoT) device profiling. The main goal is to supplement current studies that use a wide range of machine learning techniques to identify anomalous behavior in Smart Home IoT devices with an exceptionally high accuracy rate. The intended framework is positioned to play a crucial function in bolstering IoT security in the future because it is made to include several types of abnormal activity detection. Our technological motivation stems from IoT smart sensors' high processing power and advanced connectivity capabilities. Notably, these sensors have the potential to be manipulated for malicious purposes only on a single sensed data point rather than the complete collection of collected data from sensors, such as temperature, humidity, light, and voltage measurements. Such a threat lowers the detection effectiveness of many machine learning algorithms and has a substantial impact on the accuracy of aberrant behavior detection. To identify occurrences of alteration in one specific data point among the four potential data points collected by a single sensor, we compared and used different classifiers in our investigations, including the Decision Tree Classifier, KNeighbors Classifier, Support Vector Classifier (SVC), Logistic Regression, AdaBoost Classifier, Random Forest with Extreme Gradient Boost (XGBRF) Classifier, Random Forest Classifier, Light Gradient Boosting Machine (LGBM) Classifier, Gradient Boosting

S. K. Das · S. Bebortta · D. Senapati (✉)
Department of Computer Science, Ravenshaw University, Cuttack, Odisha 753003, India
e-mail: senapatidillip@gmail.com

B. Pati · C. R. Panigrahi
Department of Computer Science, Rama Devi Women's University, Bhubaneswar, Odisha 751022, India
e-mail: patibibudhendu@rdwu.ac.in

C. R. Panigrahi
e-mail: panigrahichhabi@rdwu.ac.in

D. Senapati
Department of Computer Science, University of Delhi, Delhi, Delhi 110007, India

J. Nayak et al. (eds.), *Machine Learning for Cyber Physical System: Advances and Challenges*, Intelligent Systems Reference Library 60,
https://doi.org/10.1007/978-3-031-54038-7_4

Classifier, and XGB Classifier. The results showed that the Gradient Boosting Classifier algorithm using random search attained an 85.96% detection accuracy, indicating a somewhat lower vulnerability to such changes. As a result, the Gradient Boosting Classifier algorithm with random search was the foundation for the carefully constructed suggested framework, which used four hyperparameter tuning mechanisms for comparison.

Keywords Internet of Things · Device profiling · Smart homes · Machine learning · Attack identification

4.1 Introduction

The term "Internet of Things" (IOT) refers to networks of physical objects and products that are integrated with electronics, sensors, actuators, software, and connections to allow for data exchange and communication between them. The IoT is currently the most extensively utilized technology, with projections indicating its continued dominance until 2030, there will be more than 25.4 billion linked IoT devices globally. Due of its widespread prevalence, the COVID-19 epidemic has contributed significantly to the rapid development of IoT technology.

The Internet of Things (IoT), which refers to a growing number of technical devices connected to the Internet, brings about new modern conveniences. The way we link IoT devices to our living spaces is expected to undergo a transformation thanks to the constantly expanding variety of smart, high-tech items available today. IoT helps and benefits us in practically every aspect of our life. The Internet of Things (IoT) gradually integrates into our daily lives. IoT devices have been making their way in a variety of industries recently, including residential and commercial applications.

To take advantage of the greater capacity to be aware of and control important characteristics of their houses, many individuals are setting up domestic devices and IP-enabled gadgets in their homes. However, there are numerous media stories regarding IoT devices that are installed in consumer residences and other living spaces that have security flaws that might be used by attackers. IoT device suppliers should be able to release timely fixes to address vulnerabilities, but they appear to be unable or unwilling to do so. This would be the best way to deal with IoT device vulnerabilities. A large number of IoT users lack the necessary knowledge or motivation to carry out such procedures, or they may forget about unattended IoT devices that were previously put in their network, leaving them with software that is outdated. Upcoming safety measures for IoT technology must take into account the possibility of unpatched IoT devices coexisting alongside other Internet of Things (IoT) devices during their entire lifecycle in the user's network and posing dangers.

A huge number of installed devices in advance IoT-connected smart city scenarios are widely accessible and, as a result, their physical security is of utmost significance. The main security vulnerabilities are those connected with poor physical protection, such as simple gadget disassembly, unauthorised access to device records and data, and portable storage media [13]. Because of this, despite the various conveniences and adaptability benefits they provide, they also pose a number of security risks and issues [14]. Separating the devices and forbidding connectivity to different IoT devices via a gateway is a key factor in attack prevention for IoT devices. Given these security issues, effective identification of devices is likely a preferable strategy for administering networks than device isolation.

The cause of received communications on the server that leads to determine thefts is the fundamental difficulty with device-side authorization. Using a document called a certificate, which may be faked, is one option. Fingerprinting devices may be the best option for allowing network managers to automatically detect linked IoT devices [12]. The fingerprinting procedure involves identifying a certain type of equipment from its personal network data from a distance. For the purpose of to safeguard and maintain the network, network administration has to be aware of the devices connected to the system. Network managers needed a better knowledge of the linked devices as more Internet of Things (IoT) gadgets are added to a network. Device profiling, which is a strategy to continuously perform identifying a device or detection by taking into account behavioural aspects, is comparable to device fingerprinting.

4.1.1 Motivations

Many more organisations now enable IOT devices to connect to their networks, which might put such networks at risk for security. In order to determine which devices are connected to their networks and if they are deemed legal and do not constitute a risk, organisations must be able to identify these devices.

In past studies, it has become more common to use network data to identify devices in general. Particularly, there is growing interest in the field of identifying IOT devices since it is crucial to do so in an organisational setting (particularly in terms of security).

This study aims to address the issue of identifying an IoT device by utilizing machine learning techniques to analyze its high-level network traffic data. We want to provide a mechanism for locating such a device, even if its IP address has been spoofed (which is simple to achieve), and to be able to see any unusual behaviours that would point to the device that is being used. We would want to analyse the traffic's high-level data (that is, the metadata and traffic statistics, rather than analysing the content), as we can't rely on the IP address to identify the device (since this number might be faked).

The topic that we want to tackle in this study is fundamentally a multi-class issue. We'll make use of a dataset gathered from 10 various IOT devices. The dataset includes details on these devices' network traffic. The strategy we'll employ in this study is to identify the device based on a specific traffic session or series of sessions. For each device, we will start by developing one-vs-rest classifiers, and we will keep going until we are able to discriminate between every type of device.

4.1.2 Contribution

To identify devices at the device type level, the suggested solution uses a cross-layer methodology that includes network, data connection, transport, and application data. To limit IoT devices' access to sophisticated features accessible to conventional devices such as laptops and smartphones, the fundamental concept is to analyse and recognise the distinctive behaviour patterns of IoT devices. IoT devices continue to be vulnerable as possible network weak spots despite these precautions, though. It would not stop, for instance, the monitoring and control of IoT devices in a home network, such as the use of a camera to trigger an action like opening a garage door. Additionally, the adopted regulation gives all IoT devices the same degree of capabilities without distinction.

4.2 Literature Survey

This study proposes a novel method for recognising and categorising devices that incorporates cutting-edge machine learning techniques. In particular, it offers a ground-breaking framework for big data-based traffic categorization that is extendable, distributed, scalable, and portable. The study also suggests a distributed approach for processing real-time ingress IoT flow streams that makes use of H2O.ai. This technique effectively fulfills crucial requirements such as on-demand scaling, storage capacity, computation dissemination, latency, and privacy. The study proposes the method, which categorises IoT devices based on their behavioural traffic characteristics. The input dataset is composed of flow entries extracted from the incoming network traffic for training the model. The learning algorithm set employed consists of MMetaModel, XGBoost, DRF, GBM, and GLM. The study greatly advances device identification and classification methods in the IoT space by utilising this thorough and complex methodology.

An additional.pcap file with 802,582 packets in binary format from 17 distinct devices was used to verify the efficacy of the suggested fix. The framework's examination revealed exceptional performance, obtaining a remarkable accuracy rate of 99.94%. Additionally, the solution showed good performance metrics for F1 score, Precision, and Recall. These findings demonstrate the solution's potential to successfully answer concerns about cyberattacks and open the door for the creation of autonomous defence systems. The framework has a lot of promise for battling cybersecurity threats and developing resilient defence systems due to its high accuracy and robust performance.

The solution IDentifier (SysID) solution, which specialises in IoT device fingerprinting, is introduced in this research article. With just one packet, SysID can successfully identify the device type. SysID employs machine learning and genetic algorithms to independently acquire knowledge about the unique characteristics of each IoT device, in contrast to conventional techniques that need professional supervision. This method illustrates the supremacy of rule-based algorithms, which excel in capturing distinctive header traits and precisely analysing attribute values utilised for classification. SysID distinguishes out as a flexible, network-adaptive model-based technology. The three stages of this study's development were: defining the research topic; setting up a lab environment with SHIoT (Smart Home IoT) devices; and, finally, preprocessing the data gathered and creating a classification model. Based on the traffic flow characteristics amassed over a period of 10 days for each SHIoT device, the categorization model was built. The initial dataset had 681,684 feature vectors spread across four classes, however it was discovered that this distribution was out of balance. In order to overcome this, stratification techniques were utilised, producing a dataset with 117,423 feature vectors that was then used to create further models.

It was decided that the Precision-Recall Curve (PRC) metric was superior than the Receiver Operating Characteristic (ROC) measure. The M4 model emerged as the best option after several observed models were analysed since it performed more consistently than the others. It was discovered that the characteristics of the observed traffic flow that had the most influence on the classification model were the packet length, interarrival packet timings, segments within the traffic flow, and the amount of data transferred inside the sub-stream. These characteristics were essential for correctly categorising SHIoT devices.

Federated learning (FL) is becoming increasingly important in this field as machine learning (ML) and deep learning (DL) approaches are used to discover cybersecurity vulnerabilities in IoT systems. Realistic splitting tactics, like those reported in the MedBIoT dataset, may be used in FL approaches, which call for datasets that correctly represent cyberattacks directed at IoT devices. However, the split of centralised databases must be taken into account by current FL-based systems. This component's goal is to train a federated ML model for malware detection using four different multilayer perceptrons (MLP) and autoencoder architectures.

The aggregation function is considered as a parameter in two FL methods, Mini-batch aggregation and Multi-epoch aggregation. The supervised solution is used primarily to compare the supervised method to the unsupervised approach and to carry out extensive experiments that are in-depth in nature. Three separate challenges are studied and analysed as a result of the three various methods the dataset is rebalanced.

This study proposes a unique method called HFeDI that uses horizontal federated learning with privacy protection to identify IoT devices. Three publicly accessible datasets are used to assess the effectiveness of HFeDI, with encouraging findings. When using centralised training, the 23 features along with the 2 additional features discovered by Miettinen et al. [39], were shown to offer the greatest accuracy. The output data of the feature extractor tool is enhanced through the utilization of SK resampling, a resampling method, to improve the quality of the data. By employing kaiming weight initialization, group normalisation, a loss function that incorporates weights to calculate the cross-entropy, and a straightforward averaging technique at the server, HFeDI substantially enhances the efficiency of IoT device identification. For cases involving both independent identically distributed (IID) and non-independent and identically distributed (non-IID) data, the findings show a significant improvement in accuracy, recall, precision, and F1-score. These results demonstrate the efficacy and promise of HFeDI in improving IoT device identification while preserving privacy using federated learning techniques.

We employ machine learning (ML) techniques and encrypted analysis of traffic to address the issue of identifying IoT devices based on their unique characteristics. This study utilized the dataset given by the University of New South Wales and IBM Research. A TP-Link router, which serves as a connection point to the public Internet, is equipped with the OpenWrt operating system and other essential packages, enabling the collection of traffic in pcap files to record pertinent actions. Then, in order to extract useful characteristics, these files are analysed. Exploration Evaluation is used to find the most effective classifier estimators in order to maximise training. This method aids in choosing the best models for the task at hand. Additionally, a comparative evaluation employing a variety of predictive measures is carried out to compare these classifiers to a baseline. These assessments make it possible to evaluate the performance of the classifiers in-depth.

The testing set is then used to evaluate and verify how well the chosen classifiers perform in comparison to the metrics and benchmarks that have been created. In order to accurately map encrypted data streams to the appropriate device types, this guarantees thorough study and evaluation of our ML-based encrypted traffic analysis technique.

A robust security solution called IOT SENTINEL was created expressly to address the security and privacy issues brought on by unreliable IoT devices. In order to manage and restrict the flow of traffic from susceptible devices, it makes use of software-defined networking and an advanced device-type recognition approach. A Security Gateway and an IoT Security Service provided by an IoTSSP (IoT Security Service Provider) are the two essential parts of the system. IOT SENTINEL automatically detects susceptible devices inside an IoT network and implements customised rules to restrict their communication capabilities with the purpose of minimising any harm resulting from hacked devices. The approach dramatically lessens the potential harm caused by hacked IoT devices by putting these preventative steps in place. IOT SENTINEL guarantees strong security and safety in the quickly developing IoT landscape by combining the benefits of device-type identification and software-defined networking.

This study introduces a machine learning approach for accurate IoT device categorization using network traffic analysis. The suggested method uses a recursive feature selection model to find and choose the IoT-AD-20 dataset's most important properties. In addition, the characteristics are ranked according to how crucial they are to the classification process using the random forest method. A cross-validation test is carried out to guarantee the model's dependability and prevent overfitting. When using flow-based characteristics, the results show the usefulness of the suggested approach, attaining a phenomenal 100% identification rate for all IoT devices. The detection of weak IoT devices is made possible by this precise categorization capacity, which also makes it easier to implement strict security regulations. The proposed approach demonstrates its potential to improve IoT device security and reduce possible dangers in IoT networks by utilising the strength of machine learning and careful feature selection.

With the use of a machine learning algorithm, this work pioneers the creation of an anomaly-based protection (ABP) system. It investigates how slight changes to sensed data might affect how accurate a machine learning algorithm is, additionally, it covers the process of constructing an ABP with a specific machine learning approach. The dataset for the experiment consist of 32,000 samples collected from the Intel Berkeley Research Laboratory. The remaining 12,000 samples were produced in a way that resembled anomalous behaviour, whereas the other 20,000 samples were obtained during routine operations. 24,000 samples from the complete dataset were designated for training, while 8,000 samples were reserved for testing. The ABP system was used to find instances of signal injection that were intended to compromise services, such heating or cooling in an office context, and were directed at specific detected data. Insights into the behaviour and effectiveness of the machine learning algorithm in spotting abnormalities were gathered through this research, which helped enhance anomaly detection methods for protecting crucial systems.

This paper explores the dangers of IoT traffic analysis by outlining a two-stage classification method for identifying devices and recognising their statuses. Two different datasets—self-collected packet traces and publicly accessible packet traces—are used to assess the suggested approach. It was found that each time a state of an appliance changes, a discrete sequence of packets with different sizes is sent along with it. This discovery was made by careful examination of traffic on the network caused by IoT devices in a controlled laboratory environment. This paper thoroughly investigates the effects of traffic profiling attacks on IoT devices. Notably, machine learning (ML) techniques are used to accurately and efficiently learn user actions. This research highlights the hazards and vulnerabilities present in IoT networks by thoroughly analysing IoT traffic data and applying ML approaches. The findings help to build strong security measures in the IoT ecosystem by offering insightful information on device identification, state recognition, and the possibility of hostile traffic analysis assaults. The summary of various worked carried on IoT devices profiling has been represented in Table 4.1.

4.3 Research Gap and Objectives

Network management and monitoring face additional issues as IoT devices grow. Statistical analysis can classify IoT devices. IoT rules must be executed consistently using a device type recognition framework. IoT devices may not be detectable because skilled hackers can use malware to find their MAC addresses. There is no MAC address-based system identification detection standard. This study classifies IoT devices by traffic patterns using composite controlled machine learning algorithm. The machine learning algorithms RF, k-NN, DT, NB, and SVM are capable

Table 4.1 Summary of literature surveyed on IoT devices profiling

Author(s)	Contributions	Model(s) used	Dataset used	Advantages	Limitations
Snehi and Bhandari [33]	IoT devices identification and classification	Stack Ensemble algorithm, k-fold validation, XGBoost, DRF, GBM, GLM, and MMetaModel	Sydney UNSW dataset	Scalable, extensible to security solutions, and portable	N/A
Aksoy et al. [34]	IoT device fingerprinting system	Rule-based ML algorithms	N/A	Extract features, generate fingerprints, automatically detect feature	Might be enhanced by the analysis of packet groups and GA optimisation
Cvitić et al. [35]	Identifying the research problem, establishing a laboratory environment, preprocessing data, developing a classifcation model	The IG method	10 days for each SHIoT device dataset	The M4 model not deviating significantly from other observed models	N/A
Rey et al. [36]	Identify cybersecurity vulnerabilities in IoT scenarios	The algorithm involves mini-batch aggregation and multi-epoch aggregation	N/A	N/A	N/A
Sumitra et al. [37]	A strategy for identifying IoT devices using horizontal federated learning while respecting privacy	suggested multilevel federated learning system with privacy protection	Aalto University and UNSW Sydney dataset	Increased efficiency for IID as well as non-IID identify skewed situations along with data distribution	Additional IoT device datasets will allow for greater degrees of diversity

(continued)

Table 4.1 (continued)

Author(s)	Contributions	Model(s) used	Dataset used	Advantages	Limitations
Msadek et al. [38]	ML-based secured traffic analysis for IoT device fingerprinting	Application of the sliding window technique for the analysis of encrypted traffic and various classification algorithms	The dataset is a collaboration between the University of New South Wales and IBM Research	Find the best estimators for classifiers to train optimally	Devices may have long sequences of traffic, performing tracing using a small window is impossible to manage or solve
Miettinen et al. [39]	manage security and privacy risks posed by insecure IoT devices	machine learning-based classification model	N/A	Automatically identifies and enforces rules to constrain the communications, minimizing damage resulting from their compromise	Unable to fully investigate the impact of software updates on test devices
Ullah et al. [40]	classify IoT devices based on network traffic analysis	recursive feature selection, random forest and cross-validation test	IoT-AD-20 dataset	Used to expose vulnerable IoT devices, enforce security policies	Lack of publicly accessible datasets for many devices
Lee et al. [41]	detect abnormal behaviour of IoT device, detect accuracy, build the ABP	Abnormal Dataset Generation, Dimension Reduction with PCA, training data using the k-Means technique, training with SVM	Intel Berkeley Research Lab dataset	Detection of chances of injecting a signal into targeted data being sensed to compromise service	The accuracy rate of anomalous behavior detection in IoT devices reduced when a malevolent attacker tampered with a single data point

(continued)

Table 4.1 (continued)

Author(s)	Contributions	Model(s) used	Dataset used	Advantages	Limitations
Skowron et al. [42]	Examines the hazards associated with the study of Internet of Things (IoT) communications	N/A	5-device test- bed dataset	Attacks on IoT gadgets have been suggested and examined, with high recognition accuracy and ML methods being used to learn user activities	Need to evaluate more sophisticated IoT gadgets have been suggested and examined

of identifying IoT devices. The innovative method groups novel, undiscovered IoT devices by network utilization. Network information like SSID probes, packet destination, MAC protocol fields, and broadcast packet size identify users, while device driver hardware and some hardware features are fingerprinted.

This chapter summarizes IoT device categorization research. Several studies have used application and device packet features to define systems. Miettinen et al. [11], tested 31 IoT devices. Fingerprint readers get 23 new functions. Nineteen of the 23 elements were binary, indicating domains or protocols at several protocol stacking levels, including link (LLC and ARP), network (ICMPv6, IP, and EAPoL), transport (UDP and TCP), application layer (HTTPS and HTTP), payload, and IP selections. Target IP counter, packet size, source-and-destination port class were integer-type properties. Authors employed Random Forest (RF) to classify 17 IoT devices with 95% accuracy and all of their system's devices with 50% accuracy.

Researchers created IoT device fingerprinting methods using proactive probes or anonymous data grabs. Nmap can detect devices [1]. Manufacturers classify devices using various network stacks. Nmap determines OS or device from 16 probes. Several passive fingerprinting approaches target network packet characteristics. For OS verification, P0f passively profiles and modifies TCP SYN headers and metadata [2]. Gao et al. [3] locate access sites using packet traffic wavelet estimation. Many approaches emphasize timing. Several passive and periodic authentication solutions leverage application layer protocol timing to identify devices [4]. In addition to SVM classification, RTF shows tree-based spatial finite phase machine signatures. Radhakrishnan et al. [5], categorize devices and simulate packet inter-arrival times using ANN. Formby et al. [6], develops fingerprint commercial control systems using actual running times and data and information response computation durations. Devices are categorized by TCP timestamp clock skew prevalence by Kohno et al. [7] are examined using wireless network properties. Desmond et al. [8] use 802.11 probe request packet timing analysis to find WLAN devices. Clustering was employed to create fingerprints. Radiometrics was used by Nguyen et al. [9] to passively profile identity tampering. Measurements include radio signal frequency and volume. The authors then identified the device using non-parameterized Bayesian approaches. Recent research by Xu et al. [10] used unconstrained learning and a white-list algorithm to detect wireless devices. MAC, top layer, and physical properties were prioritized.

4.4 Methodology and Evaluation

The level of complexity of the IoT market is increasing, therefore there is still plenty to learn about the many categories of IoT devices [44, 45]. The rising need for IoT technology presents various kinds of challenges for the infrastructure as it tries to sustain network services. This section outlines a method for building a structure to identify devices in networked IoT devices whenever an additional IoT device gets added, an IoT device is compromised, or as an IoT device provides erroneous data [48, 49]. New network analysis processes are needed in order to locate IoT equipment that is attached with the system. This makes it practicable to employ analytical methods for interpretation of information find typical setups that might distinguish between various device types. IoT systems are more established than conventional desktop computers since they exclusively carry out certain tasks. In order to detect IoT devices along with excellent accuracy and minimal error messages, communication analysis is advised. The proposed method will guard against different attacks on the IoT systems' activities by tracking and analysing the activity of IoT devices [50]. Figure 4.1 depicts our recommended format for sensor characteristics in an IoT network. A variety of connected devices and platforms for communication make up the testbed. IoT devices have sensors for data collection and transmission from or into the actual surroundings. Figure 4.2 shows five stages of the IoT device recognition procedure [44–47].

4.4.1 Acquisition Phase

A network management tool gathers IoT network traffic. The access point and the intelligent systems will be the two points of contact for the monitoring process [51, 52]. This method has the advantage of detecting malicious IoT device activity before the access point is accessed. Network traffic is recorded using packet capture software like Wireshark [53–55]. The origin IP, root ports, goal IP, ports at the goal, and the content of the packets are all included in Wireshark traffic. Data from IoT devices can be gathered from different payload to create the device identity. To determine device behaviour, the information from every system will be analysed.

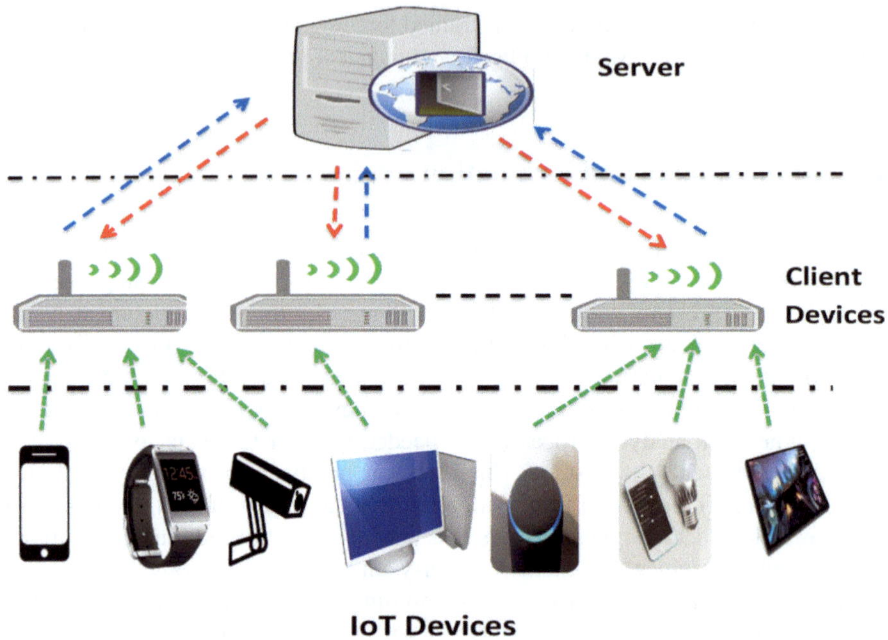

Fig. 4.1 Model of the device recognising system

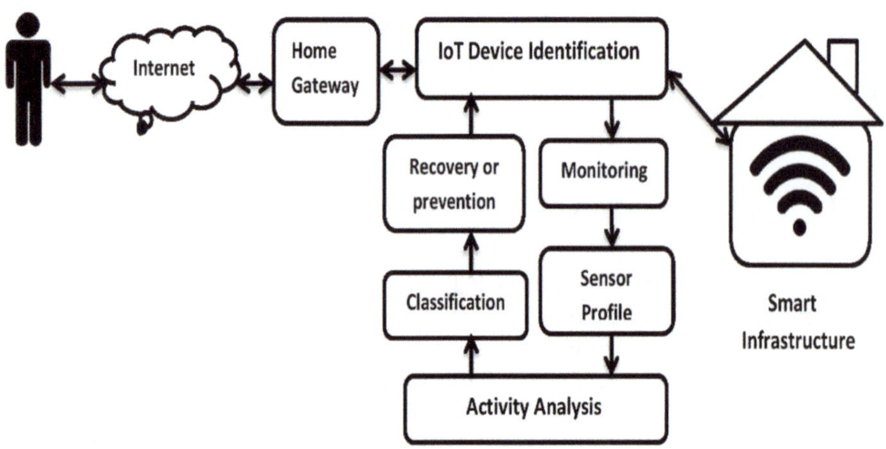

Fig. 4.2 System for profiling IoT devices in smart home infrastructure

4.4.2 Sensor Configuration

A system of data with the typical functioning of IoT sensors is described in the section for sensor description [56, 57]. The routine operation of IoT sensors has been described using machine learning techniques. It is best for determining every potential stage of typical sensor behaviour using a thorough model. To be able to categorise IoT devices throughout the system, this document emphasises the communication analysis of the sensors. By device recognition, a network manager will be able to identify malicious sensors in IoT system.

4.4.3 Analysis

To check for any discrepancies in the communications of the received IoT system, the sensor profile from the preceding stage will be utilised as a baseline [58]. For the sensor communications, a runtime profile is created, and any departure from the baseline should be regarded as abnormal. The possibility of natural behaviour occurring outside of the lower or upper bound is confirmed using the probability distribution. If the system's gathered data rate exceeds the parameters specified, it will be regarded as irregular.

4.4.4 Classification

Following the discovery of malicious in the received communications by the analysis module, the categorization functionality establishes the irregularity [59]. The classification of abnormalities enables IT, management, or consumers to more clearly identify the sort of anomaly.

4.4.5 Action (Prevention and Recovery)

To safeguard IoT networks, a number of recovery procedures may be used, such as rejecting incoming data, modifying the network configuration, de-authenticating the sensor, etc [60]. You can reset an IoT device and request that the classification unit re-authenticate itself if it is unable to recognise the IoT device. In this research, we focus on the sensor profile for network traffic assessment to detect IoT gadgets. By actively analysing network data, the attacker can utilise IoT recognition to find corrupted IoT gadgets. The network manager can find hacked devices in IoT networks with the help of sensor profile and device recognition. The supervisor of IT may utilise the sensor configuration to impose various safety standards on multiple IoT devices.

4.5 System Model

The system paradigm comprises three fundamental components: authorized devices, a central hub, and a verification server. We then explore numerous potential ways an attacker can exploit a susceptible device using the adversary model [61–64]. These consist: intercepting network traffic, introducing malicious packets to damage the network, and pressuring a device to carry out operations specific to another device type. We also provide the security requirements of the proposed approach, which relies regarding the authentication of devices, it involves the classification of devices depending on their device type. Lastly, we provide a concise overview of the machine learning techniques and algorithms employed in this thesis. The system framework employed in this study replicates a network populated with IoT devices. The following are the main parts of this system, as shown below:

1. Legitimate devices (D): By using a variety of currently used strategies, these devices have already builtup confidence inside the network [12, 39, 40]. The capabilities or security requirements of these devices are not constrained in any way.
2. Hub (A): The hub takes on the responsibility of supporting authorised devices inside the network. It performs initial trust establishment and checks the devices''s credentials. The hub also makes it easier for devices to connect to the Internet, which gives it the ability to look at different network layers''s headers.
3. Verification Server (V): The verification server is in charge of classifying devices based on observed traffic patterns. The verification server, connected to the hub's administrators as a cloud service, accepts the collected patterns of traffic from the hubs for classification. It is assumed that the hub and the verification server have a secure route of communication. Modern cryptographic methods can be used to create this channel, guaranteeing the realisation of the authenticated encryption (AE) function designated as K() [6].

4.5.1 Attack Model

By utilizing various methods such as exploiting vulnerabilities in firmware or revealing pre-shared secrets stored in databases, the adversary (M) has the ability to compromise any of the authorized devices [20, 21, 26]. The adversary can take control of a networked device that is weak by taking use of this compromised knowledge and attempting to [65].

- Inject malicious packets into the network to contaminate it.
- Record network traffic to get critical information,
- Control a gadget to carry out operations usually reserved for a distinct kind of equipment.

We assume that the attacker is unaware of the traffic patterns displayed by any lawful device that has been infiltrated [22, 23, 25]. This assumption is valid since the adversary lacks access to the authorized equipment required for capturing and analyzing traffic trends once they have obtained the compromised secrets.

4.5.2 *Security Analysis*

The authentication of devices according to their device type classification is a security requirement. The hub is responsible for verifying credentials and comparing the claimed and observed device types based on traffic patterns [66, 67]. The hub and the verification server can be conceptualized as a unified secure gateway. The following tasks are carried out via this secure gateway:

- Initial trust building: This can be done in a number of ways that are currently in use [12, 39, 40]. It is assumed that each device is given unique credentials once the user has achieved their initial level of confidence. Pre-Shared Keys (PSK) that are particular to a certain device are assigned for WPA2 or WPA3 in technologies like WiFi Protected Setup (WPS) [39]. These keys can be used to establish future security features like integrity verification or confidentiality or for subsequent authentication.
- Policy-based network access involves granting different levels of access to devices based on their known security vulnerabilities [24]. A novel classification technique has been developed to effectively differentiate between various types of IoT devices. This technique enables the implementation of certain policies at a granular level. The Common Vulnerabilities and Exposures (CVE) database provides information about known vulnerabilities in different types of devices, which can be used to modify access rules. The objective of this is to provide an additional modality during the initial and continuous authentication procedure for devices. The gadget's behavior should align with both previously observed behavior and credentials. The secure gateway maintains the login information and the unique identifiers of network traffic, and utilizes them as parameters for subsequent authentication. To get access to the network, a malicious device must not only obtain login information but also replicate the communication patterns of the compromised device. The suggested solution enhances network security by integrating three features: device type categorization, vulnerability-based policies, and traffic pattern matching. This improvement in device authentication is supported by references [68–71].

4.5.3 *Machine Learning Models*

The three categories of machine learning algorithms are supervised learning, unsupervised learning, and semi-supervised learning. Each type has unique traits and uses in a variety of industries. Let's look more closely at these categories.

4.5.3.1 Supervised Learning Algorithms

Training a model on labelled data through supervised learning entails associating the input data with the matching output labels. The objective is to discover a mapping between the features of the input and the desired output [72]. The model gains the ability to predict outcomes during training by extrapolating patterns from the labelled samples. The labelled data helps the model make precise predictions about fresh, unforeseen data. Linear regression, decision trees, support vector machines, and neural networks are examples of common supervised learning methods [70].

4.5.3.2 Unsupervised Learning Algorithms

In this method, the model is given unlabeled data without any associated output labels. Finding patterns, structures, or relationships within the data is the aim [73]. There is no specific right output to direct the learning process, unlike supervised learning. Unsupervised learning algorithms seek out hidden patterns, collect related data points, or make the data less dimensional. Unsupervised learning is frequently used in clustering algorithms like k-means and hierarchical clustering as well as dimensionality reduction methods like principal component analysis (PCA) and t-distributed stochastic neighbour embedding (t-SNE).

4.5.3.3 Semi-Supervised Learning Algorithm

This type of learning is somewhere in between supervised and unsupervised. For training, it uses a mix of labelled and unlabeled data. Acquiring labelled data can be expensive or time-consuming in many real-world settings. Through the use of unlabeled data to supplement the sparse labelled data, semi-supervised learning improves model generalisation [74, 75]. Comparing this method to pure unsupervised learning, performance may be increased. Semi-supervised learning frequently employs strategies including self-training, co-training, and multi-view learning. Every machine learning algorithm has unique advantages and uses. When labelled data is available and precise predictions are needed, supervised learning is appropriate. Unsupervised learning is useful for identifying patterns and hidden structures in massive datasets that lack explicit labelling. When there is a dearth of labelled data, semi-supervised learning is advantageous, although performance can be enhanced by using more unlabelled data. Researchers and practitioners can apply relevant strategies to various data analysis and prediction jobs across a variety of domains by knowing and utilising these numerous sorts of machine learning algorithms.

4.5.4 Machine Learning Classifiers

This section gives a thorough explanation of supervised learning and discusses five different supervised machine learning algorithms that are used to categorise the different kinds of IoT devices that are present in the network. The following are these algorithms.

4.5.4.1 Random Forest

A decision degree growing technique variant known as random forest enables random growth branches inside the chosen subspace, setting it apart from other classifiers as shown in Fig. 4.3. Constructed from a collection of random depend regression trees, the random forest approach makes predictions about the result. At each random base regression, the algorithm chooses a node and splits it to develop additional branches. Given that it integrates many trees, and since Random Forest is a combined algorithm, it is crucial to keep this in mind. Ensemble algorithms integrate a number of classifiers with various kinds in the ideal case. It is possible to think of random forest as a bootstrapping method for enhancing decision tree outcomes.

The algorithm operates in the sequence shown below. The parameter designating the sample used for bootstrapping $U^{(i)}$ employs the ith bootstrap. Despite using a modified decision tree technique, the programme learns a traditional decision tree. As the tree develops, the alteration is methodically carried out and is particular. This means that instead of performing an iterative for each conceivable value splits at

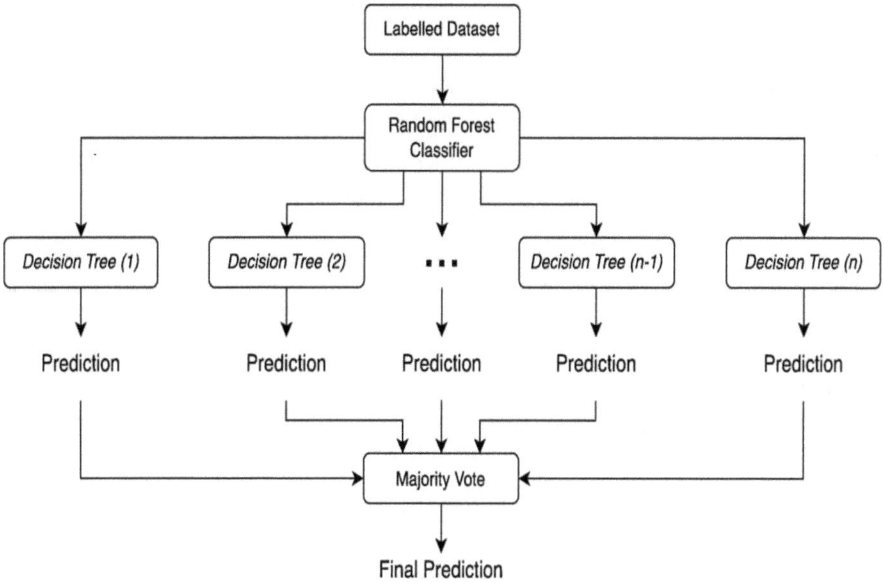

Fig. 4.3 Structure of RF classifier

every point of the decision tree, RF independently chooses a subgroup of attributes such as $f \subseteq Z$ and next divides the attributes in the subgroup (f).

The approach selects a group that is considerably less than the total number of all characteristics during implementation since the division is determined on the subgroup's best attribute. Since the data sets with big size subgroups likely to possess greater computational difficulty, choosing the number of characteristics to separate minimises the strain. Therefore, the technique learns more quickly when the properties to be taught are limited.

Algorithm 1 Random Forest

Set the training parameters as a requirement U: $= (x_1, y_1), \ldots, (x_n, y_n)$ the whole list of attributes Z, and the quantity of trees that will be present in forest Q.

```
perform RF (U, Z)
A ← 0
for i ∈ 1, ..., Q do
U^(i) ← 1 an instance of bootstrap taken from U
a_i RTL(U^(i), Z)                    (RTL = RandomizedTreeLearn)
A ← A ∪ {a_i}
close for
return A
close RF
perform RTL (U, Z)
At every point:
f ← Create a condensed collection of Z
Divide the most significant attribute in f
return RTL (Model)
close RTL
```

The subject it takes into account is that wrapping lowers the difference of decision tree method, which is how the algorithm implements the ensemble decision tree.

4.5.4.2 Support Vector Machine

A collection of supervised learning methods known as Support Vector Machine categorise data using regression analysis. In order for the learning method to assign the addition of a fresh category value as prediction output, among the learning specimen's parameters needs to be specific. As a result, SVM uses the linear characteristics to create a non-likelihood binary classifier.

SVM is flexible when used for high dimensionality problems and, in addition to classification and regression, finds outliers [15, 19]. An ideal definition following describes a learning vector parameter with minimum two different types.

$$x_i \in R^p, i = 1, \ldots, n$$

where R^p denotes the p-dimensional data space and forecast vector domain with real values and x_i stands for the training observation. A basic Support Vector Machine algorithm's pseudo-code is displayed.

Algorithm 2 (SVM)

FeatureSupportVector(FSV) = {Most Similar Feature Pair of Differing Groups}.

1: while there spots that breaches the margin limitation do
2: Figure out the offender
3: FSV = FSV ∪ offender
4: if any $\alpha_p < 0$ due to inclusion of c to S then
5: FSV = $\frac{FSV}{p}$
6: Repeat that all offences are removed
7: terminate if
8: terminate while

This technique looks for potential supported vector, designated as S which makes the assumption that the SV represents the dimension in which hyperplane's linear attributes' parameters are kept.

4.5.4.3 K-nearest Neighbor

kNN categorises data by utilising the identical distance measuring method as Linear Discriminant Algorithm and further approaches employing regression. While the technique delivers the worth of a characteristic or predictor in a regression application, it creates class members in a classification application [16]. The method was chosen for the study because it can pinpoint the most important predictor. Despite being regarded as resistant to outliers and adaptable among many other qualifying criteria, the method has significant memory requirements and is attentive to feature that are not contributed. The average space between individual data points is used by the method to form classes or clusters. The following Eq. (4.1) can be used to get the mean distance [40].

$$\varphi(x) = \frac{1}{K} \cdot \sum (x_i, y_i) \in kNN(x, L, K) y_i \tag{4.1}$$

In above Eq. (4.1), kNN (x, L, K), the letter k stands for the input attribute's K nearest neighbours in the learning set space (i).

The dominant k class determines how the algorithm applies classification and prediction, and the following Eq. (4.2) is the prediction formula [40].

$$\varphi(x) = arg\max_{c \in y} \sum (x_i, y_i) \in kNN(x, L, K)y_i \tag{4.2}$$

So, it is crucial to understand the fact which resulting class is made up the desired attribute's participants, and that the Euclidean distance is used to assign the attributes to classes. Six phases make up the algorithm's implementation. The calculation of distance according to Euclidean is first stage. The calculated n distances organised from highest to lowest in the second stage, and k is a positive integer selected based on the ordered Euclidean distances in the third phase. The fourth stage establishes and assigns k-points that match the k-distances due to closeness to the centre of the group. Finally, if $k_i > k_j$ for every $i \neq j$ is true, an feature x is added to that group for $k > 0$ and for the amount of points there in i. The kNN stages method is shown in Algorithm 3.

Algorithm 3 (kNN)

As a requirement, provide a training sample (X), the class samples (Y), with an undefined training data (x):

1: Categorised (X, Y, x)
2: for $i = 1$ to m do
3: Enter distance d (X_i, x)
4: *terminate*
5: *for* Enter Set I consists of the smallest groups with k distances. d (X_i, x)
6: Return dominant label for $\{Y_i; i \in I\}$

4.5.4.4 Logistic Regression

For the J class, the LR technique simulates with a conditionally probability for noticed instance comprising a specific group $Pr(G = j|X = x)$, where it is feasible to identify the classes of unidentified cases using below Eq. (4.3).

$$j = arg\max_{j} Pr(G = j|X = x) \tag{4.3}$$

j is the j-th member among the groupings

G; G is a collection of groups (1,…,J);

x is an distinct attribute from the group X;

X is the group of distinct attributes.

By modelling probability for x using linear functions, logistic regression makes sure that their sum stays within certain bounds [0, 1]. According to Eq. (4.4) to Eq. (4.6), the template is described with the *J—1* is the log-odds which splits each class into groups from the "basic" group J [50].

$$log \frac{Pr(G = j|X = x)}{Pr(G = j|X = x)} = \beta_j^T x_i; \, j = 1, \cdots, J-1 \tag{4.4}$$

where: β_j distinct feature's logistic value for group j [50];

$$Pr(G = j|X = x) = \frac{e^{\beta_j^T x_i}}{1 + \sum_{l=1}^{J-1} e^{\beta_j^T x_i}}; \, j = 1, \cdots, J-1 \tag{4.5}$$

$$Pr(G = j|X = x) = \frac{1}{1 + \sum_{l=1}^{J-1} e^{\beta_j^T x_i}} \tag{4.6}$$

Equation (4.4) denotes a multiclass classification model where J indicates a group and *j* {0, 1, 2,.., J–1} below the constraint that J ≥ 3. With such an approach, linear differences between areas that belong to different categories are made. Inferring Pr(G = j|X = x) = Pr(G = J|X = *x*) expresses P*r*(G = *j*|X = *x*) = P*r*(G = J|X = *x*), This additionally equates to *log odds* = 0, because these cases (x_i) are those that are within the dividing line between two groups (*j* and J). Estimating parameter β_j is necessary for the logistic regression model's adaptation. where the goal of the common statistical method is to find the greatest in terms of the probability function [17, 30].

4.5.4.5 AdaBoostclassifier

The AdaBoost [31] method acts as a meta-estimator which initially matches an estimator with the data set that was originally used, thereafter matches another copy of that estimator on an identical dataset while altering the coefficients of occurrences that have been classified incorrectly ensuring that subsequently classification focus on greater difficult instances.

The fundamental idea behind AdaBoost aims to adapt a number of poor learner-models that only marginally more accurate than independently assuming —on frequently altered copies of the data. Then, utilising a vote with a significant majority (or total), the estimation from each participant is added to form the overall projection. Every 'boosting' phase modifies the results through introducing weights $\omega_1, \omega_2, \ldots,$ ω_n for every training set. All of those weights' establishing configured with $\omega_1 = \frac{1}{N}$, therefore the initial step only instructs a weak learner on the starting set of data. The sample weights are individually changed for each subsequent iteration, and the learning process is then reapplied to the reweighted data [32].

4.5.4.6 Gradient Boosting Classifier

This approach's main idea is to build templates one after another with aiming to decrease the shortcomings of the model prior to it. However, how should we approach that? How will the error be reduced? Such is achieved by building another system on the flaws or any residuals of the previous one.

Boosting the gradient whenever the objective column is constant, the regressor is employed; whenever the issue involves a single of categorization, the gradient boosting algorithm utilised. The only difference within two of them is the "Loss factor". The objective is to add poor learners then decrease this loss factor using gradient descent techniques. Due to the fact that it relies on a loss factor, we are going to have a variety of loss factor for regression challenges like Mean Squared Error along with difficulties in classification as log-likelihood.

Let's consider X and Y as the input and goal, respectively, with N samples each. We seek to comprehend the function f(x) that transforms the input characteristics X into the required variables y. It represents the cumulative count of trees, including those that have been improved or modified. The discrepancy between the observed and predicted variables is referred to as the loss function as shown in Eq. (4.7) [50].

$$L(f) = \sum_{i=1}^{N} L(y_i, f(x_i)) \tag{4.7}$$

With regard to f, we aim to minimise the loss function L(f) as shown in Eq. (4.8) [50].

$$\widehat{f}_0(x) = arg \min_f L(f) = arg \min_f \sum_{i=1}^{N} L(y_i, f(x_i)) \tag{4.8}$$

If our gradient boosting approach is in M stages, the algorithm can add a new estimator as h_m having $1 \leq m \leq$ M to enhance the f_m as shown in Eq. (4.9).

$$\widehat{y_i} = F_{m+1}(x_i) = F_m(x_i) + h_m(x_i) \tag{4.9}$$

The steepest Descent determines $h_{\text{m}} = -p_m g_m$ for M stage gradient boosting where p_m is constant and known as step length and g_m is the gradient of loss function L(f) as shown in Eq. (4.10).

$$g_{im} = -\left[\frac{\partial L(y_i, f(x_i))}{\partial f(x_i)}\right]_{f(x_i) = f_{m-1}(x_i)} \tag{4.10}$$

The gradient refers to the rate of change of a function at a certain point. It represents the slope or steepness of the function at that point. Similarly, the same applies to M trees as shown in Eq. (4.11).

$$f_m(x) = f_{m-1}(x) + \left[arg \min_{h_m \in H} \left(\sum_{i=1}^{N} L(y_i, f_{m-1}(x_i) + h_m(x_i)) \right) \right] . x \tag{4.11}$$

The proposed solution at present is shown in Eq. (4.12),

$$f_m = f_{m-1} - \rho_m g_m \tag{4.12}$$

4.5.4.7 XGB Classifier

A scattered, configurable gradient-boosted decision tree machine learning system is called XGBoost. It provides parallel tree boost features alongside which is the best ML package for solving classification, regression, and ranked problems. To understand XGBoost, one must have a solid understanding of the principles and methods of machine learning around which controlled machine learning, collaborative learning, tree models, as well as gradient boosting are based. In controlled machine learning, a prediction model is developed using techniques to find trends in the data set with attributes and labelling, after which the framework is applied to forecast the tags on the attributes of the latest dataset.

Decision trees offer a framework that estimates the tag by examining a tree that uses if–then-else true/false attribute queries and determining the absolute minimal number of queries necessary to assess the chance of making the correct option. Utilising decision trees, a number could use regression to forecast a constant number or categorization to anticipate a classification. The simple instance underneath uses a decision tree to predict the label (home price) based on the attributes (size and number of bedrooms).

4.5.4.8 Decision Tree Classifiers

Among the algorithms utilised for controlled machine learning are detectors from decision trees. This shows that they develop a method which may forecast outcomes from pre-labeled data. Decision trees can also be used to address regression-related problems. Much of the information you learn in this lesson may be applied to prob lems with regression. Classifiers using decision trees function in a manner resembling flow diagrams. The nodes of a decision tree typically represent a point of choice that splits into two nodes. Every one of such nodes represents the result of the option, and each possibility includes a chance to evolve into a decision node. A conclusive classification will be produced as the culmination of all the various assessments. The primary node is the root or base node. Every of the decision points are referred to by the decision nodes. The result decision point is known as a leaf node.

4.5.4.9 LGBM Classifier

LightGBM is a type of gradient boosting method made with decision trees which enhance model efficiency while utilizing small data or memory. To detect the features of the histogram-dependent method, which is greatly deployed using entirety Gradient Boost Decision Tree approach, it uses 2 creative strategy: gradient-dependent first type testing and another bundles of special features. Characteristics of the LightGBM method are made by above couple of techniques. Collectively, they apply the model to perform accurate and give an edge over GBDT method.

For a training set of n occurrences, each of which is represented by a vector of dimension s in space {x1, x2, …., xn}. The -ve inequalities in the demise factor with relation to the technique outcomes shows up {g1,…., gn} within every iteration of gradient boosting. The learning instances are rated using this GOSS technique based on the absolute values of their gradients in descending order. Then, we acquire an instance subset A by keeping the top-a 100% instances with the largest gradients. Then, for the set A^c that is still present, we randomly choose a group B with dimension b $|A^c|$ that contains (1−a)× 100% cases with lower gradients [50]. The mathematical formula is represented in Eq. (4.13).

$$V_j(d) = \frac{1}{n}\left(\frac{\left(\sum_{x_i \in A_l} g_i + \left(\frac{1-a}{b}\right)\sum_{x_i \in B_l} g_i\right)^2}{n_l^j(d)} + \frac{\left(\sum_{x_i \in A_r} g_i + \left(\frac{1-a}{b}\right)\sum_{x_i \in B_r} g_i\right)^2}{n_r^j(d)}\right) \tag{4.13}$$

where

- $A_l = \{x_i\ ?\ A\colon x_{ij}\ ?\ d\}$,
- $A_r = \{x_i\ ?\ A\colon x_{ij} > d\}$,
- $B_l = \{x_i\ ?\ B\colon x_{ij}\ ?\ d\}$,
- $B_r = \{x_i\ ?\ B\colon x_{ij} > d\}$,

also value (1–a)/b is utilize which regularize the addition of gradients above B return to the scale of A^c.

4.5.4.10 Classification of Devices

Gadgets are divided into seven categories by us: smart speaker, smart electricity and lighting, smart camera, smart sensor, smart home assistant, and non-IoT gadgets. These categories were chosen because of the functionality they provide. While home assistants can carry out tasks, cameras and sensors are largely employed to gather information. We may successfully implement regulations that forbid data-gathering gadgets from doing acts that would jeopardise privacy by classifying them in this way. Furthermore, as they both acquire data with variable degrees of privacy violation, we distinguish between cameras and sensors as information-gathering devices. Comparing cameras to sensors, in particular, provides more thorough data about user privacy.

We use a threshold-based iterative classification approach to increase the process' efficiency and accuracy. The first division of the obtained dataset places 80% of it in the training category and 20% in the testing category. The training data is then used to train five separate models independently by fitting the data to each model. By using this strategy, the classification procedure is optimised in terms of its efficiency and accuracy.

In this part, we analyse the proposed framework procedure as well as the applied classification methods covered in this study. We concentrate on emphasising the crucial component that makes the suggested solution resistant to diverse assaults. We also look at other situations where the protocol may be used to provide complete system security. Depending on how each device behaves, these scenarios entail the use of various authentication procedures. We want to give a complete grasp of the protocol's capabilities and its potential for successfully securing the system by diving into these elements.

The user uses device credentials to start a new device's authentication procedure, which uses cryptographic methods to create the first connection between the device and the hub. This framework protocol uses the aforementioned categorization approaches to efficiently identify whether a device is known or unknown. The protocol handles different attack scenarios in detail:

1. Injecting malicious packets into a system by taking advantage of a weak point, which compromises the network [16].
2. Even if the network packets are encrypted, using an insecure IoT device can extract sensitive data [2].
3. Using a susceptible IoT device to perform actions usually reserved for another kind of device [43].

The protocol uses classifiers that are maintained in the database and that have been trained using fingerprints of previously authenticated devices to find such attacks. When generating predictions on datasets that are comparable to the training dataset, machine learning models demonstrate excellent accuracy. Therefore, the content of the data packets will vary if a device is compromised, and these changes may be found by comparing the fingerprints with the classifiers that are kept in the database. Since the fingerprint of the hacked device will be different from the fingerprint used to train the model in the beginning, it will be possible to spot the deviation and take the necessary action.

The hub then takes action by revoking full network access for the compromised device and only allowing limited access to the network after obtaining these results from the verifier. The hub also eliminates the hacked device's credentials from the list of authorised devices. The device must thus be reintroduced to the network and go through the authentication procedure later on.

The permanent variation in the device's fingerprint will prevent the classifiers from providing an appropriate classification upon re-authentication if the device is still affected. If an IoT camera is hacked, the classifiers in the verifier will be unable to recognize it as a camera since the fingerprints obtained from the compromised device will not correspond to any fingerprints from known IoT cameras that were utilized to train the classifiers.

4.5.5 Analysis of ML Models

This section outlines our method of classifying the device during each re-authentication, regardless of its previous authentication status with the server. This approach differs from the reliance on identifying a new device solely based on its MAC address, as demonstrated in [24–27]. Such reliance on MAC address can be easily manipulated by adversaries to authenticate themselves with the hub, as discussed in [28, 29, 36].

The database only verifies the device's MAC address to see if it is already present. However, even in the event of a falsified MAC address, the classifier in the database will not possess the capability to accurately categorize any information to the highest degree. This occurs as a result of a discrepancy between the fingerprint of the device and the fingerprint that is already stored in the database. Although the MAC address may be same, the device's unique fingerprint results in a discrepancy. Consequently, this addresses the situation where a device has been compromised, as the patterns of communication and the unique characteristics of the device would have changed from their previous state.

The MAC addresses match but the fingerprints don't, which means the device must be fake, the verifier may immediately confirm to the hub. By using this strategy, we tighten the authentication procedure and stop unauthorised devices from connecting to the network, even if they try to impersonate a device that has previously been authenticated.

This concept is founded on the notion that a machine learning model exhibits a propensity to provide highly precise forecasts when it undergoes training and testing using either the same or a similar dataset. The motivation behind acquiring traffic data of a novel device is rooted in this concept. After the device has been properly authorised, we build a model based purely on its fingerprint. This model is then saved in the database. By using this strategy, we make use of machine learning to improve the overall efficacy of the authentication process and assure accurate forecasts.

4.5.6 Dataset and Feature Selection

This section describes about the dataset, various preprocessing and feature selection approaches used in the proposed system.

4.5.6.1 The Dataset

The baby monitor, lights, motion sensor, security camera, smoke detector, socket, thermostat, TV, watch, and water sensor were among the ten IOT devices from which the dataset for this study was primarily compiled. It had previously been split into a train set, a validation set, and a test set. The dataset includes details on these devices' network activity as it was gathered over an extended period of time. A TCP connection from the SYN packet to the FIN packet is represented by each instance (example) in the dataset. The device's type categorization serves as the dependent variable. Nearly 300 characteristics and almost 400,000 cases make up the training set.

4.5.6.2 Dataset Preprocessing

This section outlines various approaches used in data preprocessing.

Handling Missing Data

It is important to note that not all of the data in the dataset was accessible for all of the offered sessions. It happens frequently to come across a dataset where not all of the data is accessible and usable for training. There are several ways to deal with this missing data, and the one we used was to get rid of the instances where it existed. We found that the number of cases with missing data is rather low and that there are still enough examples to provide successful learning. In the original dataset, missing data are denoted with a question mark. We also had to cope with the "thermostat" device having just one class represented in this test set when we got to the testing and utilising step. Therefore, we could determine its accuracy score but not AUC.

Feature Scaling

We quickly observed from the data that the different attributes have varying ranges of values. It is well knowledge that such adjustments to the feature ranges may result in less accurate findings and issues with training. As a result, we have chosen to employ the Python sklearn library's built-in MinMaxScaler. You may use this scaler to conduct min–max scaling, which will result in the dataset's values all falling inside the range of (0,1). We really saw that the test set findings were substantially more accurate and had a better AUC value after the feature scaling was done.

Feature Selection

Feature selection is one of the factors that should be taken into account in many machine learning situations. The well-known idea of "the curse of dimensionality" might make the model overfit or underperform. The feature selection idea was developed to address it. Some models, such as Decision Trees and Random Forests, do not often need feature selection. The rationale is that because of how these models are trained (the "best" feature is chosen at each split of the tree), the feature selection process is done on the fly. But in order to get better outcomes, some models could require feature selection. Given the large samples-to-features ratio in this study (the training set contains over 400,000 instances and has about 300 features), the "curse of dimensionality" shouldn't have much of an effect.

4.5.7 *Evaluation*

We have 10 IoT devices like baby monitor, lights, motion sensor, security camera, smoke detector, socket, thermostat, TV, watch, and water sensor. So according to the device categories we include here the device's data to the training set.

The Fig. 4.4 graph shows the appearances of IoT devices in training data set according to their count. Similarly using dataset, we also count the categories of test set of device categories as shown in Fig. 4.5. Y axis describe count of device and X axis shows device categories.

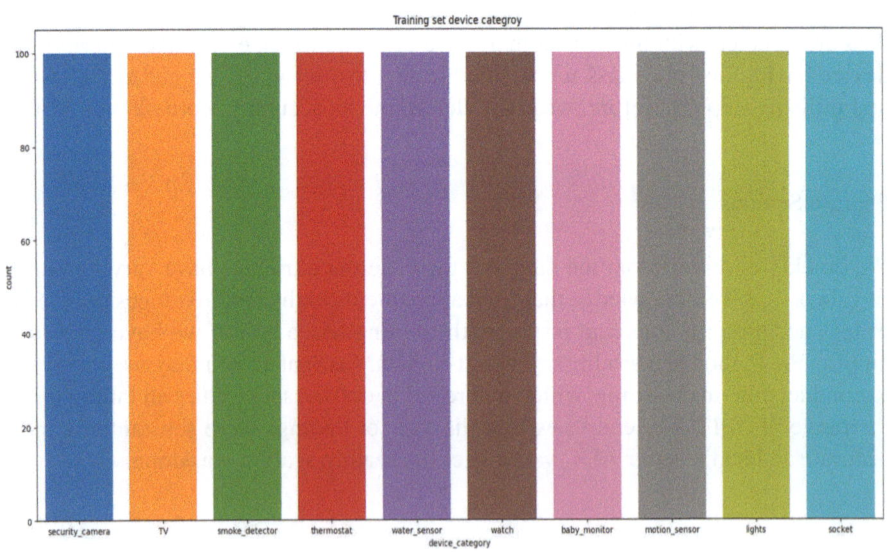

Fig. 4.4 Training set device categories

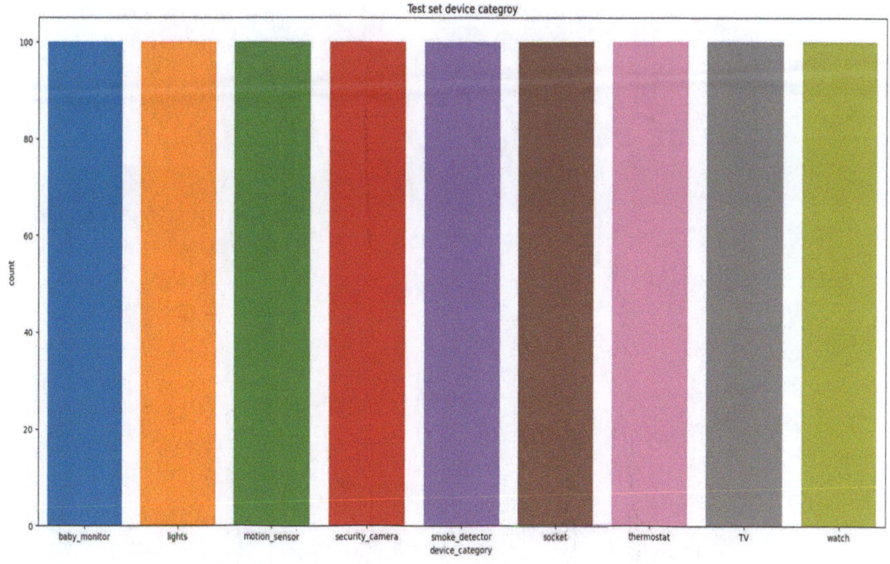

Fig. 4.5 Test set device categories

By considering training and test set device categories we have above correlation matrix and heat map shows the value of correlation between device categories. Figure 4.6 shows heat map of correlation of IoT devices.

After getting correlation matrix we move towards our next stage by defining base line model scores according to different machine learning models. Here base line model used different ML models such as AdaBoost Classifier, GradientBoostingClassifier, LGBM Classifier, XGB Classifier, SVC, DecisionTreeClassifier, KNeighborsClassifier, RandomForestClassifier, LogisticRegression, XGBRFClassifier etc.

Calculating different classifier model we get following baseline model scores as follows:

The Table 4.2 shows the respective value of classifier models in order to determine the baseline score model. From Fig. 4.7 we get the baseline model precision score and found that a top 2 model scores of Gradient Boosting Classifier is 0.859649. To see if the model becomes better, we may try adjusting the hyperparameters. So, for this here we use Random search (RS) method to improve the classifier's performance.

From Table 4.3 we get the values of different random search model of Gradient Boosting Classifier in order to improve performance of the classifier. Since RS 4 yields the greatest results, we will build the model on it i.e. RS model 4 Gradient Boosting Classifier 0.8491228070175438.

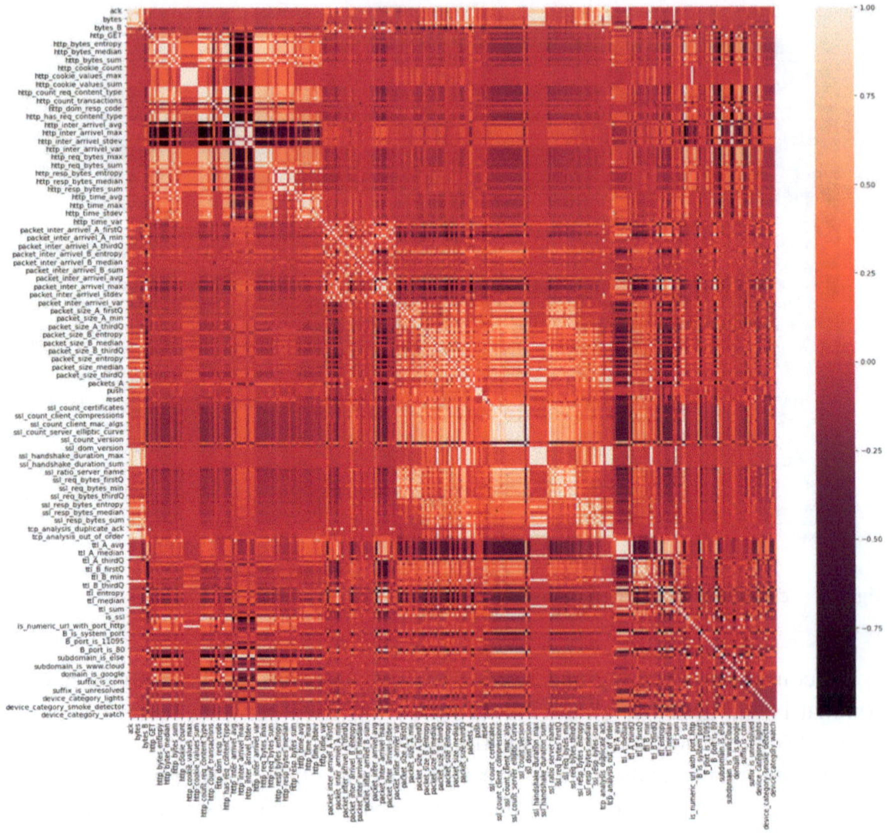

Fig. 4.6 Correlation matrix

Table 4.2 Scores for baseline ML models

Model classifier	Score
Decision tree classifier	0.815789
KNeighbors classifier	0.833333
SVC	0.791228
Logistic regression	0.840351
AdaBoost classifier	0.538596
XGBRF classifier	0.842105
Random forest classifier	0.856140
LGBM classifier	0.857895
Gradient boosting classifier	0.859649
XGB classifier	0.856140

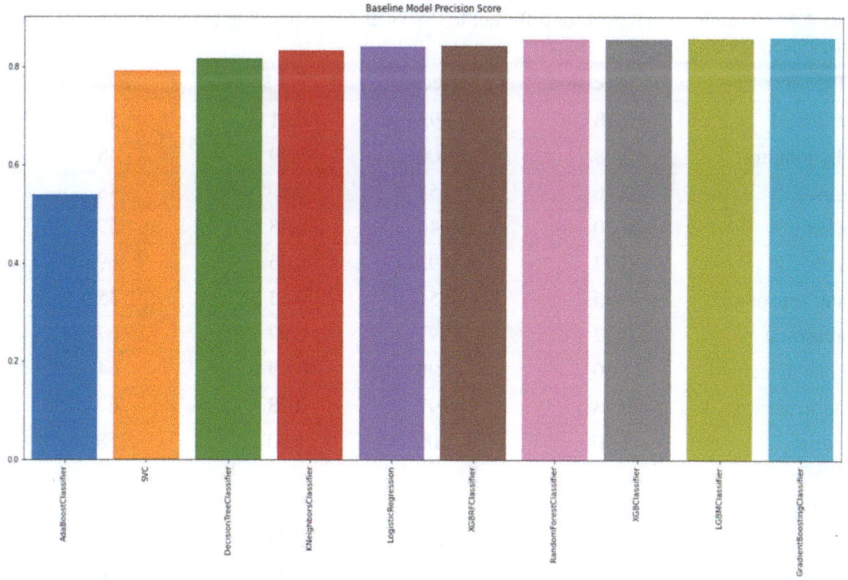

Fig. 4.7 Baseline model precision score

Table 4.3 Random search models

RS models	Score
RS model 1 Gradient boosting classifier	0.6859649122807018
RS model 2 Gradient boosting classifier	0.856140350877193
RS model 3 Gradient boosting classifier	0.8596491228070176
RS model 4 Gradient boosting classifier	0.8491228070175438

4.6 Result and Analysis

So here we getting the result of RS model 4 and considering that value further we have to calculate different confusion matrix parameter like precision, recall, f–1 score. Table 4.4 shows the supported value of different IoT device with their precision, recall, f-1 score and support score value by considering the below confusion matrix graph (Fig. 4.8). Finally, the table calculate the accuracy, macro and weighted average of classification report.

Classification report shows the accuracy of the model in order to corresponding RS model 4 value. Figure 4.8 describe the confusion matrix of 10 IoT device which is plot in basis of predicted label and true label. The higher value shows the easiness of predicating of some device label and lower value shows the difficulty of predicating value of device label.

Table 4.4 Classification report for different IoT devices corresponding to the various performance metrics used for evaluation

IoT devices	Precision	Recall	F-1 score	Support
TV	0.88	0.93	0.91	57
Baby monitor	0.98	1.00	0.99	48
Lights	0.46	0.54	0.50	48
Socket	0.50	0.46	0.48	57
Watch	0.97	0.93	0.95	69
Water sensor	0.55	0.51	0.53	35
Thermostat	0.97	0.97	0.97	67
Smoke detector	1.00	0.98	0.99	64
Motion sensor	0.98	0.97	0.98	67
Security Camera	1.00	1.00	1.00	58
Accuracy			0.85	570
Macro avg	0.83	0.83	0.83	570
Weighted avg	0.85	0.85	0.85	570

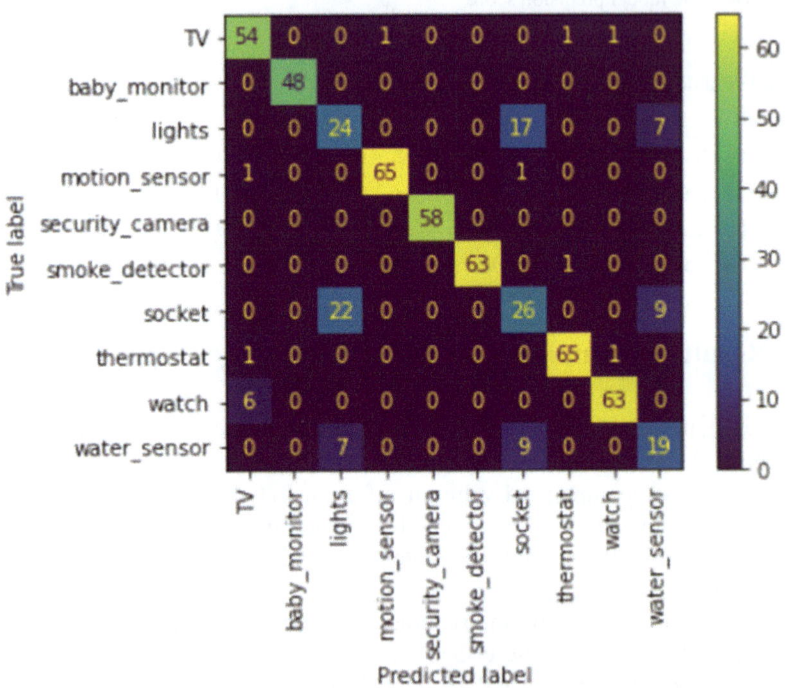

Fig. 4.8 Confusion matrix

Table 4.5 Cross validation score

	Accuracy score
Cross validation accuracy scores	0.75263158
	0.90789474
	0.87368421
	0.87368421
	0.84210526
Cross validation accuracy mean score	0.85

After getting the confusion matrix we apply the cross-validation method in order to get better accuracy of the model. Table 4.5 describe the value of cross-validation of accuracy value and the mean score cross-validation shows the exact and better accuracy of the model.

With a Cross validation accuracy of 0.85 the Overall, model does rather well, although it struggles to forecast the outcomes for lights, water sensor and sockets.

4.7 Conclusions, Challenges and Future Work

In order to profile the devices, the study established a more objective system of categorization, isolating attackers The structure brings successful traits to future IoT security, and uses mixed machine learning techniques to more reliably detect abnormal behavior from various Smart Home devices. Internet of Things sensing technology was inspired by the better processing and communication abilities offered by smart sensors. But these sensors can be used to deliberately strike a single sensed point rather than the entire data set. This hazard greatly undermines the accuracy with which machine learning algorithms can detect deviant behavior. We tested eight classifiers, including the DT, KNN, SVC, LR, AdaBoost, XGBRF, and LGBM. Therefore, the framework utilized XGB with random search and four hyperparameter tuners to thoroughly compare them all while also achieving overall attack detection accuracy of 85.96%.

However, the framework's limits must be acknowledged. A single detected data point for anomaly detection may make the system vulnerable to sophisticated assaults targeting sensor metrics, bypassing the detection algorithms. The model may also be limited in its applicability to different IoT contexts and device kinds, requiring further validation across different scenarios. The focus of this paper is on the classification and detection of IoT devices employing flow-dependent system communication assessment. Attacker may exploit IoT device categorization to uncover insecure IoT devices by performing an effective network stream of traffic assessment. The device characterization and recognition permit the network's operator to recognise rogue sensor in IoT system. We need to examine this proposed model in future research based on IoT system networks utilised with more types of IoT devices. The XGBRF showed promising anomaly detection results in IoT device profiling, but real-time machine learning model deployment challenges must be addressed. Since it relies on single observed data points for anomaly identification, the proposed system is vulnerable to targeted attacks on sensor metrics. In instances when adversaries exploit this vulnerability, they may avoid detection and compromise IoT system security. Due to the complexity of varied IoT contexts and device kinds, the anomaly detection method may be less successful, making the model difficult to modify and generalize. This approach must be validated and considered to assure its dependability and efficacy in dynamic, live situations when translating machine learning models from controlled experimental settings to real-world applications.

References

1. Lyon, G.F.: Nmap network scanning: The official Nmap project guide to network discovery and security scanning. Insecure, (2009)
2. Bebortta, S., Senapati, D., Panigrahi, C.R., Pati, B.: Adaptive performance modeling framework for QoS-aware offloading in MEC-based IIoT systems. IEEE Internet Things J. **9**(12), 10162–10171 (2021)
3. Bebortta, S., Singh, A.K., Pati, B., Senapati, D.: A robust energy optimization and data reduction scheme for iot based indoor environments using local processing framework. J. Netw. Syst. Manage. **29**, 1–28 (2021)
4. Francois, J., Abdelnur, H., State, R., Festor, O.: Ptf: Passive temporal fingerprinting. In: 12th IFIP/IEEE International symposium on integrated network management (IM 2011) and workshops, pp. 289-296. IEEE, (2011)
5. Tripathy, S.S., Imoize, A.L., Rath, M., Tripathy, N., Bebortta, S., Lee, C.C., Chen, T.Y., Ojo, S., Isabona, J., Pani, S.K.: A novel edge-computing-based framework for an intelligent smart healthcare system in smart cities. Sustainability. **15**(1), 735 (2023)
6. Bebortta, S., Senapati, D., Panigrahi, C.R., Pati, B.: An adaptive modeling and performance evaluation framework for edge-enabled green IoT systems. IEEE Trans Green. Commun. Netw. **6**(2), 836–844 (2021)
7. Senapati, D.: Generation of cubic power-law for high frequency intra-day returns: Maximum Tsallis entropy framework. Digital Signal Processing. **1**(48), 276–284 (2016)
8. Mukherjee, T., Singh, A.K., Senapati, D.: Performance evaluation of wireless communication systems over Weibull/q-lognormal shadowed fading using Tsallis' entropy framework. Wireless Pers. Commun. **106**(2), 789–803 (2019)
9. Nguyen, N.T., Zheng, G., Han, Z., Zheng, R.: Device fingerprinting to enhance wireless security using nonparametric bayesian method. In: INFOCOM, 2011 Proceedings IEEE, pp. 1404–1412. IEEE (2011)
10. Xu, Q., Zheng, R., Saad, W., Han, Z.: Device fingerprinting in wireless networks: Challenges and opportunities. IEEE Commun. Surv. & Tutor. **18**(1), 94–104 (2016)
11. Miettinen, M., Marchal, S., Hafeez, I., Asokan, N., Sadeghi, A.R., Tarkoma, S.: IoT SENTINEL: Automated device-type identification for security enforcement in IoT. In: Proceedings—International conference on distributed computing systems, pp. 2177–2184 (2017). https://doi.org/10.1109/ICDCS.2017.283
12. Nayak, G., Singh, A.K., Bhattacharjee, S., Senapati, D.: A new tight approximation towards the computation of option price. Int. J. Inf. Technol. **14**(3), 1295–1303 (2022)
13. Bertino, E., Islam, N.: Botnets and internet of things security. Computer **50**, 76–79 (2017)
14. Shah, T., Venkatesan, S.: Authentication of IoT device and IoT server using secure vaults. In: Proceedings of the 2018 17th IEEE international conference on trust, security and privacy in

computing and communications/12th IEEE international conference on big data science and engineering (TrustCom/BigDataSE), New York, NY, USA, pp. 819–824 (2018)
15. Nayak, G., Singh, A.K., Senapati, D.: Computational modeling of non-gaussian option price using non-extensive Tsallis' entropy framework. Comput. Econ. **57**(4), 1353–1371 (2021)
16. Mukherjee, T., Pati, B., Senapati, D.: Performance evaluation of composite fading channels using q-weibull distribution. In: Progress in Advanced Computing and Intelligent Engineering: Proceedings of ICACIE 2019, vol. 1, pp. 317–324. Springer Singapore, (2021)
17. Mukherjee, T., Nayak, G., Senapati, D.: Evaluation of symbol error probability using a new tight Gaussian Q approximation. Int. J. Syst., Control. Commun. **12**(1), 60–71 (2021)
18. Yi, H.-C., Huang, D.-S., Li, X., Jiang, T.-H., Li, L.-P.: A deep learning framework for robust and accurate prediction of ncrna-protein interactions using evolutionary information. Mol Ther-Nucleic Acids. **1**(11), 337–344 (2018). https://doi.org/10.1016/j.omtn.2018.03.001
19. Ling, H., Kang, W., Liang, C., Chen, H.: Combination of support vector machine and k-fold cross validation to predict compressive strength of concrete in marine environment. Constr. Build. Mater. **206**, 355–363 (2019). https://doi.org/10.1016/j.conbuildmat.2019.02.071
20. Kaur, H., Kumari, V.: Predictive modelling and analytics for diabetes using a machine learning approach. Appl Comput. Inf. 2018. https://doi.org/10.1016/j.aci.2018.12.004
21. Zhang, H., Yu, P., et al.: Development of novel prediction model for drug-induced mitochondrial toxicity by using naïve bayes classifier method. Food Chem. Toxicol. **10**, 122–129 (2017). https://doi.org/10.1016/j.fct.2017.10.021
22. Donzé, J., Bates, D.W., Schnipper, J.L.: Causes and patterns of readmissions in patients with common comorbidities: retrospective cohort study. BMJ. **347** (7171), (2013). https://doi.org/10.1136/bmj.f7171
23. Smith, D.M., Giobbie-Hurder, A., Weinberger, M., Oddone, E.Z., Henderson, W.G., Asch, D.A., et al.: Predicting non-elective hospital readmissions: a multi-site study. Department of veterans affairs cooperative study group on primary care and readmissions. J. Clin. Epidemiol. **53**(11), 1113–1118 (2000)
24. Han, J., Choi, Y., Lee, C., et al.: Expression and regulation of inhibitor of dna binding proteins id1, id2, id3, and id4 at the maternal-conceptus interface in pigs. Theriogenology **108**, 46–55 (2018). https://doi.org/10.1016/j.theriogenology.2017.11.029
25. Jiang, L., Wang, D., Cai, Z., Yan, X.: Survey of improving naive bayes for classification. In: Alhajj, R., Gao, H. et al., (eds). Lecture notes in computer science. Springer, (2007). https://doi.org/10.1007/978-3-540-73871-8_14
26. Jianga, L., Zhang, L., Yu, L., Wang, D.: Class-specific attribute weighted naive bayes. Pattern Recogn. **88**, 321–330 (2019). https://doi.org/10.1016/j.patcog.2018.11.032
27. Han, L., Li, W., Su, Z.: An assertive reasoning method for emergency response management based on knowledge elements c4.5 decision tree. Expert Syst Appl. **122**, 65–74 (2019). https://doi.org/10.1016/j.eswa.2018.12.042
28. Skriver M.V.J.K.K., Sandbæk, A., Støvring, H.: Relationship of hba1c variability, absolute changes in hba1c, and all-cause mortality in type 2 diabetes: a danish population-based prospective observational study. Epidemiology. **3**(1), 8 (2015). https://doi.org/10.1136/bmjdrc-2014-000060
29. ADA: Economic costs of diabetes in the U.S. in 2012. Diabetes Care. (2013)
30. Landwehr, N., Hall, M., Frank, E.: Logistic model trees. Mach. Learn. **59**, 161–205 (2005)
31. Freund, Y., Schapire, R.: A decision-theoretic generalization of on-line learning and an application to boosting. (1995)
32. Hastie, T., Tibshirani, R., Friedman, J.: Elements of statistical learning Ed. 2", Springer, (2009)
33. Snehi, M., Bhandari, A.: A novel distributed stack ensembled meta-learning-based optimized classification framework for real-time prolific IoT traffic streams. Arab. J. Sci. Eng. **47**(8), 9907–9930 (2022)
34. Aksoy, A., Gunes, M.H.: Automated iot device identification using network traffic. In: ICC 2019–2019 IEEE international conference on communications (ICC). IEEE, (2019)
35. Cvitić, I., et al.: Ensemble machine learning approach for classification of IoT devices in smart home. Int. J. Mach. Learn. Cybern. **12**(11), 3179–3202 (2021)

36. Rey, V., Sánchez, P.M.S., Celdrán, A.H., Bovet, G.: Federated learning for malware detection in IoT devices. Comput. Netw. **204**, 108693 (2022)
37. Shenoy, M.V.: HFedDI: A novel privacy preserving horizontal federated learning based scheme for IoT device identification. J. Netw. Comput. Appl. **214**, 103616 (2023)
38. Msadek, N., Soua, R., Engel, T.: Iot device fingerprinting: Machine learning based encrypted traffic analysis. In: 2019 IEEE wireless communications and networking conference (WCNC). IEEE, (2019)
39. Miettinen, M., et al.: Iot sentinel: Automated device-type identification for security enforcement in iot. In: 2017 IEEE 37th International Conference on Distributed Computing Systems (ICDCS). IEEE, (2017)
40. Ullah, I., Mahmoud, Q.H.: Network traffic flow based machine learning technique for IoT device identification. In: 2021 IEEE International Systems Conference (SysCon). IEEE, (2021)
41. Lee, S-Y. et al.: ProFiOt: Abnormal Behavior Profiling (ABP) of IoT devices based on a machine learning approach. In: 2017 27th International telecommunication networks and applications conference (ITNAC). IEEE, (2017)
42. Skowron, M., Janicki, A., Mazurczyk, W.: Traffic fingerprinting attacks on internet of things using machine learning. IEEE Access **8**, 20386–20400 (2020)
43. Shafagh, H., Hithnawi, A.: Poster: come closer: proximity-based authentication for the internet of things. In: Proceedings of annual international conference on mobile computing and networking, pp. 421–424. (2014)
44. Sheng, Y., Tan, K., Chen, G., Kotz, D., Campbell, A.: Detecting 802.11 mac layer spoofing using received signal strength. In Proc of IEEE INFOCOM, pp. 1768–1776. IEEE, (2008)
45. Sivanathan, A., Gharakheili, H.H., Loi, F., Radford, A., Wijenayake, C., Vishwanath, A., Sivaraman, V.: Classifying IoT devices in smart environments us-48 ing network traffic characteristics. IEEE Trans. Mob. Comput. **18**(8), 1745–1759 (2018)
46. Bebortta, S., Singh, A.K., Senapati, D.: Performance analysis of multi-access edge computing networks for heterogeneous IoT systems. Materials Today: Proceedings. **1**(58), 267–272 (2022)
47. Bebortta, S., Dalabehera, A.R., Pati, B., Panigrahi, C.R., Nanda, G.R., Sahu, B., Senapati, D.: An intelligent spatial stream processing framework for digital forensics amid the COVID-19 outbreak. Smart Health. **1**(26), 100308 (2022)
48. Bebortta, S., Tripathy, S.S., Modibbo, U.M., Ali, I.: An optimal fog-cloud offloading framework for big data optimization in heterogeneous IoT networks. Decis. Anal. Journal. **1**(8), 100295 (2023)
49. Bebortta, S., Singh, A.K., Mohanty, S., Senapati, D.: Characterization of range for smart home sensors using Tsallis' entropy framework. In: Advanced computing and intelligent engineering: proceedings of ICACIE 2018, vol. 2, pp. 265–276. Springer Singapore, Singapore (2020)
50. Bebortta, S., Singh, S.K.: An adaptive machine learning-based threat detection framework for industrial communication networks. In: 2021 10th IEEE international conference on communication systems and network technologies (CSNT), pp. 527–532. IEEE (2021)
51. Yun, J., Ahn, I.-Y., Song, J., Kim, J.: Implementation of sensing and actuation capabilities for IoT devices using oneM2M platforms. Sensors **19**(20), 4567 (2019)
52. Tripathy, S.S., Rath, M., Tripathy, N., Roy, D.S., Francis, J.S., Bebortta, S.: An intelligent health care system in fog platform with optimized performance. Sustainability. **15**(3), 1862 (2023)
53. sklearn.metrics.f1 score.: https://scikit-learn.org/stable/modules/generated/sklearn.metrics.f1_score.html. [Online; Accessed 24 Mar 2022]
54. Bebortta, S., Senapati, D., Rajput, N.K., Singh, A.K., Rathi, V.K., Pandey, H.M., Jaiswal, A.K., Qian, J., Tiwari, P.: Evidence of power-law behavior in cognitive IoT applications. Neural Comput. Appl. **32**, 16043–16055 (2020)
55. Ahmad, T., Zhang, D.: Using the internet of things in smart energy systems and networks. Sustain. Cities Soc., 102783 (2021)
56. Bebortta, S., Singh, S.K.: An opportunistic ensemble learning framework for network traffic classification in IoT environments. In: Proceedings of the seventh international conference on mathematics and computing: ICMC 2021, pp. 473–484. Springer Singapore, Singapore (2022)

57. Bebortta, S., Senapati, D.: Empirical characterization of network traffic for reliable communication in IoT devices. Secur. Cyber-Phys. Syst.: Found. Appl., 67–90 (2021)
58. Bebortta, S., Panda, M., Panda, S.: Classification of pathological disorders in children using random forest algorithm. In: 2020 International conference on emerging trends in information technology and engineering (ic-ETITE), pp. 1–6. IEEE (2020)
59. Das, S.K., Bebortta, S.: Heralding the future of federated learning framework: architecture, tools and future directions. In: 2021 11th International conference on cloud computing, data science & engineering (Confluence), pp. 698–703. IEEE (2021)
60. Bebortta, S., Senapati, D.: Characterizing the epidemiological dynamics of COVID-19 using a non-parametric framework. Curr. Sci. **122**(7), 790 (2022)
61. Mukherjee, T., Bebortta, S., Senapati, D.: Stochastic modeling of q-Lognormal fading channels over Tsallis' entropy: Evaluation of channel capacity and higher order moments. Digit. Signal Processing. **1**(133), 103856 (2023)
62. Bebortta, S., Tripathy, S.S., Basheer, S., Chowdhary, C.L.: FedEHR: A federated learning approach towards the prediction of heart diseases in IoT-based electronic health records. Diagnostics. **13**(20), 3166 (2023)
63. Bebortta, S., Rajput, N.K., Pati, B., Senapati, D.: A real-time smart waste management based on cognitive IoT framework. In: Advances in electrical and computer technologies: select proceedings of ICAECT 2019, pp. 407–414. Springer Singapore, Singapore (2020)
64. Bebortta, S., Singh, S.K.: An intelligent framework towards managing big data in internet of healthcare things. In: International conference on computational intelligence in pattern recognition, pp. 520–530. Springer Nature Singapore, Singapore (2022)
65. Bebortta, S., Singh, S.K., Rath, M., Mukherjee, T.: Dynamic framework towards sustainable and energy-efficient routing in delay tolerant IoT-based WSNs. Int. J. Syst., Control. Commun. **15**(1), 79–94 (2024)
66. Bebortta, S., Senapati, D.: Toward cost-aware computation offloading in IoT-based MEC systems. Natl. Acad. Sci. Letters. **24**, 1–4 (2023)
67. Bebortta, S., Das, S.K.: Assessing the impact of network performance on popular e-learning applications. In: 2020 Sixth international conference on e-learning (econf), pp. 61–65. IEEE, (2020)
68. Bebortta, S., Tripathy, S.S., Basheer, S., Chowdhary, C.L.: DeepMist: Towards deep learning assisted mist computing framework for managing healthcare big data. IEEE Access., (2023)
69. Singh, A.K., Senapati, D., Bebortta, S., Rajput, N.K.: A non-stationary analysis of Erlang loss model. In: Progress in advanced computing and intelligent engineering: proceedings of ICACIE 2019, vol. 1, pp. 286–294. Springer Singapore (2021)
70. Tripathy, S.S., Bebortta, S., Gadekallu, T.R.: Sustainable fog-assisted intelligent monitoring framework for consumer electronics in industry 5.0 applications. IEEE Trans. Consum. Electron. (2023)
71. Bebortta, S., Senapati, D.: A secure blockchain-based solution for harnessing the future of smart healthcare. In: InIoT-based data analytics for the healthcare industry, pp. 167–191. Academic Press, (2021)
72. Bebortta, S., Das, S.K., Chakravarty, S.: Fog-enabled intelligent network intrusion detection framework for internet of things applications. In: 2023 13th international conference on cloud computing, data science & engineering (Confluence), pp 485–490. IEEE (2023)
73. Bebortta, S., Singh, S.K.: An intelligent network intrusion detection framework for reliable UAV-based communication. In: International conference on cryptology & network security with machine learning, pp. 169–177. Springer Nature Singapore, Singapore (2022)
74. Bebortta, S., Senapati, D.: Precision healthcare in the era of IoT and big data. Comput. Intell. Aided Syst. Healthc. Domain. **14**, 91 (2023)
75. Bebortta, S., Panda, T., Singh, S.K.: An intelligent hybrid intrusion detection system for internet of things-based applications. In: 2023 International conference on network, multimedia and information technology (NMITCON) (pp. 01–06). IEEE, (2023)

Chapter 5
Application of Machine Learning to Improve Safety in the Wind Industry

Bertrand David Barouti and Seifedine Kadry

Abstract The offshore wind industry has been gaining significant attention in recent years, as the world looks to transition to more sustainable energy sources. While the industry has successfully reduced costs and increased efficiency, there is still room for improvement in terms of safety for workers. Using machine learning (ML) and deep learning (DL) technologies can significantly improve offshore wind industry safety by facilitating better accident prediction and failure prevention. The current study aims to fill a significant gap in the existing literature by developing a useful selection of machine learning models for simple implementation in the offshore wind industry. These models will then be used to inform decision-making around safety measures, such as scheduling maintenance or repairs or changing work practices to reduce risk. The development of this tool has the potential to significantly contribute to the long-term viability of the offshore wind industry and the protection of its workers. By providing accurate predictions of potential accidents and failures, the tool can enable companies to take proactive measures to prevent incidents from occurring, reducing the risk of injury or death to workers and reducing the financial cost of accidents and downtime. The chapter concludes with a summary of the present study's research challenge and the literature gaps. It highlights the importance of developing effective machine learning models and implementing stricter data records to improve safety in the offshore wind industry and the potential impact of these tools on the long-term viability of the industry. The chapter also notes that the high performance of selected models proves the reliability of the expected predictions and demonstrates the effectiveness of machine learning models for decision- making around safety in the offshore wind industry.

Keywords Undersampling · SMOTE technique · Ridge classifier · Extra trees classifier · Wind industry · Neural network · LSTM models

---

B. D. Barouti · S. Kadry (✉)
Department of Applied Data Science, Noroff University College, Kristiansand, Norway
e-mail: Seifedine.Kadry@noroff.no

J. Nayak et al. (eds.), *Machine Learning for Cyber Physical System: Advances and Challenges*, Intelligent Systems Reference Library 60,
https://doi.org/10.1007/978-3-031-54038-7_5

5.1 Introduction

"The existing literature lacks a comprehensive problem formulation addressing safety concerns in the wind industry, particularly in the context of leveraging Machine Learning (ML) and Deep Learning (DL) models. The authors aim to bridge this gap by developing and implementing ML and DL solutions to enhance safety measures within the wind industry. This involves identifying critical safety challenges, exploring relevant data sources, and, the focus of the present research, designing effective predictive models that can proactively mitigate risks and improve overall safety standards in wind energy operations."

Application of machine learning and deep Learning to Health and Safety processes in the Wind industry can contribute to the safety of the workers, reduce the cost of wind farm projects, and increase the reputation of the companies involved. Multiple factors, amongst them technology improvements driving the price down and rising investments in renewable energy sources and favorable government policies, have led to a sharp increase in wind (turbine generator) farms. In 2021 in the United States of America alone, the offshore wind pipeline grew by 13,5% over the previous year only.

As this market grows, so do the related activities, from installing wind turbines (from preparing and installing the foundations through laying cables to erect the towers and turbine) to their operation and maintenance. All these phases in the life cycle of a wind farm require human intervention, and the personnel working on these installations is exposed to hazards that can have dire consequences (from exposure to high energy sources to working at heights under challenging conditions offshore). A lot of effort has been put into optimizing the design of the wind turbines (to maximize energy output and have more reliable blades) and their maintenance, as well as improving routing and scheduling to minimize costs.

To achieve this, machine learning and deep learning have been successfully used, but the same process has not been applied with the same rigor to improve health and safety. Machine learning and deep learning can enhance personnel safety as well as promote operational excellence by eliminating wastes (the correct resources are assigned where they matter the most) and therefore also offer a financial, competitive advantage to the organization that adopts them. Machine learning, particularly for health and safety incident predictions and discovering their main contributing factors, is poorly implemented in the wind industry, where it could help prevent unnecessary incidents (including injuries), operations downtime and sub-optimal financial results.

The primary objective of this chapter is to find out the best models that can assist users in deploying predictive models in Production, amongst other classes, for the severity, cause or type of safety incidents in the wind industry. To achieve this primary objective, the following secondary objectives have been identified.

- Perform a review of the state of machine learning implementation for health and safety in the wind industry
- Identify the most common key performance indicators for health and safety in the wind industry and the data gathered by the companies operating in this sector.

- Make a comparative study to select the most suitable predictive models to fit certain types of datasets.

Physical work is the most challenging and demanding in an offshore wind farm. Wearing safety suits and climbing ladders to the turbine with heavy equipment takes a toll on workers. The work is physically demanding, with tasks involving work at heights and inside installations. Transfer and access to facilities are also physically and psychologically challenging. The risk of accidents and unstable weather leads to mental stress. The 12-hour work shifts of workers are tedious and lengthy, often involving overtime. Pressure to complete activities is very much present due to the losses incurred with any downtime of the turbines. Each second spent rectifying a wind turbine costs the wind farm company. Waiting times are also challenging in offshore wind farms due to bad weather conditions. The occurrence of delays in workflow due to modifications and the high costs involved is problematic.

Similarly, communication between offshore and onshore personnel is difficult. Also, any medical emergency has limited treatment options and long emergency routes. This brings risks to people working in offshore wind farms. Due to the complex nature of work, many technicians and other personnel operating offshore wind farms follow rotational shift schedules [5]. 12-hour shift rotations are standard at these remote sites, making working schedules hectic. Another significant challenge with offshore wind farms is downtime waiting for work during corrective maintenance. In such a situation, shift rotation time can be prolonged due to the shortage of technical staff. Furthermore, scheduling conflicts can arise if fault duration is elongated, contrary to what was set in the work orders. Nevertheless, with appropriate strategies, all these challenges can be managed efficiently. Operation and maintenance costs are among offshore wind farms' most significant cost components.

One way to reduce costs is to make maintenance activities more efficient by streamlining maintenance schedules and vessel routing. The European Committee for Standardization categorizes maintenance activities for wind power systems into corrective maintenance, preventive maintenance, and condition-based maintenance [11]. Offshore wind farm Operations and Maintenance specific challenges can be broadly listed as follows, according to [23].

- **High crew dispatch costs**: Assembling and deploying a maintenance team is quite expensive since offshore turbines are frequently deployed offshore and in remote locations where wind conditions are best.
- **High production losses**: As the scale and capacity of offshore turbines increase, the cost of downtime is growing intolerably high due to the related production losses of a failing ultra-scale offshore turbine.
- **Limited accessibility**: Access to turbines can frequently be restricted for extended periods due to harsh weather and sea conditions, ranging from a few hours to many days.

Scheduling activities, work orders and personnel rotation are all part of the operation and maintenance of offshore wind farms. The men and women working in offshore wind farms undergo many challenges to cope with their work. According to [8], the employers' offshore wind body covering offshore wind, the Total Recordable Injury Rate (TRIR) was 3.28 while the number of Lost work day injuries was 50 for 2021 and, noticeably, the number of high potential incidents in the category' civil works onshore including excavations' increased by 175% compared to 2020.

Following [20], implementing predictive indicators using predictive analytics would benefit organizations trying to determine the likelihood of incidents occurring. Although already somewhat implemented within the oil and gas sector, another high-hazard industry, the wind industry, has not yet embraced machine learning and predictive analytics. Machine Learning has been identified as a way to improve process safety [2]. Similarly, some authors argue that machine learning will contribute to learning from major accidents [26]. Although some industries, such as construction [28], automotive [4] and aerospace [19], have adopted machine learning for safety. The offshore wind industry is relatively new and could benefit from the same application of Machine Learning, specifically to improve the safety of the personnel working in this industry, from installation to operation and maintenance of those assets.

5.2 Literature Review

This section describes various literature works carried on offshore wind industry along with its challenges, challenges of traditional safety management practices, various machine learning and deep learning approaches used in the offshore wind industry along with its challenges.

5.2.1 Context

The number of offshore wind energy installations and activities has increased dramatically during the last several years. This sector offers a steady supply of clean energy, but such facilities' building, installation, operation, and maintenance are dangerous. Therefore, novel and efficient approaches to enhance safety in the sector are urgently required [3]. It has been suggested that machine learning (ML) and deep learning (DL) methods might help with this issue. The offshore wind sector may utilize these techniques to create prediction models to help them avoid accidents and keep workers safe. This study intends to remedy the wind industry's sluggish adoption of ML/DL to ensure worker safety [7].

The created application is a GUI-based model selection tool for machine learning and deep learning. With this graphical user interface, customers may choose the most appropriate prediction model for their requirements and understand the results. Weather occurrences, equipment failures, or other possible threats to safety may all be anticipated with the help of this program. Overall, ML/DL's use in the offshore wind business has the potential to improve worker safety significantly. The continuing success of this vital sector of the economy depends on our ability to create accurate predictive models that can be used to identify and eliminate possible threats to employees. The rapid expansion of wind turbine installation has introduced new hazards that must be mitigated for the sake of worker welfare and the long-term health of the business sector. Machine learning and deep learning technologies provide a possible answer to these issues by allowing the prediction and prevention of accidents and equipment breakdowns [13].

This study of the relevant literature introduces the reader to the offshore wind business, its safety problems, and the conventional safety management practices now in use. The reader is also given an overview of machine learning and deep learning and an examination of recent research on the use of ML to enhance safety in various sectors [12]. Studies on wind turbine failure prediction, structural health monitoring, and blade icing detection are examples of how ML has been used in the offshore wind sector and are included in this overview. Challenges and restrictions of using ML in the offshore wind business are explored, including data and computing resource scarcity and the difficulty of understanding and explaining ML results. Finally, the possible advantages of ML in the offshore wind business are highlighted. These advantages include enhanced safety, decreased maintenance costs, and enhanced efficiency. The review finishes with a synopsis of the research challenge addressed by the present study and the gaps in the existing literature. Challenges and restrictions of using ML in the offshore wind business are explored, including data and computing resource scarcity and the difficulty of understanding and explaining ML results.

5.2.2 *Overview of the Offshore Wind Industry and Its Safety Challenges*

Installing and operating wind turbines in coastal or offshore marine settings is the focus of the offshore wind business. Demand for renewable energy sources and government measures to minimize greenhouse gas emissions have rapidly expanded the sector in recent years [13]. However, the harsh and dynamic maritime environment, high voltage electrical systems, and complicated logistics connected with offshore installation and maintenance provide specific safety problems for the offshore wind sector.

The danger of falls from height during turbine construction, maintenance, and repair is one of the biggest safety problems in the offshore wind business. Often in inclement weather and with little access to safety equipment, workers must go to tremendous heights to complete these operations [16]. If enough precautions are not taken, this might lead to catastrophic consequences, including loss of life. High-voltage electrical systems pose a threat due to the potential for electrical dangers. Workers' safety is at stake if undersea cables, which carry power generated by wind

Table 5.1 Various works carried on the offshore wind industry

Reference	Problem definition	Findings	Key learnings
Jordan and Mitchell [13]	General overview of offshore wind industry	Expansion of offshore wind business, safety challenges, and government measures	Provides a comprehensive overview
Lian et al. [16]	Falls from height during turbine operations	Workers face the risk of falls from height during construction and maintenance	Highlights a significant safety challenge
Lian et al. [16]	High-voltage electrical systems	Potential electrical dangers from undersea cables if not properly maintained	Identifies a critical safety concern
Lian et al. [16]	Transportation and logistics challenges	Challenges in transporting large components, leading to potential accidents	Addresses logistical difficulties in the offshore wind business

turbines and transported to land, are not correctly maintained or separated. Transportation and logistics are other areas where the offshore wind business has difficulties. Offshore installations rely on ships or barge to transfer large components like blades, nacelles, and towers, which things like bad weather might delay. This increases the potential for accidents and equipment failure due to delayed or disrupted maintenance schedules [16]. Improving safety planning, identifying possible risks, and forecasting equipment breakdowns are all areas where machine learning and other cutting-edge technology might be helpful. Table 5.1 presents the analysis of various works carried out on the offshore wind industry.

5.2.3 *Review of Traditional Safety Management Practices in the Offshore Wind Industry*

The offshore wind industry's standard procedures for managing risk have always included several safeguards designed to protect employees and bystanders. Risk analyses, safety checks, contingency plans, and employee education and training are all examples of such procedures. The offshore wind industry relies heavily on risk assessments for its safety management [18]. Assessing risk entails seeing prospective threats and weighing them against their probability and potential impact. Before installing wind turbines and periodically afterwards, it is common to practice doing a risk assessment. In the offshore wind business, safety checks are also a common practice. Wind turbines undergo regular inspections by qualified workers to check for damage, wear, and other potential safety hazards. In addition to routine checks, occasional deep checks may be performed [18].

Safety management in the offshore wind sector should also include emergency response planning. Emergency procedures, including those for dealing with fires and turbine failures, are developed as part of these plans. Worker preparedness and coordination with local emergency services are essential to any disaster response strategy. As a last point, safety management in the offshore wind business relies heavily on employee training programs. Workers are educated on several safety aspects, such as PPE usage, danger identification, and emergency protocol, as part of these programs. Traditional safety management practices are crucial for protecting employees and the general public [18]. However, as the offshore wind sector grows and changes, so may the need for innovative solutions to new safety issues. Offshore wind farms might benefit from incorporating machine learning and other cutting-edge technology into their already established safety management procedures.

Traditional safety management in the offshore wind sector includes the practices above and careful adherence to industry rules and regulations (Table 5.2). For example, the International Electrotechnical Commission (IEC) and the Occupational Safety and Health Administration (OSHA) are two organizations and bodies responsible for developing such standards and rules. Wind turbines and associated equipment may be more confidently relied upon if built, installed, and maintained following these guidelines [21] (Table 5.3). Using safety equipment and protective clothing is also an important part of the conventional approach to safety management in the offshore wind sector. During turbine construction and maintenance, workers must wear personal protection equipment (PPE) such as hard helmets, safety glasses, and harnesses to reduce the likelihood of harm. Safety features like emergency stop buttons, fire suppression, and lightning protection systems may also be added to wind turbines.

The offshore wind sector still faces safety difficulties, notwithstanding the success of conventional safety management practices (Table 5.4). For instance, equipment

Table 5.2 Traditional safety management practices in offshore wind industry

Reference	Practice	Description
Maldonado-Correa et al. [18]	Risk assessments	Identifying and evaluating potential threats, weighing them against their probability and impact
	Safety checks	Regular inspections of wind turbines for damage, wear, and safety hazards by qualified workers
	Emergency response planning	Development of emergency procedures, including fire and turbine failure response plans
	Employee training programs	Educational programs covering safety aspects, such as PPE usage, danger identification, and emergency protocols

Table 5.3 Adherence to industry rules and regulations

Reference	Organization/bodies	Standards and rules
Mitchell et al. [21]	International Electrotechnical Commission (IEC)	Development of standards for building, installing, and maintaining wind turbines and equipment
	Occupational Safety and Health Administration (OSHA)	Establishment of safety guidelines and regulations for the offshore wind industry

Table 5.4 Challenges in traditional safety management practices and potential solutions

Reference	Description	Potential solutions
Olguin et al. [22]	Difficulty in forecasting safety hazards in harsh sea environment	Utilize machine learning and cutting-edge technologies to enhance predictive skills, identify potential hazards in advance, and optimize maintenance schedules for reduced accident risk

breakdowns and other safety dangers might be hard to forecast in the harsh and dynamic sea environment. By enhancing predictive skills, spotting possible hazards in advance, and optimizing maintenance schedules to reduce accident risk, machine learning and other cutting-edge technologies may assist in solving these issues [22]. Conventional safety management practices are crucial to ensure the safety of employees and the general public in the offshore wind business. However, to guarantee the sustained safety and success of offshore wind projects, the industry must stay watchful and adaptive in the face of increasing safety problems and be open to incorporating new technology and practices.

5.2.4 Introduction to Machine Learning and Deep Learning Technologies

Algorithms and models that can learn from data and make predictions or judgments based on that data are the focus of both machine learning (ML) and deep Learning (DL), two subfields of artificial intelligence (AI) that are closely connected. The offshore wind sector is only one area where ML and DL technologies are finding widespread usage. Building algorithms and models that can automatically learn from data is a critical component of ML. In other words, ML algorithms may learn from existing data, apply that knowledge to new data, and then make predictions or choices. Examples of popular ML algorithms include decision trees, SVMs, and neural networks [36].

Creating algorithms and models that perform as the human brain does is DL's goal, a subfield of ML. Deep learning algorithms use simulated neural networks to sift through data, draw conclusions, or make predictions. DL algorithms shine when analyzing multifaceted data types like photos, voice, and natural language. The capacity to swiftly and effectively analyze vast and complicated datasets is a significant benefit of ML and DL technology. This may aid with the discovery of trends and patterns that human analysts would miss, leading to better decisions across many domains [33]. ML and DL technologies have several potential applications in offshore wind, including failure prediction, maintenance optimization, and enhanced safety planning. To forecast when maintenance is needed, ML algorithms may examine data collected by sensors installed on wind turbines to look for red flags that suggest impending failure. Images and videos captured at offshore wind farms may be analyzed using DL algorithms to spot possible dangers, such as personnel doing activities in risky environments. The offshore wind business is just one area where ML and DL technologies may enhance productivity, security, and decision-making. It's safe to assume that as these technologies advance, they'll gain significance across various sectors and use cases. In Table 5.5, an overview of ML and DL approaches has been presented.

Table 5.5 Overview of Machine Learning (ML) and Deep Learning (DL) technologies

Reference	Aspect	Description
Zulu et al. [36]	Definition of ML algorithms	ML focuses on algorithms and models that learn from data to make predictions or judgments. Popular ML algorithms include decision trees, SVMs, and neural networks
Yeter et al. [33]	Definition of DL algorithms	DL, a subfield of ML, aims to create algorithms and models that mimic the human brain. DL algorithms use simulated neural networks to analyze complex data types like photos, voice, and natural language
Zulu et al. [36]	Applications in offshore wind	ML and DL technologies have potential applications in the offshore wind sector, including failure prediction, maintenance optimization, and enhanced safety planning
Zulu et al. [36]	ML in offshore wind	ML algorithms can analyze sensor data from wind turbines to predict maintenance needs and detect red flags indicating potential failures
Yeter et al. [33]	DL in offshore wind	DL algorithms can analyze images and videos from offshore wind farms to identify potential safety risks, such as personnel in hazardous environments
Zulu et al. [36]	Overall impact and future trends	ML and DL technologies are expected to play a significant role in enhancing productivity, security, and decision-making across various sectors as they continue to advance

5.2.5 Previous Studies on the Application of Machine Learning to Improve Safety in Other Industries

Several research projects have looked at how machine learning may be used to improve industry-wide safety. Some of the more notable instances include.

- Machine learning algorithms have been used to predict patient outcomes and spot hidden health hazards in the healthcare sector. In the healthcare industry, ML algorithms have been used to analyze patient data and predict, for example, hospital readmission rates and illness risk [32].
- Machine learning algorithms may detect faulty machinery and anticipate service requirements in the industrial sector. For instance, ML systems may examine sensor data from machinery to forecast when maintenance is needed to discover patterns that may suggest possible equipment breakdowns [25].
- Safer mobility is possible using ML algorithms in the transportation sector. For instance, ML systems may examine vehicle sensor data to spot red flags like unexpected stops or erratic driving that might threaten passengers [6].
- Safer building sites may be achieved via the application of ML algorithms by the construction sector. ML systems may analyze image and videos from construction site to spot employees in hazardous situations [1].

These studies show that machine learning can potentially enhance safety across many sectors. Machine learning algorithms can produce more accurate predictions and inferences by analyzing massive datasets in ways that would be impossible for human analysts. The offshore wind business is no exception, and there is a rising interest in using machine learning to increase safety. Various studies carried on the enhancement of safety in various industries using ML approaches has been represented in Table 5.6.

5.2.6 Application of ML to the Offshore Wind Industry

Several studies in recent years have looked at the possibility of using machine learning (ML) in the offshore wind business to boost efficiency, save costs, and promote safety. In particular, as will be seen below, ML has been used in predicting failures in wind turbines, checking on structural health, and identifying instances of blade icing.

- **Prediction of Wind Turbine Failures**: Several researchers have looked at the possibility of using ML algorithms to foresee breakdowns in wind turbines. For instance, SsCADA data and a long short-term memory (LSTM) neural network were used to predict gearbox breakdowns in wind turbines [32]. Li et al. [15] also employed vibration data and a stacked auto-encoder to foresee bearing failures in wind turbine generators. These results show the promise of ML in predicting and avoiding problems in offshore wind turbine components.
- **Monitoring of Structural Health**: Offshore wind turbines' structural health has also been tracked using ML. For instance, Zhu and Liu [35] analyzed spectrogram

Table 5.6 Previous studies on ML for safety improvement in various industries

Reference	Problem definition	Model used	Findings	Advantage	Limitation
Yan [32]	Healthcare—patient outcomes	ML algorithms	Predicted hospital readmission rates and illness risk	Improved accuracy in healthcare predictions	Dependent on the quality and diversity of patient data
Surucu et al. [25]	Industrial - Faulty machinery	ML systems	Anticipated service requirements and detected faulty machinery	Early identification of potential equipment breakdowns	Relies on accurate sensor data for effective detection
Gangwani and Gangwani [6]	Transportation—safer mobility	ML systems	Identified red flags in vehicle sensor data threatening passengers	Real-time monitoring for enhanced passenger safety	Limited by sensor data availability and accuracy
Adekunle et al. [1]	Construction—Safer building sites	ML systems	Detected employees in hazardous situations on construction sites	Proactive identification of safety risks	Depends on the quality and availability of visual data

imagery using a convolutional neural network (CNN) to detect fractures and other structural flaws in wind turbine blades. Similarly, the results of a study demonstrated that a proposed framework could identify blade cracks using unmanned active blade data and artificially generated images [29]. These results show the potential of ML as a technique for proactively assessing the structural health of offshore wind turbines and preventing catastrophic failure.

- **Detection of Blade Icing**: Blade icing detection is another area where ML has found use in the offshore wind sector. The probability of ice shedding from wind turbines may be increased by icing, which reduces their performance and poses safety risks. Several researchers have looked at the possibility of using ML algorithms to identify blade icing as a solution to this problem. Using deep neural networks and wavelet transformation [34] could detect icing on wind turbine blades using a classification anomaly detection system. These results show how ML may be used to identify and prevent blade icing on offshore wind turbines, increasing their efficiency and safety.

Table 5.7 represents various studies carried on Offshore wind industry using ML techniques. Overall, these studies show how ML has the potential to enhance safety and efficiency in the offshore wind sector via the detection of ice on turbine blades, the prediction of equipment failures, and the monitoring of structural health. The

Table 5.7 Application of ML to offshore wind industry studies

Reference	Problem definition	Model used	Findings	Advantage	Limitation
Yan [32]	Predicting wind turbine failures	LSTM neural network	Predicted gearbox breakdowns in wind turbines	Proactive maintenance to avoid breakdowns	Relies on the availability and quality of SCADA data
Li et al. [15]	Predicting bearing failures	Deep belief network (DBN) integrated with back-propagation (B-P) fine-tuning and layer-wise training	Foreseen bearing failures in wind turbine generator	Early identification of potential failures	Reliance on vibration data for accuracy
Zhu and Liu [35]	Monitoring structural health	CNN	Detected fractures and structural flaws in wind turbine blades	Proactively assess structural health	Quality and diversity of training data matter
Wang and Zhang [29]	Monitoring structural health	Extended cascading classifier is developed (from a set of base models: LogitBoost, Decision Tree, and SVM)	Identified blade cracks using unmanned active blade data and artificially generated images	Early detection of potential structural issues	Dependence on the proposed framework
Yuan et al. [34]	Detection of blade icing	Deep neural networks, Wavelet transformation	Detected icing on wind turbine blades using a classification anomaly detection system	Increased efficiency and safety through prevention	Accuracy depends on data quality and quantity

offshore wind sector should expect ML to become a vital resource as technology advances.

5.2.7 *Challenges of Using ML in the Offshore Wind Industry*

The offshore wind business might gain immensely by implementing ML, but several obstacles and restrictions must be overcome first. Below are mentioned some of the major barriers that have arisen due to the implementation of ML in the offshore wind sector (Table 5.8).

Table 5.8 Challenges of Using ML in the offshore wind industry

Reference	Challenge	Description	Potential solutions
Wolsink [30]	Lack of data	Scarcity of data due to the novelty of offshore wind turbines	Explore the use of sensors, drones, and other monitoring technologies to collect more data
			Encourage collaboration between industrial players and academic institutions for data sharing
Thulasinathan et al. (2022)	Computational resources	Computing resources needed for analyzing vast and complicated datasets	Investigate edge computing and other methods to optimize ML algorithms for application in contexts with limited resources
			Address personnel and infrastructure limitations through advancements in technology and collaboration
Ren et al. [24]	Interpretability and explainability	Difficulty in understanding and explaining ML algorithms, often considered "black boxes"	Develop ML models that are more interpretable and explainable, using decision trees and other methods
			Foster research and development to enhance transparency and accountability in ML algorithms
			Promote the use of interpretable models in safety–critical applications

- **Lack of Data**: The paucity of data is a major obstacle to use ML in the offshore wind business. Due to the novelty of offshore wind turbines, information on their operation and performance is typically scant. ML algorithms need enormous volumes of data to discover significant patterns and generate accurate predictions, so it might be challenging to train them properly [30]. Researchers are also considering using sensors, drones, and other monitoring technologies to solve this problem.
- **Computational Resources**: The computing resources needed to analyze vast and complicated datasets are another difficulty when using ML in the offshore wind

business. Computationally demanding ML algorithms may benefit from dedicated hardware and software. Offshore wind businesses may struggle with this since they lack the personnel or infrastructure to collect and analyze massive amounts of data (Thulasinathan et al., 2022). Edge computing and other methods are being investigated as potential solutions for optimizing ML algorithms for application in contexts with limited resources.

- **Interpretability and Explainability**: The difficulty in understanding and explaining ML algorithms is another obstacle to their widespread adoption in the offshore wind sector. Unfortunately, the decision-making mechanisms of ML algorithms are typically considered mysterious "black boxes" that defy explanation [24]. This might be a concern in safety–critical applications since it can be hard to understand why a machine-learning algorithm came to a particular conclusion. Researchers are attempting to solve this problem by creating ML models that are easier to understand and explain via the use of decision trees and other methods.

While there is much to be gained by applying ML to the offshore wind business, several obstacles must first be removed before it can be used to its full potential. Technology advancement, data collecting and administration, and cooperation between industrial players and academic institutions will be required to meet these difficulties.

5.2.8 Traditional Safety Metrics in the Wind Industry

Worker, contractor, and visitor safety is a top priority in the wind energy sector. Measuring and monitoring safety performance across industries requires the use of safety metrics. Common applications for these measures include problem area identification and creating plans to enhance worker safety. The Total Recordable Incident Rate (TRIR) is one of the most popular safety indicators used in the wind business. Injuries and illnesses are included in the total recordable incident rate (TRIR), expressed as a number per 200,000 hours worked. The TRIR is a valuable tool for comparing and contrasting firms' and sectors' safety records, and it is used by a broad range of businesses, including the wind energy sector.

The LTIR, or Lost Time Incident Rate, is another crucial indicator of wind sector safety. The number of occurrences leading to missed workdays per 200,000 hours worked constitutes the LTIR. The LTIR is often used to gauge the seriousness of accidents regarding lost time and productivity; it is a more nuanced indicator than the TRIR. The wind industry also uses the severity rate (SR), quantifying the severity of injuries and illnesses, and the near-miss rate (NMR), estimating the number of near-misses that did not result in injuries or illnesses. However, these indicators do not provide a whole picture of safety performance and have shortcomings. These indicators, for instance, do not assess the value of safety initiatives or the bearing of safety culture on safety results. In addition, these indicators can only be used as a health check of any organization or group thereof. They do not provide the necessary information to specifically assign resources to the areas that would prove most beneficial. One of the risks of using these kinds of indicators is that minor incidents (with the potential to become major accidents and fatalities) are not always recorded, thus giving an inaccurate reflection of the organization's real performance.

5.2.9 Limitations and Challenges Associated with Traditional Safety Metrics

While conventional safety criteria have been employed in the wind sector for a while, they have drawbacks and cannot guarantee a completely risk-free environment. The following are examples of restrictions and difficulties.

- Traditional safety metrics are reactive in nature [9], as they seek to analyze historical events and accidents to determine safety patterns and enhance safety precautions. This method is reactive rather than proactive and therefore does not permit the early detection of possible security flaws.
- Traditional safety measurements are typically based on partial data, which may lead to erroneous conclusions about a system's safety. Some occurrences involving safety may not be recorded, and not all safety-related tasks may have their data gathered.
- Due to the lack of a universally accepted set of safety criteria in the wind sector, it is difficult to establish meaningful comparisons between the safety records of various projects and businesses.
- Traditional safety measurements, which emphasize trailing indications like injury rates, fail to capture the whole picture of safety performance [17]. Safety culture, safety leadership, and safe practices are examples of leading indicators that are poorly recorded.
- Traditional safety measurements frequently have a narrow emphasis, considering just the risks associated with one or two areas of a wind farm's operations. Thus, there may be blind spots in terms of safety if we stick to this method.
- Traditional safety measures typically fail to engage employees in safety management due to a lack of motivation. Employees may not participate in safety initiatives because they see safety measures as a tool for management rather than a tool to enhance safety performance.
- Predicting future safety hazards is challenging since traditional safety indicators do not consider industry shifts or emerging threats. This means they may not be reliable indicators of impending danger.
- Technology to enhance safety performance is typically overlooked in conventional safety measurements. Proactive safety management is made possible by innovations like artificial intelligence and machine learning, which give real-time data on potential hazards.

In summary, the limits and difficulties of existing safety criteria render them sub-optimal for assuring the best safety for personnel in the wind industry. Finding and fixing these weaknesses is crucial for enhancing safety performance. Interest in using cutting-edge technology like machine learning to strengthen wind sector safety measures has risen in recent years. Using these innovations, we may be able to gauge safety performance better, discover new areas of concern, and devise more efficient methods for enhancing it.

5.2.10 Potential Benefits of Using Machine Learning and Deep Learning in the Wind Industry

Machine learning (ML) can enhance safety, decrease maintenance costs, and boost productivity in the offshore wind sector. Some possible gains from using ML in offshore wind are listed.

- **Improved Operations**: ML may increase safety in the offshore wind sector through predictive maintenance and early warning of possible dangers [22]. Machine learning algorithms can examine data collected by sensors and other monitoring systems to predict when a piece of machinery may break down. This may aid in preventing accidents and unscheduled downtime for offshore wind enterprises.
- **Reduced Maintenance Costs**: ML may provide more focused and effective maintenance actions in the offshore wind business, hence lowering maintenance costs. Offshore wind businesses may save money by planning for maintenance using ML algorithms to forecast when their equipment will go down. In addition, ML may be used to improve maintenance procedures by, for example, pinpointing when parts should be replaced or investigating the reason for equipment breakdowns [21].
- **Increased Efficiency**: ML may positively impact productivity by streamlining operations and decreasing downtime in the offshore wind sector. By analyzing variables like wind speed and direction in real-time, ML algorithms may improve the performance of wind turbines, for instance [22]. This may aid offshore wind farms in optimizing energy output while minimizing losses due to inclement weather.
- **Incident prevention**: With the use of machine learning algorithms, safety parameters can be tracked in real-time, allowing for immediate responses to any emerging threats. Potential risks may be detected and handled before they become an issue, leading to more proactive safety management methods [31].
- **Better safety culture**: Using machine learning to monitor safety parameters helps businesses foster an environment where security is a top priority and is rigorously administered. The result may be a more productive workforce and a safer workplace [14].

- **Improved decision-making**: The data-driven insights that machine learning algorithms offer can help people make better decisions, which could lead to improved safety management strategies. Decision-makers may enhance safety management practices by utilizing machine learning to assess safety measurements and pinpoint problem areas [27].

Table 5.9 provides an overview of the application of ML and DL approaches in Wind Industry along with its advantages.

When applied to the offshore wind sector, ML has the potential to significantly enhance safety, decrease maintenance costs, and boost productivity. Offshore wind enterprises will need to spend money on data collecting and administration and

Table 5.9 Potential benefits of ML and deep learning in the wind industry

Reference	Problem definition	Findings	Advantage	Limitation
Olguin et al. [22]	Improved operations	Predictive maintenance and early warning of potential dangers	Enhanced safety through preventive measures	Dependence on sensor data for accuracy
Mitchell et al. [21]	Reduced maintenance costs	More focused and effective maintenance actions, lower maintenance costs	Cost savings through optimized maintenance planning	Relies on accurate prediction of equipment failures
Olguin et al. [22]	Increased efficiency	Streamlined operations, decreased downtime, improved wind turbine performance	Enhanced productivity through optimized energy output	Sensitivity to real-time data accuracy
Xu and Saleh [31]	Incident prevention	Real-time tracking of safety parameters, detection, and handling of potential risks	Proactive safety management methods with immediate responses	Accuracy and timeliness of data crucial
Le Coze and Antonsen [14]	Better safety culture	Fostering a safety-focused environment, resulting in a more productive and safer workplace	Improved workforce productivity and safer work environment	Requires cultural adaptation and acceptance
Taherdoost, [27]	Improved decision-making	Enhanced decision-making, improved safety management strategies	Better-informed decision-makers leading to improved safety practices	Dependence on quality and relevance of data

acquire the skills to properly install and oversee ML algorithms if they want to reap the advantages of this technology.

5.2.11 Summary of the Gaps in the Current Literature and the Research Problem

Research shows that the offshore wind sector might benefit greatly from using machine learning and deep learning techniques to enhance safety. There is a shortage of widespread usage of ML/DL for worker safety in the sector, despite some research looking at its potential for forecasting wind turbine failures, monitoring structural health, and identifying blade icing. The literature has also brought to light many difficulties and restrictions connected to using Machine Learning in the offshore wind business, including a shortage of data and computing resources and questions about the interpretability and explainability of Machine Learning algorithms (Table 5.10). In addition, while there has been progress in the wind business in using machine learning, there has not been as much study of machine learning's potential impact on safety metrics. Most prior research has been on the application of machine learning to the problems of failure prediction, structural health monitoring, and blade icing detection in wind turbines. However, a more significant investigation into the potential of machine learning to enhance safety metrics in the wind business is required.

The present study seeks to remedy the knowledge gap caused by the offshore wind industry's inconsistent use of ML/DL to ensure worker safety. The researcher has performed a comparative evaluation of commonly available and used machine learning models. It then establishes guidelines to select the best model (performance) for a given data set. The study aims to improve offshore wind sector safety by facilitating better use of ML and DL technologies in accident prediction and failure prevention. Because of the enormous stakes in human and environmental safety in the offshore wind sector, this research topic is all the more pressing [22]. The number of wind turbines built in offshore areas has increased significantly in recent years, indicating the industry's rapid development. However, new hazards and difficulties have emerged alongside this expansion, and they must be resolved for the sake of worker welfare and the sector's long-term health. It is crucial to address the current lack of widespread adoption of ML/DL, given the potential advantages of doing so for increasing safety in the offshore wind sector. The present study addresses a significant gap in the literature, and the creation of a useful ML/DL tool to enhance offshore wind industry safety has the potential to significantly contribute to the long-term viability of the industry and the protection of its workers.

Table 5.10 Summary of gaps in current literature and research problem

Problem	Findings	Advantage	Gaps
Lack of ML/DL adoption for worker safety in offshore wind	Limited usage of ML/DL for worker safety despite potential benefits	Establishing guidelines for model selection based on performance	Inconsistent use of ML/DL in offshore wind industry
Gaps and challenges in ML application in offshore wind	Difficulties and restrictions in ML application: shortage of data, computing resources, interpretability issues	Identification of challenges and gaps in the literature	Limited research on ML's impact on safety metrics in offshore wind sector
Insufficient exploration of ML's impact on safety metrics	Limited research on ML's impact on safety metrics in offshore wind sector	Addresses a significant gap in the literature	Lack of exploration of ML's potential impact on safety metrics
Research objective and methodology	Aims to improve offshore wind sector safety by facilitating better use of ML/DL in accident prediction and prevention	Comparative evaluation of ML models for effective safety enhancement	Limited focus on safety metrics improvement in prior ML research in the sector
Research objective and methodology	Aims to improve offshore wind sector safety by facilitating better use of ML/DL in accident prediction and prevention	Establishing guidelines for model selection based on performance	Inconsistent use of ML/DL in offshore wind industry
Urgency of addressing lack of ML/DL adoption	Emphasizes the urgency of addressing the lack of ML/DL adoption for safety in the offshore wind sector	Potential significant contribution to the long-term viability of the industry	Urgency emphasized due to human and environmental safety stakes

5.3 Data Processing

In this chapter, after performing detailed analysis and text cleaning, we evaluated the dataset in three ways. As our dataset is highly imbalanced, (1) we tested our models on the original dataset that is imbalanced, (2) we tested our models on undersampled datasets, and (3) we tested our models on oversampled datasets. Sampling techniques details are mentioned.

5.3.1 Data Description

The dataset has been provided by an Offshore wind company as an extract from the incident reporting system (This tool serves to document incidents, near-misses, and observations, facilitating the analysis of reported occurrences. It helps identify underlying causes and implement corrective measures to enhance safety performance. However, it should be noted that the tool does not offer predictive or prescriptive analysis based on the gathered data.). The dataset is composed of 2892 rows and 12 columns and represent the data collected from January 2020 to December 2021.

5.3.2 Columns Description

The following is the description of various columns used in the dataset.

- Incident case ID: unique identifier of the case
- Unit_location: Location (wind farm) where the incident happened
- Location_details: More information about the where the incident happened
- Work_activity: Activity being carried out at the moment of the incident
- Equipment_involved: what equipment was involved at the moment of the incident
- Vehicle_involved: what vehicle was involved at the moment of the incident
- Accident_level: from I to IV, is the actual severity of the accident (I being the least severe and IV the most severe)
- Potential Accident_level: Registers how severe the incident could have been (depending on other factors)
- Incident_category: represent the safety category under which the incident falls, from Observation to Serious Injury
- Relation to Company: relationship between the personnel involved in the incident and the company reporting the incident
- Cause: the triggering event or condition resulting in the incident
- Description: description as of the incident as written by the reporter of the incident.

5.3.3 Undersampling Technique and SMOTE Technique

This section describes various approaches used in balancing the dataset.

5.3.3.1 Undersampling Technique

Undersampling is a technique used to address the class imbalance in a dataset, which occurs when one class has significantly more samples than the other(s). This can lead to poor performance in machine learning algorithms as they tend to be biased towards the majority class. Undersampling aims to balance the class distribution by randomly removing instances from the majority class to match the number of instances in the minority class.

Advantages of Undersampling

- It reduces the size of the dataset, which can help reduce the computational time and resources needed for training a model.
- It can improve the performance of machine learning algorithms on imbalanced datasets by pro- viding a more balanced class distribution.

Disadvantages of undersampling

- It may lead to the loss of potentially important information as instances from the majority class are removed.
- There is a risk of underfitting, as the reduced dataset may not represent the overall population.

To balance the dataset, we use the resampling technique to upsample both minority classes (1 and 2) to match the number of instances in the majority class (0).

5.3.3.2 SMOTE Technique

Synthetic Minority Over-sampling Technique (SMOTE) is an advanced oversampling method that generates synthetic samples for the minority class to balance the class distribution. SMOTE works by selecting samples from the minority class and generating new synthetic instances based on the feature space similarities between the selected samples and their k-nearest neighbours.

The SMOTE algorithm involves the following steps:

- For each minority class instance, find its k-nearest neighbours in the feature space.
- Choose one of the k-nearest neighbours randomly.
- Generate a new synthetic instance by interpolating the feature values of the selected instance and its chosen neighbours.
- Repeat the process until the desired number of synthetic instances is created.

Advantages of SMOTE

- It creates a balanced dataset without losing important information, unlike undersampling.
- The synthetic instances can help improve the performance of machine learning algorithms on imbalanced datasets.
- It reduces the risk of overfitting, as the synthetic instances are generated based on the feature space similarities.

Disadvantages of SMOTE

- It may increase computational time and resources as the dataset size increases with the addition of synthetic instances.

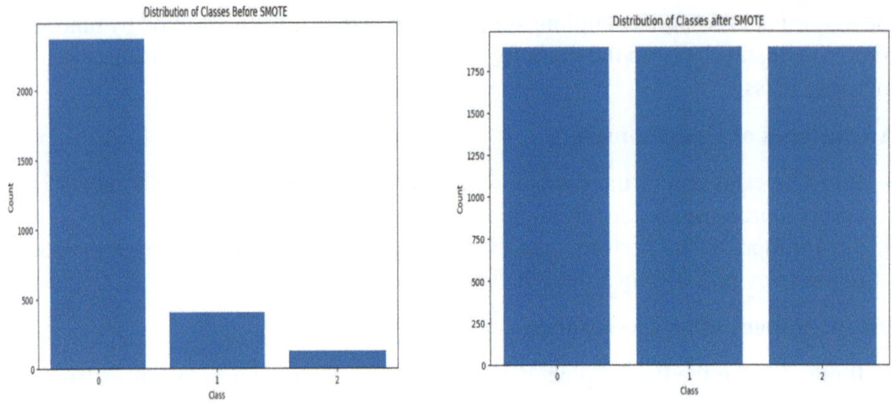

(a) (b)

Fig. 5.1 Distribution of classes **a** Before SMOTE **b** After SMOTE

- SMOTE can generate noisy samples if the minority class instances are too close to the majority class instances in the feature space, which may decrease the model's performance.

Figure 5.1 shows the impact of use of SMOTE on your datasets by displaying class distribution before and after applying SMOTE.

5.4 Models

This section describes various models used in the enhancement of safety in wind industry.

5.4.1 Machine Learning Models

Figure 5.2 shows the Machine Learning based model with integration of SMOTE.

5.4.1.1 Logistic Regression

Logistic Regression is a linear model used for binary classification tasks. However, it can be extended to handle multi-class problems using the one-vs-rest (OvR) or the one-vs-one (OvO) approach. In the OvR approach, a separate logistic regression model is trained for each class, with the target label being the class itself versus all other classes combined. In the OvO approach, a model is trained for each pair of

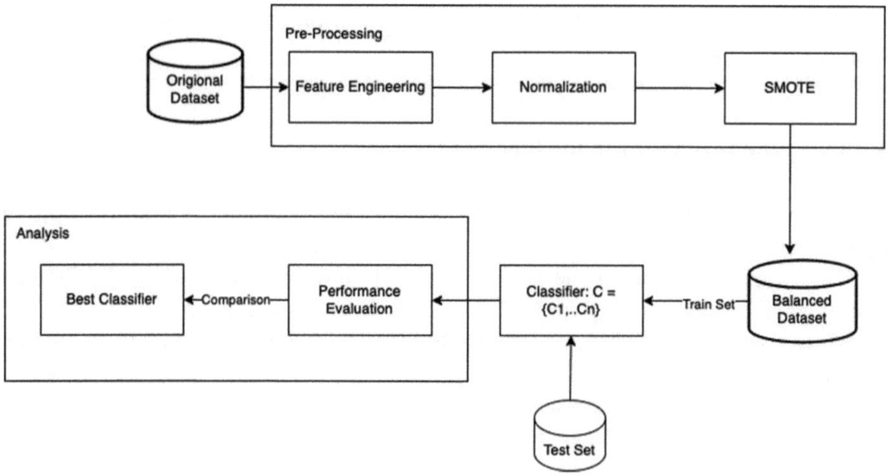

Fig. 5.2 Machine learning based model with integration of SMOTE

classes. During prediction, the class with the highest probability among all models is assigned to the instance. Logistic Regression is simple, easy to interpret, and works well when the features and target relationship is approximately linear.

5.4.1.2 Ridge Classifier

Ridge Classifier is a linear classification model that uses Ridge Regression (L2 regularization) to find the optimal weights for the features. It can handle multi-class problems using the one-vs-rest approach, similar to Logistic Regression. For each class, a Ridge Classifier model is trained to separate that class from the rest. The class with the highest decision function score is then assigned to the instance. Ridge Classifier can handle multicollinearity in the feature space and is less sensitive to overfitting than unregularized linear models.to overfitting than unregularized linear models.

5.4.1.3 K-Nearest Neighbours Classifier (KNN)

K-Nearest Neighbors Classifier is a non-parametric, instance-based learning algorithm that can be used for multi-class classification problems. It works by finding the k-nearest neighbours in the feature space for a given instance and assigning the majority class label among those neighbours. In the case of multi-class problems, KNN assigns the class with the highest frequency among the k-nearest neighbours. KNN is a lazy learner, meaning it doesn't build an explicit model during training; instead, it memorizes the training instances for making predictions. The algorithm is simple, easy to understand, and works well for problems with complex decision boundaries.

5.4.1.4 Support Vector Classifier (SVC)

Support Vector Classifier is a powerful classification algorithm that finds the optimal hyper-plane that separates the classes in the feature spaceSVC or multi-class problem SVCV.C. typically uses the one-vs-one approach. It trains a separate model for each pair of classes, resulting in n*(n–1)/2 classifiers for n classes. During prediction, each classifier votes for the class it was trained to identify, and the class with the most vote SVCs assigned to the instan SVCV.C. can handle non-linear problems using kernel functions such as the Radial Basis Function (RBF) kernel. It is robust to overfitting and works well for high-dimensional data and complex decision boundaries.

5.4.1.5 Decision Tree Classifier

A Decision Tree Classifier is a non-linear, hierarchical machine learning model that recursively partitions the input feature space into subsets based on each node's most significant feature(s). For multi-class problems, the decision tree learns to make decisions by constructing branches, with each branch representing a decision based on the feature values. The leaf nodes of the tree correspond to the class labels. During prediction, an instance traverses the tree along the branches, following the decisions made by the nodes, until it reaches a leaf node representing the predicted class. Decision trees are interpretable, easy to visualize, and can effectively handle non-linear relationships between features and target variables.

5.4.1.6 Random Forest Classifier

A Random Forest Classifier is an ensemble learning algorithm that constructs multiple decision trees and combines their predictions using a majority voting mechanism. For multi-class problems, each decision tree in the random forest is trained on a bootstrapped sample of the dataset and uses a random subset of features at each split. This strategy introduces diversity among the trees, reducing overfitting and improving generalization. The predicted class, for an instance, is the one with the majority vote among all trees. Random forests are robust to overfitting, handle non-linear relationships well, and often perform better than individual decision trees.

5.4.1.7 Bagging Classifier

A Bagging Classifier (Bootstrap Aggregating) is another ensemble learning technique that combines multiple base models, often decision trees, to improve the stability and accuracy of the predictions. For multi-class problems, the bagging CClassifier trains multiple base models, each on a bootstrapped sample of the dataset, and combines their predictions using majority voting. The algorithm reduces the variance of the base models by averaging their predictions, leading to better generalization and performance. Bagging classifiers work well with non-linear, high-variance base models and can effectively handle non-linear relationships between features and target variables.

5.4.1.8 Extra Trees Classifier

Extra Trees Classifier (Extremely Randomized Trees) is an ensemble learning method similar to the Random Forest Classifier but with a key difference in the tree construction process. For multi-class problems, both methods build multiple decision trees and use majority voting for predictions. However, in Extra Trees Classifier, the candidate feature splits are chosen randomly rather than searching for the optimal split as in Random Forest. This additional layer of randomness often results in better generalization and faster training times. Extra Trees Classifier is robust to overfitting and can effectively handle non-linear relationships between features and target variables, often achieving comparable or even better performance than Random Forest Classifier. The Classifier is robust to overfitting and can effectively manage non-linear relationships between features and target variables, often achieving comparable or even better performance than Random Forest Classifier.

5.4.1.9 AdaBoost Classifier

AdaBoost (Adaptive Boosting) is an ensemble learning technique that combines multiple weak learners, often decision trees, to build a strong classifier. For multi-class problems, AdaBoost uses a one-vs-one or one-vs-all approach to train multiple binary classifiers. The algorithm initially assigns equal weights to instances but adapts the weights of misclassified instances during each iteration, increasing their importance. Subsequent weak learners focus more on these problematic instances, aiming to classify them correctly. The final prediction is obtained by combining the weighted predictions of all weak learners through a weighted majority vote. AdaBoost is effective in handling non-linear relationships and is less susceptible to overfitting, often achieving better performance than individual weak learners.

5.4.1.10 Gradient Boosting Classifier

Gradient Boosting Classifier is an ensemble learning method that builds multiple weak learners, usually decision trees, sequentially by fitting them to the residuals of the previous learner's predictions. For multi-class problems, gradient boosting employs a one-vs-all approach, training binary classifiers for each class. Each CClassifier is fitted to the negative gradient of the logarithmic loss function, focusing on reducing the misclassification error. The final prediction is made using a weighted combination of the classifiers' decisions, with the weights determined by the classifiers' performance. Gradient Boosting is highly adaptable and can handle non-linear relationships effectively, often achieving improved performance compared to single decision trees and other boosting methods.

5.4.1.11 CatBoost Classifier

CatBoost (Category Boosting) is a gradient boosting-based algorithm designed specifically to handle categorical features effectively in multi-class problems. Like Gradient Boosting, CatBoost trains a sequence of decision trees, fitting each tree to the residuals of the previous one. CatBoost uses an ordered boosting approach that reduces overfitting by introducing randomness into the tree construction process. It also employs an efficient one-hot encoding technique called "one-hot max", which significantly speeds up the training process for categorical variables. CatBoost is robust to overfitting, handles non-linear relationships well, and often outperforms other gradient-boosting-based algorithms on datasets with categorical features.

5.4.1.12 LGBM Classifier

LGBM (Light Gradient Boosting Machine) Classifier is a gradient boosting-based algorithm that employs a unique tree construction method called "Gradient-based One-Side Sampling" (GOSS) and "Exclusive Feature Bundling" (EFB). For multi-class problems, LGBM trains binary classifiers for each class using a one-vs-all approach. The GOSS method focuses on instances with large gradients, reducing the size of the data used for tree construction and speeding up the training process. EFB bundles mutually exclusive features to reduce the number of features used in the learning process, further enhancing training efficiency. LGBM is highly scalable, capable of handling large datasets, and often achieves better performance than other gradient-boosting-based algorithms.

5.4.1.13 XGB (eXtreme Gradient Boosting) Classifier

XBG is a gradient boosting-based algorithm that aims to optimize both the model performance and computational efficiency. For multi-class problems, XGB uses a one-vs-all approach, training binary classifiers for each class. XGB employs a unique regularization term in its objective function, which controls the complexity of the trees and reduces overfitting. It also uses advanced techniques, such as column block and cache-aware access patterns, to improve the training speed. XGB is highly scalable, robust to overfitting, and can handle non-linear relationships effectively, often outperforming other gradient-boosting-based algorithms in terms of accuracy and training efficiency.

In the first step, we tested all machine algorithms with the three (3) above-mentioned datasets variations. As the model result shows that all algorithms perform best with the original dataset rather than undersample or oversample dataset, so in the next step, we tested all machine learning algorithms on the original dataset with multiple hyperparameters for each model. To perform hyperparameter tuning and find optimal parameters for each model. For each model, a parameter grid is defined to specify the hyperparameters and their possible values for tuning. These hyperparameters are used to find the best-performing models through a process called hyperparameter optimization, which includes techniques Grid Search and Random Search. By exploring various combinations of these hyperparameters, we aim to identify the best configurations for each Classifier to maximize their performance on a given multi-class classification problem.

In the next step, we tested three models, Random Forest, XGBoost, and AdaBoost performance using bootstrap sampling technique to avoid any model overfitting. Bootstrap sampling is a resampling technique that involves creating multiple datasets by randomly drawing samples from the original dataset with replacements, maintaining the same size as the original dataset. This diversity in training sets helps improve model performance and robustness by reducing overfitting and increasing the overall predictive power of the ensemble.

5.4.1.14 Neural Networks

Figure 5.3 shows neural network model with integration of SMOTE.

We have built and evaluated two variations of neural networks to check each model architecture performance on our dataset and choose the one that best fits our dataset. We have tested model performance with original data and SMOTE oversampled data. In the first model, we used original data with labels; in the second model, we used original data with one hot encoded label; and in the third model, we used oversample data with one hot encoded label. In the second and third scenarios, the model architecture is the same.

Neural Network Model-1

The model is a Sequential neural network model consisting of five layers. The first layer is a dense layer with 50 neurons, using the input dimension equal to the number of features in the training data, a ReLU activation function, and He uniform initializer for the weights. The following three layers are dense layers with 100, 150, and 40 neurons, respectively, all with ReLU activation functions and He uniform initializers. The final layer is a dense layer with a single neuron and a linear activation function for regression tasks. The moSGD is compiled using the Stochastic Gradient DesceSGDG.D.) optimizer with a learning rate of 0.001 and momentum MSE0.9. The loss function used is the mean squared error (MSES.E.). The model also employs early stopping and the ReduceLROnPlateau callbacks to prevent overfitting and to

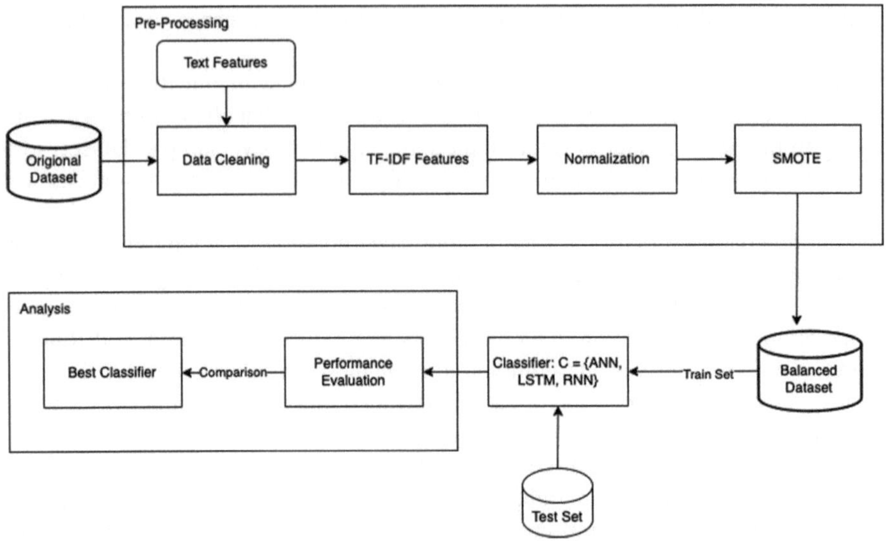

Fig. 5.3 Neural network model with integration of SMOTE

adjust the learning rate dynamically based on the validation loss. The model is trained on the dataset using a batch size of 32 and a total of 100 epochs, with the training history recorded to analyze the model's performance.

Neural Network Model-2

The Sequential neural network model comprises three (3) dense layers, with dropout and batch normalization layers in between. The first dense layer has ten (10) neurons, a ReLU activation function, He uniform initializer, L2 regularization, and a unit norm kernel constraint. Following this layer, there's a 20% dropout layer and a batch normalization layer. The second dense layer also has ten (10) neurons, a ReLU activation function, He uniform initializer, L2 regularization, and a unit norm kernel constraint, followed by a 50% dropout layer and a batch normalization layer. The final dense layer has three (3) neurons, a SoftMax activation function for multi-class classification, L2 regularization, and a unit norm kernel constraint. The model is compiled using the Stochastic Gradient Descent (SGD) optimizer with a learning rate of 0.001 and momentum of 0.9. The loss function used is categorical cross-entropy, and the metric is categorical accuracy. Early stopping and ReduceLROnPlateau callbacks are employed to prevent overfitting and dynamically adjust the learning rate based on validation loss. Custom Metrics class is used to record performance during training. The model is trained on the dataset for 100 epochs with a batch size of 32, with training history recorded to analyze the model's performance.

5.4.2 Deep Learning Models

Figure 5.4 shows Deep Learning based model with integration of SMOTE.

We have evaluated sequential-based deep neural networks Long Short-Term Memory (LSTM). The model's performances were evaluated on the original dataset without sampling. We have built three variations of models.

5.4.2.1 LSTM Models

LSTM Model Architecture-1

We have implemented Bidirectional LSTM neural network architecture using Keras. The model takes input sequences and embeds them using a pre-trained embedding matrix, with the embedding layer set to non-trainable. A bidirectional LSTM layer with 128 units is applied to the embedded inputs, followed by a global max-pooling layer to reduce the sequence dimension. The subsequent layers include a series of dense layers (with 128, 64, 32, and 10 units) interspersed with dropout layers (with dropout rates of 0.5). Each dense layer uses a ReLU activation function, except for the final dense layer, which has three (3) units and a SoftMax activation function for multi-class classification. The model is compiled using Stochastic Gradient Descent (SGD) optimizer with a learning rate of 0.001 and momentum of 0.9, and categorical cross-entropy as the loss function, with accuracy as the evaluation metric.

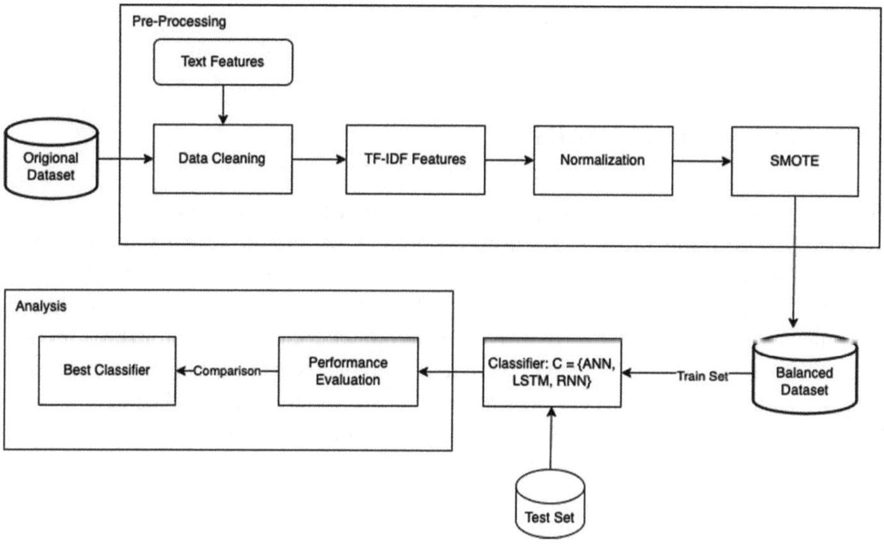

Fig. 5.4 Deep Learning based model with integration of SMOTE

LSTM Model Architecture-2

We built a single-input fully connected neural network using Keras in this model architecture. The model has a dense layer with ten (10) units and a ReLU activation function, followed by dropout (0.2) and batch normalization layers. Another dense layer with ten (10) units and a ReLU activation function is connected next, followed by another dropout (0.5) and batch normalization layers. The output layer has three (3) units and a SoftMax activation function for multi-class classification. L2 regularization with a parameter of 1e-4 and a unit norm constraint are applied to the kernel weights of the dense layers. The model is compiled using the Stochastic Gradient Descent (SGD) optimizer with a learning rate of 0.001 and momentum of 0.9, using categorical cross-entropy as the loss function and accuracy as the evaluation metric.

LSTM Model Architecture-3

In this architecture, we build a multi-input neural network architecture that combines a bidirectional LSTM and a fully connected network using Keras. The first input is passed through an embedding layer with a pre-trained embedding. This is followed by a bidirectional LSTM layer with 128 units, a global max-pooling layer, and a series of dense and dropout layers. The second input is passed through a fully connected network composed of dense, dropout, and batch normalization layers. The two branches of the model are concatenated and then connected to a dense layer with ten (10) units and a ReLU activation function, followed by an output layer with three (3) units and a SoftMax activation function for multi-class classification. The model is compiled with the Stochastic Gradient Descent (SGD) optimizer with a learning rate of 0.001 and momentum of 0.9, using categorical cross-entropy as the loss function and accuracy as the evaluation metric.

5.5 Analysis and Results

To understand the difference in performance (accuracy) of the models, two sets of models will be built, the first set done to predict "Accident severity level", where we have only four (4) classes (I, II, II and IV). In contrast, the second set that predicts "Cause" comprises 40 classes. The first table in each of the following sections is related to the "severity level" attribute, whereas the second table is related to the "cause" attribute. We have evaluated machine learning, feed-forward neural networks, and deep neural networks models' performance for the "Accident severity level" attribute. Detailed results of all models are given below.

5.5.1 Machine Learning Models Results with the Original Dataset

Table 5.11. shows the performance metrics of various classification models on a given dataset. F1- score, one of the key metrics, is the harmonic mean of precision and recall, and it ranges from 0 to 1, with 1 being the best possible score. The F1-score is particularly useful when the class distribution is imbalanced, as it accounts for false positives and negatives. Based on the F1-Score, the top three models are Extra Trees Classifier (0.9448554), Decision Tree Classifier (0.9415844), and Logistic Regression (0.9428073). These models balance precision and recall well and offer relatively high test accuracy. When comparing the remaining models, Ridge Classifier, SVC, and AdaBoost Classifier perform similarly in F1-Score. KNeighbors Classifier, Random Forest Classifier, Bagging Classifier, GradientBoosting Classifier, CatBoost Classifier, LG "MClassifier, and XGBClassifier have slightly lower F1-Scores. It is important to note that some models, such as Random Forest Classifier, Bagging Classifier, and KNeighbors Classifier, exhibit a more significant difference between train and test accuracy, which may indicate overfitting.

Table 5.12 shows the performance metrics of 13 classification models on the dataset. The models are also evaluated based on other metrics such as training accuracy, test accuracy, precision, recall, and multi-class log loss. Upon analyzing the results, the Ridge Classifier has the highest test accuracy (0.375) and F1-score (0.3378) among all the models. Although the DecisionTree Classifier and Extra Trees Classifier exhibit perfect training accuracy, they fail to generalize well to the test dataset, indicating overfitting. On the other hand, models like KNeighbors Classifier and AdaBoost Classifier perform poorly in F1-score, suggesting they may not be suitable for this problem. It is essential to consider the balance between training and test accuracy when selecting the best model for a given task and other metrics like F1-score to ensure that the chosen model performs well on unseen data and provides a good trade-off between precision and recall. The classifiers on the original data set perform worse when there are many classes, as is the case for the "cause" compared to the "severity" level.

5.5.2 Machine Learning Models Result for Undersampled Dataset

In the case of "cause", we are not dealing with the majority class (here the "accident severity level" and using undersampling with our minority class "Cause"), which already contains many classes, will create even more unbalance in our dataset and warp the results of our models.

Table 5.13 presents the results of several machine learning models' performance on an undersampled dataset. The evaluation metrics include train and test accuracy, precision, recall, F1-score, and multi-class log loss. The F1-score is particularly

Table 5.11 "Severity level" ML models results with the original dataset

Method	Train accuracy	Test accuracy	Precision	Recall	F1 score	Multi-class log loss
Logistic regression	0.9956766	0.9430052	0.9432653	0.9430052	0.9428073	0.1299078
Ridge classifier	0.997406	0.9395509	0.9379522	0.9395509	0.9381799	1
K-Neighbors classifier	0.9213143	0.8981002	0.8995582	0.8981002	0.8794643	2.285096
SVC	0.9783831	0.9395509	0.9415076	0.9395509	0.9351979	0.1425599
Decision tree classifier	1	0.9412781	0.9423546	0.9412781	0.9415844	2.0281838
Random forest classifier	0.997406	0.9360967	0.9340535	0.9360967	0.9323823	0.5341985
Bagging classifier	0.998703	0.9395509	0.9380643	0.9395509	0.9378462	0.2753485
Extra trees classifier	0.9995677	0.9481865	0.9502953	0.9481865	0.9448554	0.1759012
AdaBoost classifier	0.9468223	0.9395509	0.9391244	0.9395509	0.9360264	0.7060568
Gradient boosting Classifier	0.9580631	0.9430052	0.9430966	0.9430052	0.9396935	0.1319889
CatBoost classifier	0.9969736	0.9464594	0.9453536	0.9464594	0.944641	0.136145
LGBM classifier	1	0.9343696	0.9354349	0.9343696	0.9341319	0.2004699
XGB classifier	0.9753567	0.9343696	0.933646	0.9343696	0.932742	0.1416465

important in this context, as it provides a balanced measure of a model's performance, combining precision and recall. A higher F1-score indicates better performance. Upon analyzing the results, we can observe that the Extra Trees Classifier, Random Forest Classifier, and LGBM Classifier perform exceptionally well with F1-scores of 0.944082, 0.9319914, and 0.9426346, respectively. These models also demonstrate high test accuracy, precision, and recall, meaning they generalize well to unseen data. On the other hand, the Ridge Classifier and SVC models show very low F1 scores (0.0032996 and 0.0343382), meaning they perform poorly on this dataset. Low precision and recall suggest that these models may not be appropriate for the problem, and alternative models or approaches should be considered. Overall, the models' F-1 scores provide valuable insights into their ability to classify instances in the undersampled dataset effectively.

Table 5.12 "Cause" ML models results with the original dataset

Method	Train accuracy	Test accuracy	Precision	Recall	F1 score	Multi-class log loss
Logistic regression	0.5331325	0.3245192	0.307417	0.3245192	0.2528263	2.4832409
Ridge classifier	0.9542169	0.375	0.3614396	0.375	0.3378273	1
K-Neighbors classifier	0.5018072	0.1274038	0.1464346	0.1274038	0.1274087	23.9376348
SVC	0.2608434	0.2427885	0.1111221	0.2427885	0.1135863	2.7916195
Decision tree classifier	0.9945783	0.2427885	0.2235353	0.2427885	0.2291328	26.15316
Random forest classifier	0.9849398	0.2644231	0.2357217	0.2644231	0.2208062	13.6258541
Bagging classifier	0.9879518	0.3173077	0.299805	0.3173077	0.2641107	7.8793365
Extra trees classifier	0.9945783	0.3149038	0.426557	0.3149038	0.2377867	5.7349244
AdaBoost classifier	0.2451807	0.2331731	0.078945	0.2331731	0.0945303	3.3139672
Gradient boosting classifier	0.9560241	0.2572115	0.2640974	0.2572115	0.2310995	3.6162203
CatBoost classifier	0.6355422	0.34375	0.3241839	0.34375	0.276179	2.3950854
LGBM classifier	0.9939759	0.3389423	0.3346331	0.3389423	0.2876187	4.3011373
XGB classifier	0.523494	0.3149038	0.2506353	0.3149038	0.2464629	2.6144844

5.5.3 Machine Learning Models Results for Oversampling (SMOTE) Dataset

Table 5.14 presents the performance metrics of various classification models trained on a dataset. The metrics include train and test accuracy, precision, recall, F1-score, and multi-class log loss. The higher the F1-score, the better the model's performance. From the results, Random Forest Classifier, Extra Trees Classifier, and LGBM Classifier have the highest F1-scores of 0.9319914, 0.944082, and 0.9426346, respectively. These models perform well in terms of both precision and recall. On the other hand, Ridge Classifier and SVC have the lowest F1-scores of 0.0032996 and 0.0343382, respectively, indicating poor performance.

Table 5.13 "Severity level" ML models results with undersampled dataset

Method	Train accuracy	Test accuracy	Precision	Recall	F1-score	Multi-class log loss
Logistic regression	0.9015152	0.8860104	0.9092712	0.8860104	0.8938488	0.2813272
Ridge classifier	0.9989429	0.0414508	0.0017182	0.0414508	0.0032996	1
K-Neighbors classifier	0.9894292	0.5854922	0.747511	0.5854922	0.629735	4.3075004
SVC	0.3826638	0.1398964	0.019571	0.1398964	0.0343382	1.1047293
Decision tree classifier	1	0.9170984	0.9332522	0.9170984	0.9219016	2.8633183
Random forest classifier	1	0.9326425	0.933475	0.9326425	0.9319914	0.1888901
Bagging classifier	0.9978858	0.9188256	0.9409246	0.9188256	0.9247187	0.2472905
Extra trees classifier	1	0.9464594	0.9453092	0.9464594	0.944082	0.1900062
AdaBoost classifier	0.9651163	0.8946459	0.9399037	0.8946459	0.9055924	0.7384375
Gradient boosting classifier	0.9711064	0.9067358	0.9423256	0.9067358	0.9154937	0.2288864
Cat boost classifier	0.9899577	0.9084629	0.9396547	0.9084629	0.9164308	0.2168714
LGBM classifier	0.9971811	0.9412781	0.945825	0.9412781	0.9426346	0.1840918
XGB classifier	0.978506	0.9067358	0.9423301	0.9067358	0.9154501	0.1698508

Table 5.15 shows the performance of various machine learning classifiers regarding Train Accuracy, Test Accuracy, Precision, Recall, F1-Score, and Multi-Class Log loss. It is important to note that the classifiers used are from different categories, such as linear models (Logistic Regression, Ridge Classifier), tree-based models (Decision Tree Classifier, Random Forest Classifier, Extra Trees Classifier), and boosting models (AdaBoost Classifier, Gradient Boosting Classifier, CatBoost Classifier, LGBM Classifier, XGB Classifier).

Based on the F1-Scores, we can rank the classifiers as follows:

1. Extra Trees Classifier: 0.2747412
2. GradientBoosting Classifier: 0.2616233
3. CatBoost Classifier: 0.2594522
4. LGBM Classifier: 0.2518778
5. XGB Classifier: 0.2134572

Table 5.14 "Severity level" ML models results on oversampling (SMOTE) dataset

Method	Train accuracy	Test accuracy	Precision	Recall	F1-score	Multi-class log loss
Logistic regression	0.7924595	0.9205527	0.9173562	0.9205527	0.9164171	0.2599838
Ridge classifier	0.9076815	0.8186528	0.6701925	0.8186528	0.7370208	1
K-Neighbors classifier	0.8883016	0.1450777	0.8383258	0.1450777	0.0447925	29.3673995
SVC	0.3717407	0.0414508	0.0017182	0.0414508	0.0032996	1.0986123
DecisionTree classifier	1	0.8929188	0.9170455	0.8929188	0.902171	3.6984527
Random forest classifier	0.9996476	0.9188256	0.9220757	0.9188256	0.9190351	0.4381397
Bagging classifier	0.9982382	0.8963731	0.9319102	0.8963731	0.9041309	0.3239752
Extra trees classifier	1	0.9378238	0.9388557	0.9378238	0.9382024	0.3195194
AdaBoost classifier	0.8638125	0.9136442	0.9416438	0.9136442	0.9208222	0.8097375
Gradient boosting classifier	0.8981677	0.9050086	0.9276934	0.9050086	0.9113011	0.2174353
CatBoost classifier	0.9880197	0.9136442	0.9319687	0.9136442	0.9188692	0.1851309
LGBM classifier	1	0.925734	0.9402764	0.925734	0.9298257	0.1874874
XGB classifier	0.9670543	0.9153713	0.92837	0.9153713	0.9195072	0.163982

6. Random Forest Classifier: 0.2327086
7. Bagging Classifier: 0.2072686
8. Decision Tree Classifier: 0.1733749
9. AdaBoost Classifier: 0.0605113
10. Ridge Classifier: 0.0035892
11. KNeighbors Classifier: 0.0021914
12. Logistic Regression: 0.0009536
13. SVC: 0.000046

From the results, the Extra Trees Classifier has the highest F1-Score, followed by GradientBoosting-Classifier and CatBoost Classifier. These classifiers perform the best in terms of balancing precision and recall and are the top-performing classifiers. On the other hand, classifiers such as Logistic Regression, KNeighbors Classifier

Table 5.15 "Cause" models results on oversampling (SMOTE) dataset

Method	Train accuracy	Test accuracy	Precision	Recall	F1 score	Multi-class log loss
Logistic regression	0.1038918	0.0168269	0.0004907	0.0168269	0.0009536	3.4982925
Ridge classifier	0.9873846	0.0432692	0.0018722	0.0432692	0.0035892	1
KNeighbors classifier	0.8858015	0.0336538	0.0011326	0.0336538	0.0021914	33.3764137
SVC	0.0843503	0.0048077	0.0000231	0.0048077	0.000046	3.4578128
Decision tree classifier	0.9992579	0.1706731	0.2405388	0.1706731	0.1733749	28.6439372
Random forest classifier	0.9987632	0.2331731	0.2616756	0.2331731	0.2327086	12.7582593
Bagging classifier	0.9974439	0.2043269	0.3656235	0.2043269	0.2072686	11.5539962
Extra trees classifier	0.9992579	0.2908654	0.3065551	0.2908654	0.2747412	5.4413212
AdaBoost classifier	0.1936016	0.0697115	0.1739184	0.0697115	0.0605113	3.3131799
Gradient boosting classifier	0.9825198	0.2620192	0.2673246	0.2620192	0.2616233	2.8014981
CatBoost classifier	0.9823549	0.2740385	0.3118662	0.2740385	0.2594522	2.5746419
LGBM classifier	0.9992579	0.2596154	0.3287398	0.2596154	0.2518778	4.0591406
XGB classifier	0.9771603	0.2235577	0.3370266	0.2235577	0.2134572	2.6713203

and SVC have very low F1-Scores, indicating that they are not performing well in balancing precision and recall.

When analyzing these results, it is also essential to consider other performance metrics, such as Test Accuracy and Multi-Class Log loss, to comprehensively understand the classifiers' performance. Test Accuracy represents the proportion of correct predictions, while Multi-Class Log loss measures the classifiers' prediction probabilities' quality. It becomes apparent here that the difference in the number of classes for the "Cause" attribute leads to poor performance of the classifier models.

5.5.4 Neural Network Model Results

Table 5.16 presents the performance metrics of three neural network models on a dataset. The metrics include test accuracy, precision, recall, and F1-score. Precision measures the proportion of true positive predictions among all positive predictions made by a model. A higher precision indicates that the model correctly identifies more positive instances and minimizes false positives. From the results, Model-2 has the highest precision of 0.924324, followed by Model-1 with a precision of 0.922280, while Model-3 has a significantly lower precision of 0.288690. Model-2 is the most accurate in identifying positive instances without making too many false-positive predictions. It is, however, essential to consider other performance metrics like recall and F1-score when evaluating the overall performance of a model. Model-1, with an F1-score of 0.922280, demonstrates a balanced performance between precision and recall. In contrast, Model-3 has a low F1-score of 0.212022, indicating poor performance in terms of both precision and recall.

Table 5.17 shows the performance metrics of three different neural network feed-forward classification models on imbalanced dataset. Model-2 has the highest test accuracy (0.201923) and F1-score (0.010500) among the three models. Although the F1-score is low, Model-2 outperforms Model-1 and Model-3 in terms of precision and recall, which suggests it is the best choice among these options. Model-1, on the other hand, exhibits very low values for all metrics, indicating that it is not a suitable choice for this problem. Model-3 performs marginally better than Model-1, but its F1-score is still lower than Model-2's. Again, It becomes apparent that the difference in the number of classes for the "Cause" attribute leads to poor performance of the Neural Networks models.

Table 5.16 "Severity level" neural network model results

Method	Train accuracy	Test accuracy	Precision	Recall	F1 score
Model-1	0.953	0.922280	0.922280	0.922280	0.922280
Model-2	0.942	0.886010	0.924324	0.886010	0.904762
Model 3	0.333	0.167530	0.288690	0.167530	0.212022

Table 5.17 "Cause" neural network model results

Method	Train accuracy	Test accuracy	Precision	Recall	F1 score
Model-1	0.006	0.0048	0.0049	0.029	0.006
Model-2	0.2349	0.201923	0.006310	0.031250	0.010500
Model 3	0.0312	0.019231	0.017362	0.029990	0.013617

5.5.5 Deep Neural Network Results

Table 5.18 show varying performance levels across different evaluation metrics for three (3) deep learning models.

Model-1 has a training accuracy of 0.81 and a test accuracy of 0.818653, with precision, recall, and F1-score being the same value of 0.818653. This indicates a well-generalized balanced performance in identifying true positives and false positives but overall lower accuracy than the other models. Model-2 has the highest training accuracy of 0.94 and a test accuracy of 0.894646, with a precision of 0.916814, recall of 0.894646, and F1-score of 0.905594. These results suggest that Model-2 is the best-performing model among the three, achieving a good balance between precision and recall while maintaining high accuracy. However, the difference between training and test accuracy implies potential overfitting. Model-3 has a training accuracy of 0.91 and a test accuracy of 0.873921, with a precision of 0.890845, recall of 0.873921, and F1-score of 0.882302. While Model-3's performance is slightly lower than Model-2, it shows less overfitting, indicating a more generalizable model. In conclusion, Model-2 performs best in accuracy and F1-score, but Model-3 might be more reliable when considering overfitting concerns.

Table 5.19 shows the performance metrics of three different classification models on a dataset. Model-3 has the highest test accuracy (0.240385) and F1-score (0.019339) among the three models. Although the F1-score is relatively low, Model-3 outperforms Model-1 and Model-2 in terms of precision, recall, and test accuracy, which suggests it is the best choice among these options. Model-1 and Model-2 exhibit similar performance across all metrics, with only marginal differences in their F1-scores.

Table 5.18 "Severity" deep neural network results

Method	Train accuracy	Test accuracy	Precision	Recall	F1 score
Model-1	0.810000	0.818653	0.818653	0.818653	0.818653
Model-2	0.940000	0.894646	0.916814	0.894646	0.905594
Model 3	0.910000	0.873921	0.890845	0.873921	0.882302

Table 5.19 "Cause" deep neural network results

Method	Train accuracy	Test accuracy	Precision	Recall	F1-score
Model-1	0.006	0.0048	0.0049	0.029	0.006
Model-2	0.2349	0.201923	0.006310	0.031250	0.010500
Model-3	0.0312	0.019231	0.017362	0.029990	0.013617

5.5.6 Analaysis of Results

- **Original Data**
  - **Train Accuracy**: Models such as DecisionTree Classifier, Random Forest Classifier, and Extra Trees Classifier achieved a perfect 1.0 on training accuracy, which might suggest overfitting.
  - **Test Accuracy**: Logistic Regression had the highest test accuracy, closely followed by Ridge Classifier and Random Forest Classifier.
  - **F1-Score**: Random Forest Classifier led with the highest F1-Score, indicating a good balance between precision and recall.
- **Sampling Data**
  - **Varied Train and Test Accuracy**: There were discrepancies in the train and test accuracies across models, with some, like SVC, showing a large drop, which may indicate overfitting.
  - **F1-Score**: The F1-Scores are generally lower than those observed with the original data, which could suggest that the sampling technique might not be improving the model's ability to generalize.
- **SMOTE Data**
  - **Improved Test Accuracy**: The use of SMOTE appears to have improved the test accuracy for models like Logistic Regression and Ridge Classifier compared to the original data.
  - **F1-Score Improvement**: Models generally showed improved F1-Scores with SMOTE, suggesting better performance on the minority class.
- **Hyperparameter Tuning**
  - **Enhanced Performance**: Hyperparameter tuning likely enhanced model performance metrics across the board, although specific details were not noted.

Machine learning and deep learning models perform well on the imbalanced dataset but poorly on undersampled and oversampled datasets, which could be attributed to a few factors. Firstly, when undersampling is applied, important information might be lost as instances from the majority class are removed, leading to underfitting. Secondly, oversampling techniques, especially when synthetic instances are generated, can introduce noise or artificial patterns that do not represent the underlying relationship between features and the target variable, causing the model to overfit the synthetic data.

In contrast, models might perform better on the original imbalanced dataset if they can successfully learn the patterns and relationships in the data, despite the class imbalance. In such cases, it is essential to consider alternative techniques, such as cost-sensitive learning or ensemble methods, to handle imbalanced datasets effectively without compromising model performance.

Overall, when comparing all models with data variations, Ridge Classifier models perform better than all other models for "cause" on the original imbalanced dataset with an F1-score of 0.334. In contrast, the Extratree Classifier models perform best than all other models on the original imbalanced dataset having an F1-score of 0.9448554 for our majority class" ("accident severity level"). The above results demonstrate that it is possible to use high-performance machine learning to predict accident severity levels even with an imbalanced dataset, which is common when the datasets are obtained from real-life sources. It also strongly highlights the necessity, in the context of data related to safety and incidents, to implement strict policies for recording the information required to apply machine learning for predicting incidents. Predicting causes will allow organizations to implement such models, as studied above, to prevent the occurrence of incidents by targeting the causes and removing the conditions for the incident to happen.

5.6 Conclusion

Construction, installation, operation, and maintenance activities are all potentially dangerous parts of the wind industry, making ensuring the safety of the workers a necessity. Reliable safety measurements are crucial for worker well-being and the wind farm's operation. The inability of traditional safety measurements (reactive in nature) to record and interpret data might prevent potential safety concerns from being identified and mitigated. Using machine learning methods to advance safety measurements and enhance wind sector safety shows encouraging results.

In summary, for original data, models like Random Forest Classifier and Logistic Regression exhibited strong performance, with high F1-Scores indicating a good balance between precision and recall. When implementing SMOTE, a technique designed to mitigate class imbalance, there was an observable improvement in test accuracy and F1-Scores for several models, suggesting enhanced generalization capabilities. However, the use of sampling data did not consistently enhance model performance, with some models displaying decreased F1-Scores, which might indicate an ineffective sampling strategy or a need for more advanced techniques.

The process of hyperparameter tuning with original features was also explored, which generally improves model performance. This comprehensive evaluation showcases the importance of dataset pre-processing techniques like SMOTE and hyperparameter tuning in improving model performance, especially in scenarios dealing with imbalanced data. The chapter underscores the necessity to tailor these techniques to the specific data and problem at hand to ensure the most effective model performance. Hyperparameter tuning and the application of SMOTE appear to have a positive effect on model performance, particularly in addressing class imbalance as indicated by the F1-Scores. The integration of SMOTE into the pre-processing pipeline has led to a noticeable improvement in the model's F1-Score greater than 0.90 for all models. This enhancement in the F1-Score, which reflects a more balanced precision and recall, is indicative of the model's improved capability to classify the minority class accurately. The sampling data did not consistently improve model performance, indicating that the technique used may not have been optimal for the dataset in question or that the models may require more sophisticated sampling strategies. It's crucial to consider these findings within the context of the data and problem domain, and further model validation is recommended to ensure robustness and generalization of the results.

The study also highlighted which types of models, ExtraTree Classifier and Ridge Classifier, have the best performance for, respectively, the majority class in the imbalanced dataset and a minority class in an imbalanced dataset. Classifiers performed better than Neural Network and Deep Neural Network in the study context. Given that those models are reasonably easy to implement in Production, it should help pave the way for wider adoption of machine learning models to improve the safety of the personnel working in the wind industry. The present study demonstrates that machine learning models selection and implementation can be implemented widely in the wind industry. It also shows that the high performance of selected models can prove the reliability of the expected predictions and therefore be an effective tool for decision-making when taking measures to improve health and safety.

Few studies look at applying machine learning to safety indicators in the wind business, which is the key gap in the current literature. Existing research has dealt chiefly with establishing generic predictive models for wind turbines or predicting or detecting particular occurrences. As a result, additional study is required to build individualized machine learning models that may be used to enhance safety metrics in the wind business. There is also a shortage of studies that combine information from many sources to enhance safety measures, which is a significant research gap. Most previous research has concentrated on collecting data from sensors or maintenance records, but additional information, such as weather data, is needed to produce more all-encompassing safety metrics. Research on using machine learning models for safety metrics in the wind sector is also needed. There is a need to examine the efficacy of these models in real-world contexts since much of the previous research has concentrated on constructing models in laboratory settings or utilizing simulated data. In sum, this research intends to fill a need in the existing literature by providing a plan for using machine learning to improve wind sector safety measures. The proposed system will use data from a wide variety of sources and will be tested in real-world scenarios to see how well it performs.

References

1. Adekunle, S.A. et al.: Machine learning algorithm application in the construction industry—a review. Lecture Notes in Civil Engineering, pp. 263–271 (2023). https://doi.org/10.1007/978-3-031-35399-4_21
2. Alcides, J., et al.: Making the links among environmental protection, process safety, and industry 4.0. en. Process. Saf. Environ. Prot. **117**, 372–382 (2018). https://doi.org/10.1016/j.psep.2018.05.017

3. Bagherian, M.A. et al.: Classification and analysis of optimization techniques for integrated energy systems utilizing renewable energy sources: a review for CHP and CCHP systems. Processes **9**(2), 339 (2021)
4. Borg, M. et al.: Safely entering the deep: a review of verification and validation for machine learning and a challenge elicitation in the automotive industry. (2018)
5. Bowles, M.: What is offshore life really like? en. In: Quanta part of QCS Staffing 17. Accessed 11 Jul 2022. http://www.quanta-cs.com/blogs/2018-7/what-is-offshorelife-really-like
6. Gangwani, D., Gangwani, P.: Applications of machine learning and Artificial Intelligence in intelligent transportation system: A review. Lecture notes in electrical engineering, pp. 203–216 (2021). https://doi.org/10.1007/978-981-16-3067-5_16
7. Goodfellow, I., Bengio, Y., Courville, A.: Deep learning. en. MIT Press, (2016)
8. Gplusoffshorewindcom.: Health and safety statistics. en. (2022). Available at https://www.gplusoffshorewind.com/work-rogramme/workstreams/statistics
9. Herrera, I.A.: Proactive safety performance indicators. (2012)
10. Ims, J.B.: Risk-based health-aware control of Åsgard subsea gas compression station. en. Master's thesis, NTNU (2018)
11. Irawan, C.A. et al.: Optimization of maintenance routing and scheduling for offshore wind farms. en. Eur. J. Oper. Res **256**(1), 76–89 (2017). https://doi.org/10.1016/j.ejor.2016.05.059
12. Jaen-Cuellar, A.Y. et al.: Advances in fault condition monitoring for solar photovoltaic and wind turbine energy generation: A review. en. Energies **15** (15), 5404 (2022)
13. Jordan, M.I., Mitchell, T.M.: Machine learning: Trends, perspectives, and prospects. en. Science **349** (6245), 255–260 (2015)
14. Le Coze, J.-C., Antonsen, S.: Safety in a digital age: Old and new problems—algorithms, machine learning, Big Data and artificial intelligence. In: Safety in the digital age, pp. 1–9. https://doi.org/10.1007/978-3-031-32633-2_1
15. Li, Y. et al.: Wind turbine fault diagnosis based on transfer learning and convolutional autoencoder with small-scale data. Renew Energy **171**. https://doi.org/10.1016/j.renene.2021.01.143
16. Lian, J. et al.: Health monitoring and safety evaluation of the offshore wind turbine structure: a review and discussion of future development. en. Sustain. **11**(2), 494 (2019)
17. Luo, T.: Safety climate: Current status of the research and future prospects. J. Saf. Sci. Resil. **1**(2), 106–119 (2020). ISSN: 2666–4496. https://doi.org/10.1016/j.jnlssr.2020.09.001. https://www.sciencedirect.com/science/article/pii/S2666449620300268
18. Maldonado-Correa, J. et al.: Using SCADA data for wind turbine condition monitoring: A systematic literature review. en. Energies **13**(12), 3132 (2020)
19. Mangortey, E. et al.: Application of machine learning techniques to parameter selection for flight risk identification. pt. In: AIAA Scitech 2020 Forum, p. 1850 (2020)
20. Mills, T., Turner, M., Pettinger, C.: Advancing predictive indicators to prevent construction accidents. en. In: Towards better safety, health, well-being, and life in construction. Central University of Technology, Free State, pp. 459–466 (2017)
21. Mitchell, D. et al.: A review: Challenges and opportunities for artificial intelligence and robotics in the offshore wind sector. en. Energy and AI, 100146 (2022)
22. Olguin, E.J. et al.: Microalgae-based biorefineries: Challenges and future trends to produce carbohydrate enriched biomass, high-added value products and bioactive compounds. en. Biology **11**(8)
23. Papadopoulos, P., Coit, D.W., Ezzat, A.A.: Seizing opportunity: maintenance optimization in offshore wind farms considering accessibility, production, and crew dispatch. en. IEEE Trans. Sustain. Energy **13**(1), 111–121 (2022). https://doi.org/10.1109/TSTE.2021.3104982
24. Ren, Z. et al.: Offshore wind turbine operations and maintenance: A state-of-the-art review. en. Renew. Sustain. Energy Rev. **144**, 110886 (2021)
25. Surucu, O., Gadsden, S., Yawney, J.: Condition monitoring using machine learning: A review of theory, applications, and recent advances. Expert Syst. Appl. **221**, 119738 (2023). https://doi.org/10.1016/j.eswa.2023.119738

26. Tamascelli, N. et al.: Learning from major accidents: A machine learning approach. Comput Chem Eng **162**, 107786 (2022). ISSN: 0098–1354. https://doi.org/10.1016/j.compchemeng.2022.107786. https://www.sciencedirect.com/science/article/pii/S0098135422001272
27. Taherdoost, H.: Deep learning and neural networks: Decision-making implications. Symmetry **15**(9), 1723 (2023). https://doi.org/10.3390/sym15091723
28. Tixier, A.J.P., et al.: Application of machine learning to construction injury prediction. en. Autom. Constr. **69**, 102–114 (2016)
29. Wang, L., Zhang, Z.: Automatic detection of wind turbine blade surface cracks based on UAV-taken images. In: IEEE Transactions on Industrial Electronics, vol. 64, no.9, pp. 7293–7303 (2017)
30. Wolsink, M.: Co-production in distributed generation: renewable energy and creating space for fitting infrastructure within landscapes. en. Landsc Res **43**(4), 542–561 (2018)
31. Xu, Z., Saleh, J.H.: Machine learning for reliability engineering and safety applications: review of current status and future opportunities. (2020). ArXiv, abs/2008.08221
32. Yan, J.: Integrated smart sensor networks with adaptive real-time modeling capabilities. en. (Doctoral dissertation, Iowa State University). (2020)
33. Yeter, B., Garbatov, Y., Soares, C.G.: Life-extension classification of offshore wind assets using unsupervised machine learning. en. Reliab Eng Syst Saf **219**, 108229 (2022)
34. Yuan, B. et al.: WaveletFCNN: A deep time series classification model for wind turbine blade icing detection, (2019)
35. Zhu, Y., Liu, X.: A lightweight CNN for wind turbine blade defect detection based on spectrograms. Machines **11**(1), (2023). ISSN: 2075–1702. https://doi.org/10.3390/machines11010099. https://www.mdpi.com/2075-1702/11/1/99
36. Zulu, M.L.T., Carpanen, R.P., Tiako, R.: A comprehensive review: study of artificial intelligence optimization technique applications in a hybrid microgrid at times of fault outbreaks. Energies **16**(4), (2023)

Chapter 6
Malware Attack Detection in Vehicle Cyber Physical System for Planning and Control Using Deep Learning

Challa Ravi Kishore and H. S. Behera

Abstract Cyber-Physical Systems (CPS), which comprise smart health, smart transportation, smart grids, etc., are designed to turn traditionally separated automated critical infrastructure into modernized linked intelligent systems by interconnecting human, system, and physical resources. CPS is also expected to have a significant positive impact on the economy and society. Complexity, dynamic variability, and heterogeneity are the features of CPS, which are produced as an outcome of relationships between cyber and physical subsystems. In addition to the established and crucial safety and reliability criteria for conventional critical systems, these features create major obstacles. Within these cyber-physical systems and crucial infrastructures, for instance, connected autonomous vehicles (CAVs) may be considered. By 2025, it is anticipated that 95 per cent of new vehicles will be equipped with vehicle to vehicle (V2V), vehicle to infrastructure (V2I), and other telecommunications capabilities. To prevent CAVs on the road against unintended or harmful intrusion, innovative and automated procedures are required to ensure public safety. In addition, large-scale and complicated CPSs make it difficult to monitor and identify cyber physical threats. Solutions for CPS have included the use of Artificial Intelligence (AI) and Machine Learning (ML) techniques, which have proven successful in a wide range of other domains like automation, robotics, prediction, etc. This research suggests a Deep Learning (DL) -based Convolutional Neural Network (CNN) model for attack detection and evaluates it using the most recent V2X dataset, According to the simulation results, in this research CNN exhibits superior performance compared to the most advanced ML approaches such as Random Forest (RF), Adaptive Boosting (AdaBoost), Gradient Boosting (GBoost), Bagging and Extreme Gradient Boosting

C. R. Kishore (✉)
Department of Computer Science and Engineering, Aditya Institute of Technology and Management (AITAM), Tekkali, Andhra Pradesh 532201, India
e-mail: cauchy9@gmail.com

H. S. Behera
Department of Information Technology, Veer Surendra Sai University of Technology, Burla, Odisha 768018, India

J. Nayak et al. (eds.), *Machine Learning for Cyber Physical System: Advances and Challenges*, Intelligent Systems Reference Library 60,
https://doi.org/10.1007/978-3-031-54038-7_6

(XGBoost) and achieves an outstanding level of accuracy in the application of anomaly detection.

Keywords CPS · CAVs · IoV · ML · CNN · DL · VCPS

6.1 Introduction

The Intelligent Transportation System (ITS) is gaining popularity in the corporate and academic communities as a real-world example of CPSs and the Internet of Things (IoT) [1, 2]. The integration of CPS is increasingly becoming a fundamental component within the context of the digital society. The transmission of data occurs via CPS services, which facilitate communication between actual hardware and computer networks. Safeguarding against cyber threats is becoming more difficult due to an increasing number of harmful activities that compromise sensitive data and lead to malfunctioning gadgets. In the ITS environment, vehicles are able to interact with another vehicle (V2V) and with Road Side Units (RSUs) by using the most advanced gadgets and intelligent network technology. The reliability of V2V communications is sometimes compromised by factors such as the limited availability of neighboring automobiles or inadequate communication with on-board sensors (OBS). Assume that a vehicle has seen an automobile crash with a serious fire inside a remote tunnel during the late hours of the night. However, the vehicle is unable to communicate this critically important message due to the absence of adjacent vehicles capable of receiving and subsequently relaying this information. The occurrence of a disaster may arise when vehicles arrive suddenly and without prior knowledge of the circumstances [3].

The connection between vehicles and roadside units (V2R) is of the highest priority in the overall information network for addressing these challenges, resulting in demand for the installation of sensors on highways. Furthermore, the use of V2I connections facilitates the distribution of vehicle related information and geographically appropriate information [4]. Cybercriminals are specifically focusing on exploiting sensitive gadgets in order to deploy malware and subsequently gain unauthorized control over those devices. The widespread use of internet-connected devices is growing across several domains of human existence, such as the IoT, Internet of Vehicles (IoV), wearable gadgets, and smart phones, which are susceptible to malware infections [5]. The increasing numbers of malware infected devices are shown in Fig. 6.1, which is reported by AV-test (a reputable research organization specializing in IT security) [6]. The cyber security researcher encounters unexpected challenges in effectively detecting newly developed malwares.

The current developments of modern vehicle networks are continuously generating significant research limitations in connection with the security aspects of ITS management [7]. The insecure communications among different organizations within an ITS give rise to certain security problems [8]. The connectivity between vehicles

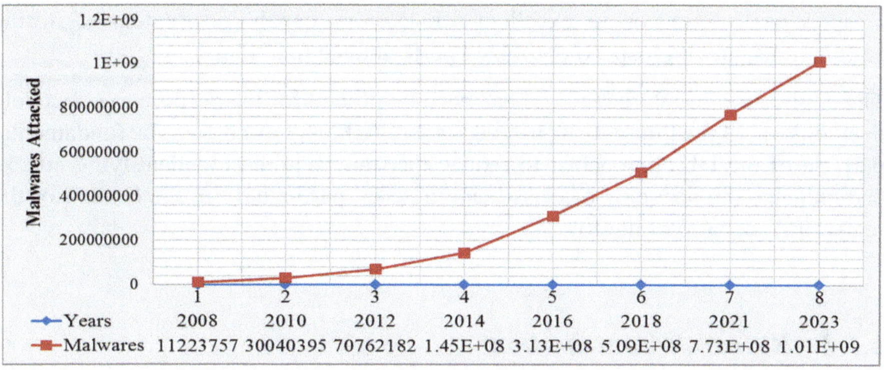

Fig. 6.1 Growth of malware programs affecting sensitive devices

and RSUs is notably inconsistent [9]. The use of encryption technology in standalone is inadequate in ensuring the authenticity of messages and lacks the ability to protect against numerous kinds of potential intruders. Conventional intrusion detection methods that depend on ML, statistical evaluation have some limits in efficiently handling the continuously growing amount of data. DL methodologies include essential capabilities that make them appropriate for overcoming these challenges.

6.1.1 Motivation

Emerging technologies in the area of Vehicles Cyber Physical System (VCPS) have been utilized, which is speeding up the evolution of the IoV. The vehicle reveals a lack of emphasis on network security which leads to enforces restrictions on storage space, functions in a complex application environment, and dependent on numerous dispersed nodes and sensor networks. As a result, challenging safety regulations are necessary. These above limitations make the VCPS environment more vulnerable to cyber-attacks, which in turn threaten the whole IoV ecosystem. The primary challenges that must be resolved are as follows:

- The IoV performs the functions of attacking behavior, monitoring and analyzing network data, categorizing normal and abnormal behavior, and identifying unusual activities like threats on the network. This technology has emerged as a key component in the defense of the IoV network.
- Current prominent research is focusing on integrating ML algorithms with more conventional Intrusion Detection Systems (IDS). The massive amount of time required for training ML-based intrusion detection algorithms is a key issue. This is because large amounts of previous network data must be analyzed.
- For analyzing recent complicated VCPS network data, particularly in the complex vehicle networking environment, DL technology in the VCPS environment is

appropriate based on its excellent self-learning capabilities, extensive storing functions, and high-speed optimized performance functions.

The application of DL techniques has been employed for the purpose to effectively overcome all of the limitations observed in the VCPS environment. The fundamental purpose of the DL approach is to reduce the time required for identifying attacks and enhance the accuracy of classification tasks, particularly in the context of the real-world scenario of the IoV.

6.1.2 Research Contribution

An increasing variety of attacks and various attacker types provide major challenges for research in the areas of misbehavior and intrusion detection in VCPS network. The high fluctuations in vehicle network architecture have a significant influence on network, routing, and security factors. In this study, the use of DL methods to the classification problems of malware attacks on the vehicular network has been done. The CPS based model is an architectural framework that, when combined with ubiquitous sensors and technologies for communication offers several advantages to the ITS and its operations. After receiving signals from a vehicle, the outermost computing devices on an RSU deploy the DL algorithms for protecting ITS communications in a significant way. Therefore, this research proposes a CNN model that can effectively address those challenges. The proposed approach employs the exceptional learning capabilities of complex CNNs for analyzing malware samples. This methodology also demonstrates better effectiveness in terms of both accuracy and the time required for detecting new types of malwares. Additionally, the model has an outstanding ability to accurately identify various malware types.

This study highlights many significant contributions, including:

1. DL-based CNN technique has been suggested as a smart IDS security system for VCPS network.
2. The proposed intelligent IDS model employs the averaging strategy for feature selection in order to enhance the performance of the IDS. This model intends to investigate the features and attacks inside VCPS network for the purposes of vehicular network monitoring.
3. The attack detection and accuracy probability of the suggested intelligent IDS model has been enhanced in relation to the F1-Score for vehicular network traffic based on VCPS.
4. The evaluation of the suggested intelligent IDS model is performed employing a variety of performance standards. The effectiveness of proposed intelligent IDS approach is evaluated by comparing it with many cutting-edge ensemble ML algorithms, including RF, AdaBoost, GBoost, Bagging, and XGBoost, specifically in the context of VCPS. The approaches those were suggested demonstrate superior performance compared to conventional methodologies when evaluated on the VCPS-based vehicle V2X_train dataset.

This article is structured in the following approach. Section 6.2 outlines the research investigations that have been conducted on the analysis of network traffic in IOVs using ML and DL techniques for the purpose of detecting malware. Additionally, it discusses the many contemporary models of AI-based IDSs that have been developed specifically for the IoV. Sections 6.3 and 6.4 include a mathematical modeling of DL-based CNN approaches as well as an overview of the experimental setup. Furthermore, these sections provide details on the dataset used including precise information pertaining to the classes and instances. The results of the suggested methodology's analysis in comparison with other models are discussed in Sect. 6.5. The critical discussion is further discussed in Sect. 6.6, while the conclusion of the study is discussed in Sect. 6.7.

6.2 Related Work

This section summarizes some relevant research addressing the implementation of an IDS based on ML and DL methodologies for IoV security. Multiple researchers have employed a wide range of technologies for IoV security optimization and attack detection systems with, spending a significant amount of time on a wide range of challenges. The primary goal of classifying malware attacks is to organize the study of vehicle communication, which provides an indicator for evaluating the effectiveness of ITS and the procedures involved in managing the security of the VCPS network. The authors investigate various types of ML and DL techniques for the purpose of traffic analysis. Additionally, the authors explore AI methodologies that are used to identify and analyze malware behaviors.

There are various studies that provide novel practical solutions on IoVs or research about their implementation using ML and DL models. Yang et al. [10] suggested a tree-structure ML model-based IDS solution for detecting both CAN bus and external assaults in IoV. To address the inadequate amount of data for certain minority populations, authors preprocessed their data using the Synthetic Minority Oversampling Technique (SMOTE). To improve accuracy, the authors proposed the stacking technique as a specific type of ensemble learning. To identify cyber-attacks in IoV, Ullah et al. [11] integrated the gated recurrent unit (GRU) and long short term memory (LSTM) models for DL-based approaches. The strategy relies on a number of preprocessing techniques, including randomization, feature filtering, and normalization. These techniques are used on the datasets to make the LSTM-GRU model more effective.

Firdausi et al. [12] employed dynamic statistical classification on both benign and infected data to investigate malware. To perform classification tasks, authors gathered a total of 220 malware samples and 250 benign samples during their investigation. Various classifiers including Support Vector Machines (SVMs), k-Nearest Neighbors (KNN), Multi-Layer Perceptron (MLP), Decision Trees (DT), and Naive Bayes (NB) were trained using the dataset. The highest level of accuracy 96.8% has been achieved by the use of DT. Rana et al. [13] have been implemented ML methods to analyze a dataset concerning Android applications and their permission access. The highest level of accuracy is achieved with the use of the KNN algorithm with an accuracy rate of 96%. Additionally, the SVM algorithm achieved an accuracy rate of 93%.

Kan et al. [14] introduced a novel approach for detecting lightweight PC malware, intended to address the computational time complexity associated with DL approaches. The fundamental design of this model is based on the CNN approach. This method has the capability to autonomously acquire features from the provided input, which is presented as a series of grouped instructions. An accuracy rate 95% has been achieved on a dataset consisting of 70,000 data points. Alzaylaee et al. [15] introduced DL-Droid, a malware detection model that employs DL techniques. Malicious Android apps are detected by the use of input generations in the process of dynamic analysis. The collection consists of 30,000 instances, including both malware and benign activities. Furthermore, experimental procedures comprise the implementation of hybrid attributes, which are formed by combining dynamic and static features. With respect to the dynamic attributes, the model exhibited a maximum accuracy of 97.8%. In contrast, the hybrid exhibits an impressive accuracy rate of 99.6%.

Xu et al. [16] introduced a malware detection framework that employs a Graph Neural Network (GNN). This research involves the conversion of the graph structure of the Android application into vectors, followed by the classification of malware families using a model. A malware detection accuracy of 99.6% has been achieved, while a classification accuracy of 98.7% has been reached. Gao et al. [17] established a model called GDroid, which utilizes a Graph Convolutional Network (GCN) for the purpose of classifying malware. This study intended to provide a graphical illustration of the interconnections between the various components of the Android platform by way of a heterogeneous graph. There were less than one percent of false positives and the accuracy is 98.99%. Table 6.1 highlights more research on IoV systems that use IDS for the detection of malware attacks.

Based on the Table 6.1 shown in the literature review section, it can be concluded that the efficacy of an IoV-IDS enabled by AI is mostly dependent upon the utilization of a suitable dataset for training purposes. ML models could have been developed with only a certain amount of data in order to get improved results. When dealing with bigger datasets, ML model may not be appropriate unless the data is automatically labeled. Due to the high costs and time requirements associated with the process of labeling, DL algorithms are seen more advantageous for handling bigger datasets. These methodologies aim to identify and extract valuable patterns from raw datasets. To enhance the effectiveness of VCPS-IDS in anomaly detection, it is essential to consistently update it with newly acquired data derived from the monitoring of network traffic. The utilization of extensive datasets and the complex architecture of DL algorithms will result in a more demanding learning process in terms of time frame and computational resources. There seems to be a tradeoff between model complexity and the level of structure achieved by DL methods. The more in-depth the approach, the more complex the model, and the more time and resources will be needed to solve the problem. As a result, this drawback will be eventually resolved by the intelligent selection of important characteristics for model training.

Table 6.1 Summary on related studies

Primary AI approach	Smart model	Comparison methods	Dataset used	Attack type detection	Performance measures	Year	References
ML	FS-Stacking	KNN, SVM, DT, RF, ET, XGBoost, Stacking	CICIDS2017	DoS, Bruteforce, portscan, Botnet, Web attack	Acc: 99.82% F1-Score: 0.997	2019	[10]
ML	RF	NB, LR, SVM, GBDT, XGBoost	NSL-KDD UNSW-NB15	Infiltration Attack, DDoS	Acc: 99.95%	2019	[18]
ML	K-mean	SVM	HCRL	RPM, Fuzzy, DoS	Acc: 99.98%	2021	[19]
DL	CNN	No	Synthetic dataset (experimental)	Not Mentioned specific type	Acc: 99.87%	2019	[20]
DL	DCNN	CNN, NN, SVM	Synthetic dataset (experimental)	DDoS	Acc: 100%	2020	[21]
DL	DCNN	DT, NB, KNN, ANN, SVM, LSTM	Synthetic dataset (experimental)	Fuzzy, DoS, RPM, GEAR	F1-Score: 99.95%:	2020	[22]
DL	LSTM	NB, SVM, ANN, DT	UNSW-NB15 car hacking dataset	RPM, GEAR, Fuzzy, DoS	Acc: 98.00% (UNSW-NB15) Acc: 99.00%(car hacking dataset)	2020	[23]
DL	GHSOM	No	Synthetic dataset (experimental)	Sybil attack, False information attack	Acc: 99.69%	2019	[24]

Where, Acc = Accuracy

The review of the existing literature reveals that many researchers have developed various methodologies, encompassing statistical and ML techniques, to enhance the effectiveness of malware detection strategies in the IoV. However, these approaches exhibit certain limitations. For instance, statistical methods struggle to adapt to the dynamic nature of IoV, posing challenges in defining appropriate evaluation threshold values. Moreover, non-parametric statistical techniques are not well suited for real-time applications due to their computational complexity. ML algorithms including DT, LR, ANN, KNN, and RF have considered for malware detection. In sensitive domains demanding high accuracy and performance, such as IoV, alternative solutions may be more promising than ML deployment. These algorithms encounter difficulties when dealing with complex and extensive datasets, resulting in processing slowdowns and limitations in effectively anticipating novel anomalous patterns. Therefore, it is the necessity to create a DL-based CNN model capable of handling substantial datasets. The objective of this study is to provide a potential solution for detecting malware in the context of IoV. CNNs offer the capability to identify anomalies in sensor data, enabling the detection of deviations from anticipated patterns. This ability holds significant value for applications related to fault diagnosis, security, and safety within the IoV domain.

6.3 Methodologies

This section outlines an in-depth explanation of the basic concepts behind ensemble learning approaches and explains the architectural development of the proposed CNN model.

6.3.1 RF

The RF method is well recognized as a prominent ensemble learning technique used in the field of ML for the purposes of both classification and regression applications. The introduction of this concept may be attributed to Leo Breiman and Adele Cutler in the year 2001.The RF algorithm generates an ensemble of DTs. Every DT is trained on a randomly selected subset of the data, which is chosen using replacement. This phenomenon is often referred to as bootstrapping. The process of bootstrapping increases variability within the dataset for every individual tree. At each split in the forest, a random subset of characteristics is chosen for each DT. This approach aids in minimizing the association between trees and enhances the adaptability of the model.

Every DT is developed autonomously by dividing nodes using specific criteria, usually Gini impurity for classification tasks and mean squared error for regression tasks. The tree continue further splitting until it reaches a stopping condition, which may include a maximum depth or a minimum amount of samples at a leaf node.

Once the training process for all DT is completed, these trees are then used to provide predictions. In the context of classification problems, each tree within the ensemble contributes a "vote" towards a certain class, and the class that receives the highest number of votes is then designated as the predicted class. In regression tasks, the final prediction is obtained by averaging the predictions of all trees. RF functionality is described by Eq. (6.1).

$$\hat{Y} = \text{mod}e(f_1(x), f_2(x), \ldots, f_n(x)) \tag{6.1}$$

where, $\hat{Y}$ = Final prediction of RF, $f_n(x)$ = Prediction of nth DT

6.3.2 AdaBoost

AdaBoost, also known as Adaptive Boosting, is a widely used ensemble learning algorithm that is mostly employed for binary classification problems. However, it may also be expanded to include multi-class classification and regression applications. AdaBoost is a ML algorithm that combines the collective predictions of numerous weak learners, often in the form of DT, in order to construct a robust classifier. The AdaBoost technique assigns more weight to data that have been misclassified by previous weak learners, hence enabling the algorithm to prioritize its attention on the more challenging instances. AdaBoost constructs a robust classifier by repeatedly training weak learners and modifying the weights assigned to the training samples. AdaBoost functionality is described by Eq. (6.2).

$$\hat{Y} = sign\left(\sum_{k=1}^{K} \alpha_k . h_k(x)\right) \tag{6.2}$$

where, $\alpha_k = \frac{1}{2}\ln\left(\frac{1-\varepsilon_k}{\varepsilon_k}\right)$ weight importance of k-th weak learner, $h_k(x)$ = prediction of kth weak learner with input x, ε_k = weight error of weak learner.

6.3.3 GBoost

Gradient Boosting is a very effective ML methodology used for the purposes of both classification and regression assignments. The algorithm in concern is an example of the ensemble learning category, similar to AdaBoost. However, it constructs a robust predictive model by using a distinct approach to aggregating the predictions of weak learners, often DT. The fundamental concept behind Gradient Boosting is the sequential construction of a robust model by repeated emphasis on the errors caused by preceding models. Every subsequent weak learner is taught with the objective of addressing the errors made by the ensemble up to that particular point in time.

GBoost functionality is described by Eq. (6.3).

$$\hat{Y} = \sum_{k=1}^{K} \eta . h_k(x) \tag{6.3}$$

where, $h_k(x)$ = prediction of kth weak learner with input x, η = learning rate hyper parameter controlling the step size during each update.

6.3.4 *Bagging*

The Bagging Classifier, also known as the Bootstrap Aggregating Classifier, is a popular ensemble ML method often used for classification purposes. The primary objective of this approach is to enhance the accuracy and reliability of classifiers by using a combination of many base learners, generally DT. This is achieved through the use of bootstrapping and aggregation techniques. Bagging is a simple but efficient technique for minimizing variance and avoiding the risk of over fitting in models. During the classification process, while generating predictions on a new data point, each base classifier contributes to the decision by "voting" for a certain class label. The final forecast is determined by assigning the class label that receives the highest number of votes. To clarify, Bagging is a technique that combines forecasts using a majority voting mechanism. Bagging functionality is described by Eq. (6.4).

$$\hat{Y}(x) = \text{mod}e(C_1(x), C_2(x), \ldots\ldots\ldots, C_n(x)) \tag{6.4}$$

where, $\hat{Y}(x)$ = Final prediction of RF, $C_n(x)$ = Prediction of nth DT.

6.3.5 *XGBoost*

XGBoost, also known as Extreme Gradient Boosting is a ML method that has excellent efficiency and scalability. It is often used for problems involving classification and regression. XGBoost is a variant of the GBoost technique that is renowned for its computational efficiency, high predictive accuracy, and adeptness in managing complex structured datasets. XGBoost was invented by Tianqi Chen and has gained significant popularity in both ML contests and practical domains. XGBoost mostly employs DT as its weak learners, while it is also capable of supporting several kinds of base models. The depth of the trees is limited and regulated by hyper parameters such as maximum depth, minimum weight, and minimum leaf weight. The XGBoost algorithm incorporates L1 (Lasso) and L2 (Ridge) regularization terms in order to manage the complexity of the model and reduce the risk of over fitting. The XGBoost algorithm constructs an ensemble of DT in an iterative manner, where the predictions of each tree are included into the ensemble with a corresponding weight. The final forecast is derived by calculating the aggregate of these predictions with each prediction being assigned a specific weight. XGBoost functionality is described by Eq. (6.5).

$$Obj(\theta) = \sum_{i=1}^{N} L(y_i, p_i) + \sum_{k=1}^{T} \Omega(f_k) \tag{6.5}$$

where, $L(y_i, p_i)$ = Loss function with y_i, p_i denoting actual target value and predicted value from weak learner respectively, $\Omega(f_k)$ = regularization term for kth trees.

6.3.6 CNN

The CNN model, additionally referred to as the convolutional neural network model, was introduced by Lecun in 1998. This particular model belongs to the category of feed-forward neural networks, which has shown superior performance in the domains of Natural Language Processing (NLP), larger complex dataset and image processing. The use of local perception and CNN weight sharing has the potential to significantly reduce the number of parameters, allowing for the projection of a diverse range of characteristics via the DL process, which in response enhances the accuracy of the learning model. Convolution-layer, pooling-layer, and at last fully-connection-layer constitute the main component of this CNN model. The computations at each convolutional layer are made up of a unique convolutional kernel. The data characteristics were recovered after the convolutional operation performed by each convolutional layer. However, it seems that the extracted features have very large dimensions. A max pooling layer was attached after this convolutional layer to deal with this complexity and reduce the network's training cost. Therefore, the features' dimensions are constrained by this layer. The last component of the neural network architecture is the fully connected layer, which plays a crucial role in linking the features obtained and determining the classification outcomes using the neural network classifier. The framework of the CNN model is shown in Fig. 6.2.

Table 6.2 explains that the traditional ML techniques have some limits due to challenges in extracting accurate features, such as the curse of dimensionality, computing constraints, and the need for domain expertise. Deep neural networks are a specific kind of machine learning technique that uses several layers of networks. In addition, deep learning addresses the issue of representation by constructing several elementary features to capture a sophisticated idea. As the quantity of training data increases, the effectiveness of the deep learning classifier intensifies. Deep learning models address intricate issues by using sophisticated and expansive models, aided by hardware acceleration to save computational time.

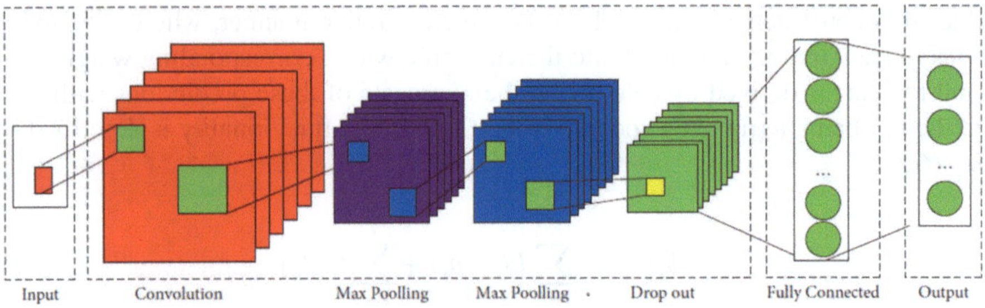

Fig. 6.2 General architecture of CNN model

6.3.7 Proposed Methodology

The primary goal of this study is to create a network IDS to detect attacks involving malware in vehicle networks. Several virus attacks could have been launched on automotive networks by cyber-assailants using wireless interfaces. Therefore, it is important to implement the suggested IDS in both private and public transit networks. The proposed IDS have the potential to effectively recognize abnormal signals inside internal vehicle networks, thus generating warnings. This is achieved by the IDS being integrated into the Controller Area Network (CAN-bus) system. The gateways on external networks could be equipped with the suggested IDS to detect and discard any malicious packets intended for the vehicles. This research introduces unique IDS based on CNN for the purpose of detecting different forms of malware infections in VCPS systems. Figure 6.3 depicts the layer architecture of proposed model. Figure 6.4 represents a detailed overview of the proposed IDS structure.

The deep architecture of CNN is used for intrusion detection is composed of four important layers (2 convolution layers and 2 pooling layers). The network consists of two convolutional layers that train 128 and 256 convolution kernels respectively with a size of 5×5. The deep design incorporates a fully connected layer which includes the use of two individual neurons for the purpose of classification. Two pooling layers are used to implement average pooling with a factor of 2. The challenge of intrusion detection could possibly be seen as a classification task; therefore the sigmoid function has been integrated into the deep architecture. Table 6.3 represents the parameter setup for the proposed model. The algorithm of proposed framework represented in Table 6.4.

The suggested architecture is shown in Fig. 6.4, and its specific steps are as follows:

1. The current research employed the real-time IoV V2X dataset. The dataset is used for the purpose of investigating many characteristics, including the source vehicle address, destination vehicle address, types of network services, network connection status, message type, and duration of connections.

Table 6.2 Summary on basic methodologies

Methodology	Advantages	Disadvantages
RF [25]	The ensemble structure of RF makes it less susceptible to over fitting. It has strong generalization capabilities to novel, unseen data, which is essential for identifying future malware risks in VCPS	RF models are usually trained on fixed datasets, and it might be difficult to update the model in real-time. In a VCPS environment, the process of adjusting the model to accommodate shifting patterns may need the use of more advanced approaches due to continuous streaming of data
AdaBoost [26]	Adaboost is proficient at handling unbalanced datasets. This is advantageous in the context of VCPS malware detection, since harmful occurrences may be smaller in number than non-malicious ones	Adaboost is primarily designed for batch learning and may not be well-suited for streaming data or online learning applications. The VCPS contexts characterized by constantly fluctuating data, this constraint might be seen as a disadvantage
GBoost [27]	Gradient Boosting is flexible enough to handle a wide range of classification issues since it can be adjusted to operate with different loss functions. Its adaptability makes it ideal for targeted VCPS malware detection operations	Determining the optimal setting of hyper parameters could need some expertise and valuable time
Bagging [28]	Bagging is a very efficient technique for dealing with unbalanced datasets that are often encountered in VCPS malware detection. It mitigates the influence of cases from the minority class by using various subsets of the data throughout the training process	Although bagging enables parallelization, training several models may still be almost demanding on resources, particularly in environments with limited processing capabilities as VCPS devices
XGBoost [29]	XGBoost is very efficient in terms of CPU resources and is capable of processing big datasets, making it well-suited for VCPS contexts that incorporate a substantial volume of data provided by numerous devices	XGBoost has several hyper parameters that need fine-tuning in order to get optimum performance. Discovering the optimal configuration of hyper parameters may be challenging and may need proficiency in model optimization
CNN [30]	Data spatial structure can potentially be captured using CNNs. For VCPS malware detection, this expertise is useful for seeing intricate spatial correlations between features, which might lead to better detection of advanced malware activities	For CNN training to be successful, a considerable quantity of labelled data is usually necessary. If there aren't enough malware occurrences, it could be difficult to get varying and representative datasets to use for detecting malware on the VCPS environment

```
Layer (type)                    Output Shape
=========================================================
conv2d_8 (Conv2D)               (None, 64, 2, 1)

max_pooling2d_8 (MaxPoolin      (None, 64, 2, 1)
g2D)

flatten_8 (Flatten)             (None, 128)

dense_43 (Dense)                (None, 64)

dense_44 (Dense)                (None, 1 )

=========================================================
```

Fig. 6.3 Layer architecture of proposed model

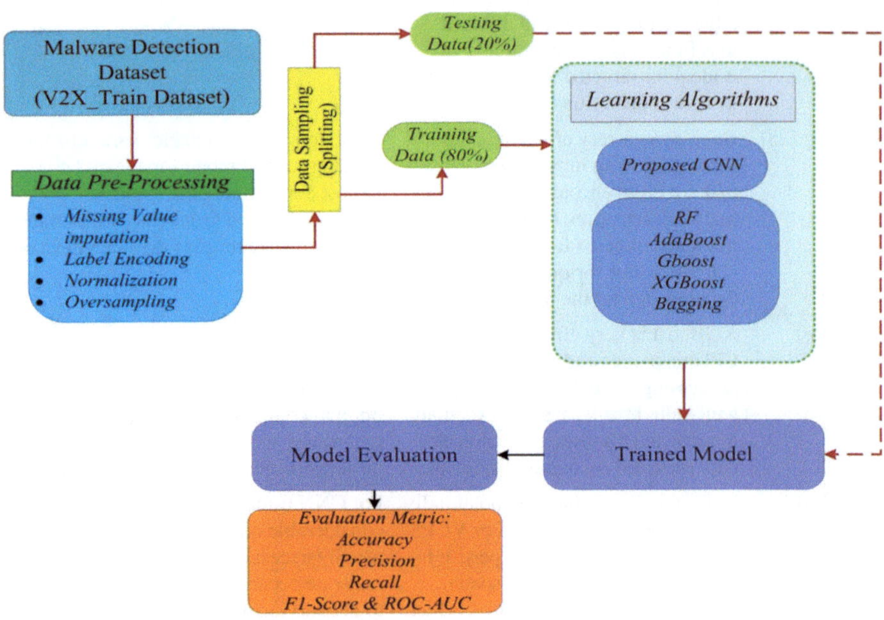

Fig. 6.4 Architecture of proposed Model

Table 6.3 Parameters set up for proposed framework

Proposed model	Parameters setting during simulation
CNN	1. No. of Conv-2D = 2 2. No. of Filters = (32,64) 3. Filter size = 2 * 2 4. Stride = 1, 5. Activation Function = Relu 6. No. of Pooling Layer = 2 7. No. of Batch Normalization Layer = 1 8. Optimizer = Adam (Learning Rate = 0.01) 9. No of Hidden Layer in FCN = 1 10. No of Neurons in Hidden Layer = 128 11. Epochs = 50 12. Batch size = 150 13. Output Layer Activation Function = Sigmoid

2. Based on the specified data processing techniques, the data undergo a series of procedures including preprocessing, handling missing values, numericalization, normalization, and oversampling.
3. After the preprocessing stage, the data is divided into training and validation sets, with a ratio of 80% for training and 20% for validation, relative to the whole dataset.
4. During this training phase, all ensemble techniques such as RF, AdaBoost, GBoost, XGBoost and Bagging are learned using training data.
5. For proposed CNN, the processed training data is sent to the convolution layer in order to extract features, which are then outputted by a two-dimensional convolution operation. In order to decrease feature dimensions, expedite convergence, and mitigate the risk of network over fitting, a pooling layer is used alongside each convolution layer. This pooling layer serves to eliminate redundant features. Subsequently, the whole of local features are combined via a fully connected layer to provide an extensive feature. Ultimately, the leaky rectified linear unit (ReLU) activation function is used in the hidden layer. The sigmoid activation function is often used in the output layer for classification purposes.
6. Following the completion of training on all of the models that are being considered, the test samples are used in order to assess the effectiveness of each model. Accuracy, precision, recall, F1-score, and ROC-AUC were some of the performance measures that were used in the evaluating the performance of each of the models.

6.4 Experimental Setup and Dataset Overview

This section provides a comprehensive description of the dataset, features, data preparation technique, simulation settings, and performance metrics for both the proposed ML algorithms and other associated algorithms.

Table 6.4 Algorithm for proposed framework

1. Initialization of input data which is represented as 3D matrix form (width * height *channels)

   Input: $X \in R^{w*h*c}$, $(w = width, h = height, c = channels)$

2. Implementing Convolutional operation to the inputs using a set of filters which produce feature map represents in equation (6).

   Convolution: $CO^{[l]}_{i,j,k} = \sum_{s=0}^{f_w-1} \sum_{t=0}^{f_h-1} \sum_{c=0}^{c-1} X_{i+s,i+t,c} * W^{[l]}_{s,t,c,k} + b^{[l]}_k$ (6)

   Where $CO^{[l]}_{i,j,k}$ is the value at position (i, j) of the $k-th$ feature map in layer l.
   f_w and f_h are the width and height of the filter.
   $W^{[l]}_{s,t,c,k}$ is the weight of the filter at position *(s, t)* in channel *c* for the $k-th$ feature map in layer l. $b^{[l]}_k$ is the bias for the $k-th$ feature map in layer l.

3. Implementing pooling operation which down sample the feature maps to reduce spatial dimensions and computational complexity shown in equation (7).

   Pooling: $PO^{[l]}_{i,j,k} = max_{m=0}^{f-1} max_{n=0}^{f-1} A^{[l]}_{(f.i+m),(f.j+n),k}$ (7)

   Where, $PO^{[l]}_{i,j,k}$ is the value at position (i,j) of the $k-th$ pooled feature map in layer l.
   f is the pooling window size.

4. Define the flatten layer which Flatten the pooled feature maps into a 1D vector to feed into a fully connected layer represented in equation (8).

   Flatten: $fl^{[l]} = Flatten\left(PO^{[l]}\right)$ (8)

   Where, fl denotes the flatten layer.

5. Define the fully connected (FC) layer which Connect every neuron in the flattened vector to every neuron in the subsequent layer represented in equation (9).

   FC operation: $Z^{[l+1]} = W^{[l+1]} * fl^{[l]} + b^{[l+1]}$ (9)

   Where, $Z^{[l+1]}$ is the output of the fully connected layer. $W^{[l+1]}$ are the weights connecting the layers, $b^{[l+1]}$ are the biases

6. Apply activation function for both hidden layer and output layer of fully connected layer.

   $Af^{[l+1]}$ = *Relu* $[Z^{[l+1]}]$ *(for hidden layers)*
   $Af^{[l+1]}$ = *Sigmoid* $[Z^{[l+1]}]$ *(for output layer classification task)*
   Where, Af denotes the activation function.

7. Compute final output and loss.

   Output: $y_{Pred} = Af^{[L]}$
   Loss: $L = (y_{Pred}, y_{true})$

8. Perform Backpropagation. Calculate the gradients of the loss with respect to the CNN parameters. Update the model's parameters (weight and biases) to minimize the loss.
9. Do repeat steps 2- 8 for multiple epochs to train the network.

6.4.1 Overview of Dataset

The V2X dataset [31] comprises a compilation of V2X (Vehicle-to-Everything) communications intended for the purposes of categorization, prioritization, and detection of spam messages. The dataset consists of 1,000 messages that exhibit diverse characteristics such as message varieties, content, priority, and spam classifications. The communications are sourced from various destination vehicles or broadcasted to all vehicles. The included message varieties are traffic updates, emergency alerts, weather notifications, danger warnings, and road works information and spam communications. The classification of communications is based on their priority, which is divided into three categories: high, medium, and low. Messages of high importance often pertain to urgent matters or critical circumstances that need rapid action. Messages of medium priority include updates on traffic conditions, notifications about ongoing road works activities, and warnings pertaining to potential hazards. Low-priority communications include unsolicited or promotional information, such as spam. The dataset includes a binary label, denoted as "spam," which serves to identify whether a given message has been identified as spam. Spam status is indicated by a binary label that may be either 1 for spam type or 0 for not spam type.

6.4.2 Data Preparation

This section offers a thorough explanation of the data preprocessing methods employed for all the models being examined. The IoV generates its network traffic through numerous sensors, resulting in a wide array of data properties, encompassing both numerical and categorical values. As a result, it is essential to preprocess this data to create the desired detection system.

6.4.2.1 Missing Value Imputation

Imputing missing values in IoV data requires careful consideration, as this data often includes a variety of data types, such as numerical and categorical variables, and may have specific characteristics related to vehicular and sensor data. In this study the missing values handled by using the mean, median, or mode imputation.

6.4.2.2 Label Encoding

For the analysis of IoV network data, employing a label encoding technique is imperative to convert categorical variables into numeric formats. This is essential due to the heterogeneous nature of IoV network traffic, which encompasses both numeric and categorical attributes requiring conversion for analysis and processing. The objective for this is because the suggested detection method has a high level of efficiency in handling numerical characteristics.

6.4.2.3 Normalization

The IoV network is equipped with a range of electronic sensors, which operate both autonomously and in conjunction with human actions. These sensors play a critical role in collecting and transmitting real-time data within the IoV system. However, the data generated by these sensors vary significantly in magnitude. To facilitate pattern analysis, enhance convergence, and reduce training time, the proposed detection system utilizes the Min–Max normalization technique.

6.4.2.4 Oversampling

Oversampling is a technique used in ML to address class imbalance, and it can be particularly relevant in the context of IoV network data. Class imbalance occurs when one class is significantly underrepresented compared to another class.in this study oversampling has been applied to mitigate the imbalance issue.

6.4.3 Simulation Environment

The research has been carried out by using the Python notebook on GPU servers provided by Google Colab, utilizing the Keras and TensorFlow frameworks. In this experimental study, the hardware configuration consisted of an Intel Core i7 CPU operating at a frequency of 2.20 GHz, 16 GB of random-access memory (RAM), the Windows 10 operating system (64-bit), and an NVIDIA GeForce GTX 1050 graphics processing unit (GPU). Several software packages in the PYTHON programming language, including Imblearn, Pandas and Numpy packages are used for the purpose of conducting additional data analysis. Furthermore, the visualization of data is facilitated by including Matplotlib and Mlxtend. Additionally, the analysis of the data is conducted using the Sklearn framework. The Keras and TensorFlow libraries were used in this study, with Keras being a library specifically designed for neural networks. In comparison, TensorFlow is an open-source framework for ML that can be used for a wide range of applications.

6.4.4 Performance Measures

This article presents a comparative analysis of CNN-based technique with measures those are specified as follows. These measures are represented in Eqs. (6.10)–(6.13) respectively.

$$Precision = TP/(TP + FP) \tag{6.10}$$

$$Accuracy = (TP + TN)/(TP + TN + FP + FN) \tag{6.11}$$

$$Recall = TP/(TP + FN) \tag{6.12}$$

$$F1 - measure = 2 * Precision * Recall/(Precision + Recall) \tag{6.13}$$

where, "true positive" (TP) refers to the number of requests accurately identified as having harmful behaviors and "false positive" (FP) refers to the number of applications incorrectly identified as normal. By contrast, true negative (TN) refers to the number of apps that are correctly labeled as normal, while false negative (FN) refers to the number of applications that are incorrectly labelled as malware. Generally, a greater level of precision, accuracy, and recall corresponds to an enhanced identifying outcome. The effectiveness of the identification strategy may be better explained by the higher F1 score, which combines the outcomes of precision and recall.

6.5 Result Analysis

This study aims to evaluate the entire effectiveness of the specified models. The analysis originates with examining advanced ML metrics and concludes by explaining the performance of the DL based CNN model. A wide variety of evaluation measures, such as recall, accuracy, precision, ROC-AUC and F1-score are used to illustrate the results. The accuracy obtained represents a metric for the measurement of the overall performance of the suggested approach. In addition; more emphasis has given on the use of the F1-Score metric across all methodologies due to its ability to facilitate the attainment of harmonized precision-recall equilibrium. The research exhibits a non-uniform and highly uneven distribution of class labels. The F1-Score is a relevant metric to appropriately evaluate performance. Table 6.5 provides a detailed description of the parameters used during the training of the other ML models.

Table 6.6 presents a comparative analysis of advanced ensemble learning methodologies, including RF, AdaBoost, GBoost, XGBoost, Bagging, and the suggested CNN. Both GBoost and XGBoost algorithms provide superior performance compared to other advanced ensemble learning algorithms, as demonstrated by their outstanding F1-Score of 98.95%. In contrast, RF technique has a suboptimal

Table 6.5 Parameter setup for considered ML models

Model	Parameters
RF	*n_estimators = 100, criterion = 'gini', min_samples_split = 2, min_samples_leaf = 1, max_features = 'sqrt', bootstrap = True, oob_score = False*
AdaBoost	*n_estimators = 50, learning_rate = 1.0, algorithm = 'SAMME.R'*
GBoost	*n_estimators = 100, learning_rate = 0.01*
XGBoost	*n_estimators = 100, learning_rate = 1.0*
Bagging	*n_estimators = 10, max_samples = 1.0, max_features = 1.0, bootstrap = True, bootstrap_features = False*

F1-Score. The CNN model under consideration has outstanding outcomes metrics including an accuracy rate of 99.64%, precision of 100%, recall of 99.30%, F1-score of 99.64%, and ROC AUC of 99.65%.

The ROC graph is frequently developed in order for evaluating the performance of the classification. The ROC curve is defined by the representation of the sensitivity test on the y-axis and the 1—false positive rate (or specificity) on the x-axis. The evaluation of classifier performance is often regarded as an efficient procedure. In general, when the area under the receiver operating characteristic (ROC) curve is 0.5, it indicates a lack of classification. This suggests that the classifier's capacity to accurately identify intrusions based on the detection of attack existence or absence is dependent on the specific circumstances employed. The range including values from 0.7 to 0.8 is often referred to as the acceptable range, while the range spanning from 0.8 to 0.9 is typically labelled as the good range. Performance over 0.9 is generally regarded as outstanding. Figure 6.5 illustrates the area under the receiver operating characteristic curve (AUC-ROC) study for several advanced ensemble learning and DL based CNN model. In this work, it can be observed that DL approach exhibits significant dominance over advanced ensemble learning techniques.

This study presents the confusion matrix of several classifiers, including XGBoost, RF, Bagging, AdaBoost, GBoost, and CNN. The employment of a confusion matrix was performed to evaluation the effectiveness of the classification algorithms employed. The confusion matrix for binary classification is shown in Fig. 6.6. The

Table 6.6 Comparison analysis of proposed method with other considered approaches

Evaluation metrics	RF	AdaBoost	GBoost	XGBoost	Bagging	Proposed CNN
Accuracy (%)	84.34	96.08	98.93	98.93	98.92	99.64
Precision (%)	82.35	92.85	98.61	98.61	98.60	100
Recall (%)	88.11	100	99.30	99.30	99.28	99.30
F1-measure (%)	85.13	96.29	98.95	98.95	99.93	99.64
ROC-AUC (%)	84.27	96.01	98.92	98.92	98.90	99.65

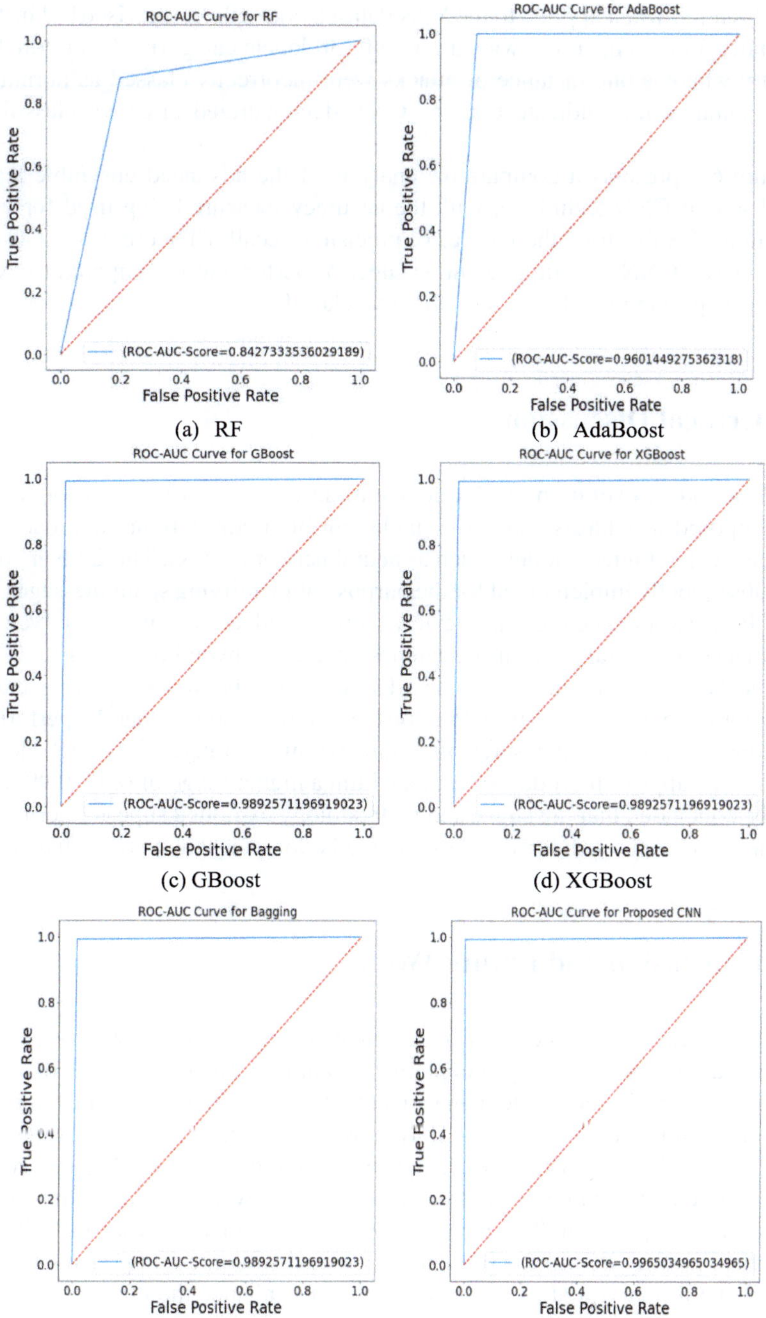

Fig. 6.5 Analysis of ROC-AUC comparison of **a** RF, **b** AdaBoost, **c** GBoost, **d** XGBoost, **e** Bagging, **f** Proposed CNN

CNN classifier, when applied to the V2X dataset, correctly categorised142 instances as assaults. In a comparable way, a total of 138 labels categorized as normal were correctly whereas one instance of attacks were incorrectly classed as normal. The experimental results indicate that the CNN demonstrated effective classification ability.

Figure 6.7 presents a comparative analysis of the advanced ensemble learning and DL-based CNN technique, with the accuracy measure being used for evaluation. Figure 6.8 illustrate the metrics of precision, recall, F1 score, and ROC-AUC. The proposed CNN technique exhibits superior performance compared to existing techniques, proving itself as a very effective classifier.

6.6 Critical Discussion

Over a period of evolutionary development, an extensive variety of methods has been proposed to address this issue, including both conventional approaches and advanced technological models, such as neural networks (NN). The CNN model has been subsequently implemented for the purpose of identifying spam messages inside the VCPS network. The literature review section produces a complete survey of the performance of several previous researches on attacks involving malware. Table 6.7 presents the previous study outcomes derived from the various dataset that was chosen for the purpose of testing. Previous research has shown that the performance of various classification measures often falls within the range of 0.8–0.9. However, the proposed approach produced results within a higher range of 0.97–0.99. Hence, it can be concluded that the suggested methodology exhibited superior performance in comparison to the previous studies and methodologies that were evaluated.

6.7 Conclusion and Future Work

A VCPS often incorporates a diverse range of advanced innovations including autonomous cars, wireless payment platforms, administrative software, communication tools, and real-time traffic management systems. Different organizations, such as cyber-criminals and hacktivists may have diverse reasons for causing disruption inside VCPS. In past decades, incidents have been reported where roadside boards, surveillance cameras, and emergency sirens were subject to unauthorized access and manipulation. The identification of malicious software in VCPS is of the highest priority due to the presence of several software components inside the system. The primary goal of this research was to enhance the efficacy of malware detection framework via the use of CNN. The study demonstrated that the detection model based on CNN yielded impressive outcomes. The effectiveness of the suggested approach against advanced attacks such as those highlighted in new case studies will be assessed in future research. The use of DL models in VCPS will

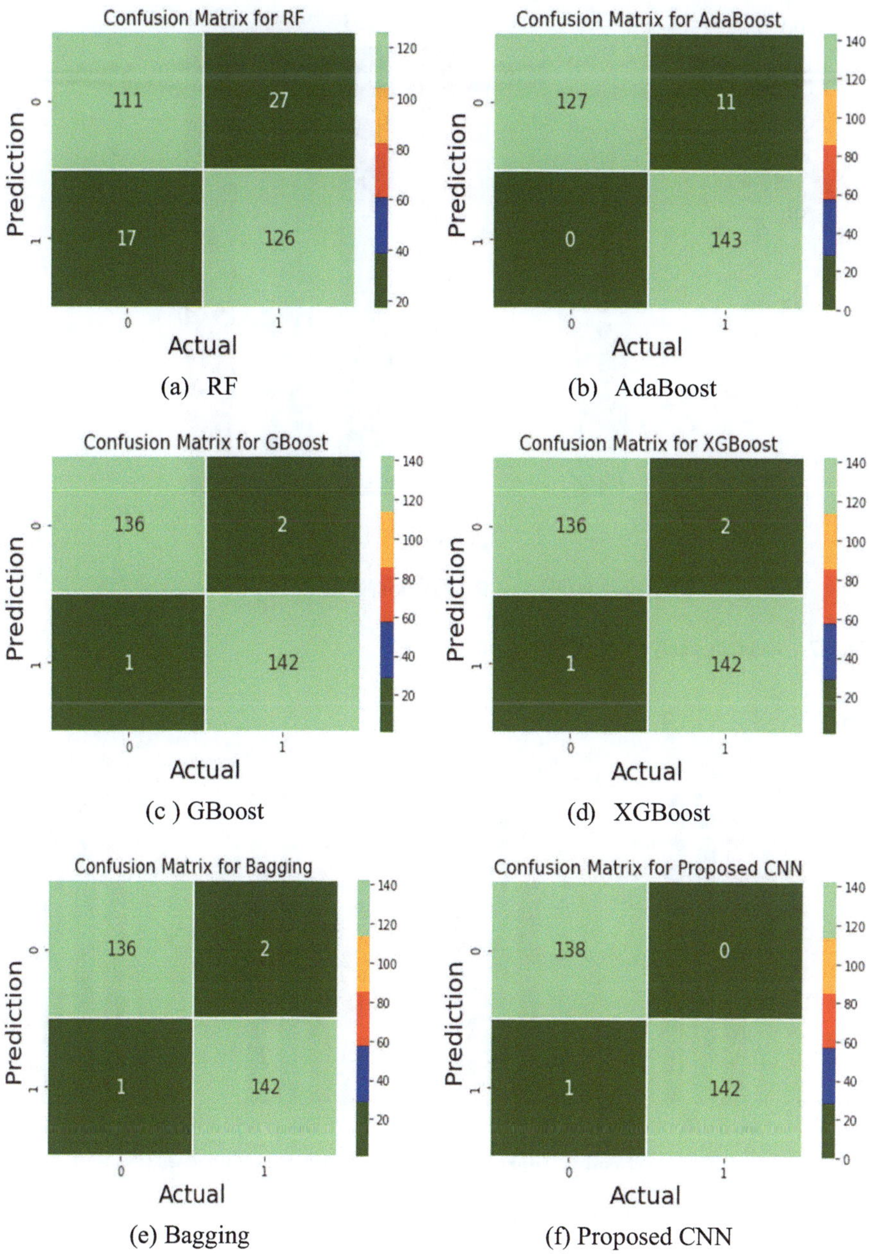

Fig. 6.6 Analysis of confusion matrix for **a** RF, **b** AdaBoost, **c** GBoost, **d** XGBoost, **e** Bagging, **f** Proposed CNN

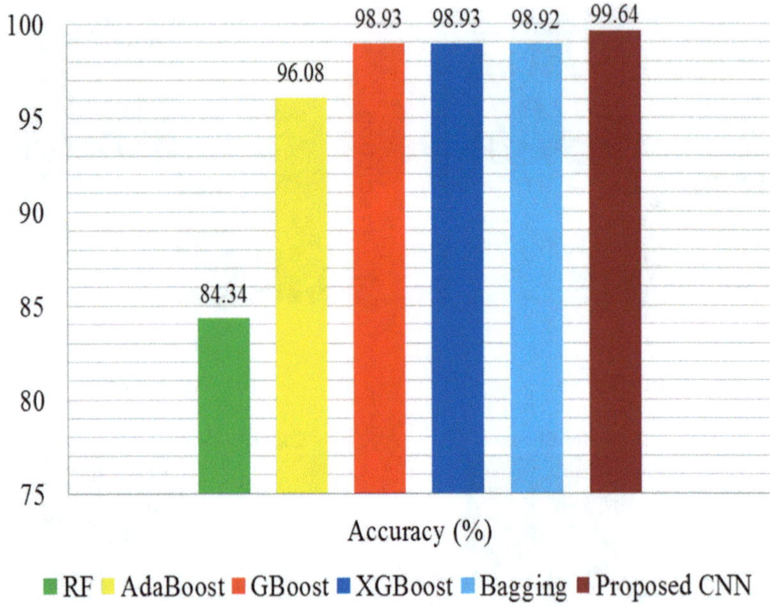

Fig. 6.7 Accuracy analysis of classification models

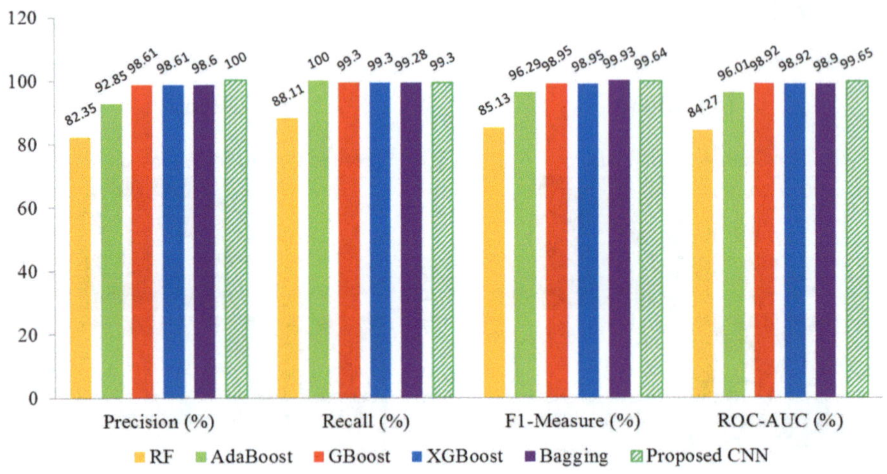

Fig. 6.8 Analyses of precision, recall, F1-score and ROC-AUC

also be discussed, with a focus on the adaptation of model hyper parameters for optimization in further research.

Table 6.7 Previous studies compared with proposed model

Year	Dataset used	Model used	Accuracy (%)	References
2019	Own experimental malware dataset	Improved Naïve Bayes	98.00	[32]
2014	Android malware genome project data samples	Deep neural network (DNN)	90	[33]
2017	Android malware genome project data samples	CNN	90	[34]
2016	Own experimental malware dataset	XGBoost	97	[35]
2020	Malimg malware dataset	ResNext	98.32	[36]
2021	Malimg malware dataset	ResNet50 with Adam	99.05	[37]
2023	V2X malware dataset	Improved CNN	99.64	Proposed model

References

1. Chen, Z., Boyi, W., Lichen, Z.: Research on cyber-physical systems based on software definition. In: Proceedings of the IEEE 12th International Conference on Software Engineering and Service Science (ICSESS) (2021)
2. Alam, K.M., Saini, M., Saddik, A.E.: Toward social internet of vehicles: concept, architecture, and applications. IEEE Access **3**, 343–357 (2015)
3. Piran, M.J., Murthy, G.R., Babu, G.P.: Vehicular ad hoc and sensor networks; principles and challenges. Int. J Ad hoc Sensor Ubiquit. Comput. **2**(2), 38–49
4. Prakash, R., Malviya, H., Naudiyal, A., Singh, R., Gehlot, A.: An approach to inter-vehicle and vehicle-to-roadside communication for safety measures. In: Intelligent Communication, Control and Devices, 624. Advances in Intelligent Systems and Computing (2018)
5. Kumar, S., Dohare, U., Kumar, K., Dora, D.P., Qureshi, K.N., Kharel, R.: Cybersecurity measures for geocasting in vehicular cyber physical system environments. IEEE Internet Things J. **6**(4), 5916–5926 (2018)
6. https://www.av-test.org/en/statistics/malware/. Accessed 11 Nov 2023
7. Lv, Z., Lloret, J., Song, H.: Guest editorial software defined Internet of vehicles. IEEE Trans. Intell. Transp. Syst. **22**, 3504–3510 (2021)
8. Maleh, Y., Ezzati, A., Qasmaoui, Y., Mbida, M.: A global hybrid intrusion detection system for wireless sensor networks. Proc. Comput. Sci. **52**(1), 1047–1052 (2015)
9. Kaiwartya, O., Abdullah, A.H., Cao, Y., Altameem, A., Prasad, M., Lin, C.-T., Liu, X.: Internet of vehicles: motivation, layered architecture, network model, challenges, and future aspects. IEEE Access **4**, 5356–5373 (2016)
10. Yang, L., Moubayed, A., Hamieh, I., Shami, A.: Tree-based intelligent intrusion detection system in internet of vehicles. In: 2019 IEEE Global Communications Conference (GLOBECOM), pp. 1–6 (2019)
11. Ullah, S., Khan, M., Ahmad, J., Jamal, S., Huma, Z., Hassan, M., Pitropakis, N., Buchanan, W.: HDL-IDS: a hybrid deep learning architecture for intrusion detection in the Internet of Vehicles. Sensors **22**(4), 1340 (2022)
12. Firdausi, I., Lim, C., Erwin, A., Nugroho, A.: Analysis of machine learning techniques used in behavior-based malware detection. In: Proceedings of the International Conference on Advances in Computing, Control and Telecommunication Technologies, Jakarta, Indonesia, 2–3 December 2010

13. Rana, J.S., Gudla, C., Sung, A.H.: Evaluating machine learning models for android malware detection: a comparison study. In: Proceedings of the 2018 VII International Conference on Network, Communication and Computing, New York, NY, USA, 14–16 December 2018
14. Kan, Z., Wang, H., Xu, G., Guo, Y., Chen, X.: Towards light-weight deep learning based malware detection. In: Proceedings of the IEEE 42nd Annual Computer Software and Applications Conference (COMPSAC), Tokyo, Japan, 23–27 July 2018
15. Alzaylaee, M.K., Yerima, S.Y., Sezer, S.: DL-droid: deep learning based android malware detection using real devices. Comput. Secur. **89**, 101663 (2020)
16. Gao, H., Cheng, S., Zhang, W.: GDroid: android malware detection and classification with graph convolutional network. Comput. Secur. **106**, 102264 (2021)
17. Xu, P., Eckert, C., Zarras, A.: Detecting and categorizing Android malware with graph neural networks. In: Proceedings of the 36th Annual ACM Symposium on Applied Computing (SAC '21), New York, NY, USA, 22–26 March 2021, pp. 409–412
18. Gao, Y., Wu, H., Song, B., Jin, Y., Luo, X., Zeng, X.: A distributed network intrusion detection system for distributed denial of service attacks in vehicular ad hoc network. IEEE Access **7**, 154560–154571 (2019)
19. D'Angelo, G., Castiglione, A., Palmieri, F.: A cluster-based multidimensional approach for detecting attacks on connected vehicles. IEEE Internet Things J. **8**(16), 12518–12527 (2021)
20. Peng, R., Li, W., Yang, T., Huafeng, K.: An internet of vehicles intrusion detection system based on a convolutional neural network. In: 2019 IEEE International Conference on Parallel & Distributed Processing with Applications, Big Data & Cloud Computing, Sustainable Computing & Communications, Social Computing & Networking (ISPA/BDCloud/SocialCom/SustainCom), pp. 1595–1599. IEEE (2019)
21. Nie, L., Ning, Z., Wang, X., Hu, X., Cheng, J., Li, Y.: Data-driven intrusion detection for intelligent internet of vehicles: a deep convolutional neural network-based method. IEEE Trans. Netw. Sci. Eng. **7**(4), 2219–2230 (2020)
22. Song, H.M., Woo, J., Kim, H.K.: In-vehicle network intrusion detection using deep convolutional neural network. Vehicul. Commun. **21**, 100198 (2020)
23. Ashraf, J., Bakhshi, A.D., Moustafa, N., Khurshid, H., Javed, A., Beheshti, A.: Novel deep learning-enabled LSTM autoencoder architecture for discovering anomalous events from intelligent transportation systems. IEEE Trans. Intell. Transp. Syst. **22**(7), 4507–4518 (2020)
24. Liang, J., Chen, J., Zhu, Y., Yu, R.: A novel intrusion detection system for vehicular ad hoc networks (VANETs) based on differences of traffic flow and position. Appl. Soft Comput. **75**, 712–727 (2019)
25. Breiman, L.: Random forests. Mach. Learn. **45**, 5–32 (2001)
26. Ying, C., et al.: Advance and prospects of AdaBoost algorithm. Acta Automat. Sin. **39**(6), 745–758 (2013)
27. Shastri, S., et al.: GBoost: a novel grading-AdaBoost ensemble approach for automatic identification of erythemato-squamous disease. Int. J. Inf. Technol. **13**, 959–971 (2021)
28. Alzubi, J.A.: Diversity based improved bagging algorithm. In: Proceedings of the International Conference on Engineering & MIS 2015 (2015)
29. Ramraj, S., et al.: Experimenting XGBoost algorithm for prediction and classification of different datasets. Int. J. Control Theory Appl. **9**(40), 651–662 (2016)
30. Jogin, M., et al.: Feature extraction using convolution neural networks (CNN) and deep learning. In: 2018 3rd IEEE International Conference on Recent Trends in Electronics, Information & Communication Technology (RTEICT). IEEE (2018)
31. https://ieee-dataport.org/documents/v2x-message-classification-prioritization-and-spam-detection-dataset
32. Kumar, R., Zhang, X., Wang, W., Khan, R.U., Kumar, J., Sharif, A.: A multimodal malware detection technique for android IoT devices using various features. IEEE Access **7**, 64411–64430 (2019)
33. Yu, W., Ge, L., Xu, G., Fu, Z.: Towards neural network based malware detection on android mobile devices. In: Cybersecurity Systems for Human Cognition Augmentation, pp. 99–117. Springer (2014)

34. McLaughlin, N., Doupé, A., Ahn, G.J., del Rincon, J.M., Kang, B.J., Yerima, S., Miller, P., Sezer, S., Safaei, Y., Trickel, E., Zhao, Z.: Deep android malware detection. In: Proceedings of the Seventh ACM on Conference on Data and Application Security and Privacy—CODASPY '17, pp. 301–308 (2017)
35. Fereidooni, H., Conti, M., Yao, D., Sperduti, A.: ANASTASIA: android malware detection using static analysis of applications. In: 2016 8th IFIP International Conference on New Technologies, Mobility and Security (NTMS), pp. 1–5. IEEE (2016)
36. Go, J.H., Jan, T., Mohanty, M., Patel, O.P., Puthal, D., Prasad, M.: Visualization approach for Malware classification with ResNeXt. In: 2020 IEEE Congresson Evolutionary Computation (CEC). IEEE, pp. 1–7 (2020)
37. Sudhakar, Kumar, S.: MCFT-CNN: Malware classification with fine-tune convolution neural networks using traditional and transfer learning in Internet of Things. Future Gener. Comput. Syst. **125**, 334–351 (2021). https://doi.org/10.1016/j.future.2021.06.029

[illegible]

Chapter 7
Unraveling What is at Stake in the Intelligence of Autonomous Cars

Dioneia Motta Monte-Serrat and Carlo Cattani

Abstract The integration of physical and cybernetic systems introduces new functionalities that modify the configuration of autonomous driving vehicles. The vehicle's driving behavior is subject to respond differently than the driver expects, causing accidents. Innovation in cybernetic systems is based on still immature information. To achieve socially responsible innovation, it is necessary to dispel the uncertainties of the *black box* of new technologies. We use an argumentative method to show that there is a pattern, a unique structure, that appears repeatedly in the cognitive linguistic process of both human beings and intelligent systems. From this, we aim not only show that this pattern guarantees coherence to the decision-making performed by cognitive computing, but also that it reveals what is at stake in the intelligence of autonomous cars and in the biases of the *black box* of AI. Therefore, by clarifying the dynamic of the unique cognitive linguistic process, as a common process for individuals and machines, it is possible to manage the interpretive activity of cyber-physical systems and the way they decide, providing safe and sustainable autonomous cars.

Keyword Cyber-physical systems · Cognitive process · Interpretive activity · Dynamic process · Autonomous cars

7.1 Introduction

The intelligence, or better saying, the intelligent decisions of autonomous cars, unite computational and physical resources, reconfiguring them to acquire autonomy, efficiency, and functionality. There are still major challenges to be overcome in relation

D. M. Monte-Serrat
Computing and Mathematics Department, Law Department, USP, Unaerp, Brazil

C. Cattani (✉)
Engineering School (DEIM), Tuscia University, Viterbo, Italy
e-mail: cattani@unitus.it

J. Nayak et al. (eds.), *Machine Learning for Cyber Physical System: Advances and Challenges*, Intelligent Systems Reference Library 60,
https://doi.org/10.1007/978-3-031-54038-7_7

to the safety of the automotive sector, revealing that scientific and engineering principles need to be deepened in terms of integrating cybernetic and physical elements. This chapter unravels the application of autonomous systems innovations confronting the fundamentals of the human cognitive linguistic process to inspire the formulation of algorithms and models of cyber-physical systems. To argue about the existence of a unique structure present in the foundations of the human cognitive-linguistic process that can be applied to intelligent systems, Chaim Perelman's argumentative method is used [1], which, instead of logical reasoning, makes use of a regressive reasoning that considers variability of situations and special values. We clarify to cyber systems' researchers and developers, who deal with language and cognition, that one cannot ignore the dynamic process through which human language and cognition are expressed. This dynamic process is unique, integrating cybernetic and physical elements.

In computational intelligence, the mechanisms of control and detection of context elements are intertwined to reconfigure the machine's cognition. The *interconnection* of these elements is still precarious because it does not imitate the human cognitive linguistic process to the satisfaction. This chapter breaks new ground by suggesting that, in addition to designing tasks that guide decision-making in autonomous systems, it is necessary to consider the fundamentals of the human cognitive linguistic *process*. Cognitive ability, when considered a 'process' encompasses the 'dynamic' aspect, which is subject to reconfiguration at different spatial and temporal scales. Overcoming this spatial and temporal difficulty means optimizing the autonomous system, preventing the degradation of its performance and the robustness of its design. It is important to highlight that the dynamic cognitive process is not limited to the influence of the logical sequence of tasks previously established in the system's cognitive core, but also responds to the unpredictability of the environment. The temporal extension of cognition, both in humans and in intelligent systems [2], has the role of making the system overcome the recurrent limited capacity to manage uncertainties arising from accidental events during its operation [3].

Pointing solutions do not solve the endemic problems of autonomous systems. There is a need to intervene in the core of the machine's cognitive system, providing it with fundamental elements and information for the generation of its cognitive activity. Under an argumentative method, we discuss the foundations of the dynamics of the human cognitive linguistic process, in order to abstract basic principles that can guide the autonomous system's core design. In this way, all technicians and researchers become aware of how they must act to improve the performance of autonomous systems, so that synchronous computational and physical processes are integrated with asynchronous computational processes. The fundamental principles demonstrated in this chapter, therefore, not only have the potential to encourage the development of tools and architectures that improve the functioning of autonomous cars, but also raise awareness among technicians and researchers about how they should act to ameliorate the performance of these systems.

In the quest to establish new principles for intelligent systems technicians to design and implement the algorithmic core of autonomous systems, we resorted to the perspectives of other branches of science, such as linguistics and neurolinguistics. This interdisciplinary approach to self-driving cars eases the challenge for computer scientists to unravel what is at stake in the autonomous systems they are developing. Machine learning, ML, algorithms, neural networks, intelligent systems are used interchangeably in this chapter to represent branches of artificial intelligence (AI). AI is generically understood as a branch of computer science that makes use of data and task sequences to mimic and outperform human behavior, as is the case with autonomous systems. It is expected that the ideas outlined in this chapter will help in the interface between the cyber world and the physical world, innovating and perfectly integrating characteristics and behaviors of autonomous cars. By understanding the dynamics of the cognitive linguistic process, it will be more manageable for the technician who designs the algorithm (system core) to identify the origin of failures and biases that cause accidents in autonomous systems.

It is important to emphasize that this chapter does not dwell on specific elements that are at play at the time of an autonomous car error or accident. Fundamentals and principles of human cognition intended to guide the decision-making of autonomous cars to be successful are discussed here. To that end, we describe some challenges encountered in highly automated systems, which caused damage to people and property. The misuse or inappropriate design of robotic consciousness of some systems intended to react to the world around them, led to failures to make meaningful predictions [3]. These negative events led us to suggest a unifying theory for the design and implementation of cybernetic and physical resources. This theory, based on the functioning of human cognition, can be applied to the core of intelligent systems in various domains of Artificial Intelligence.

This Chapter unravels what is at stake in the intelligence of autonomous cars. To this end, it opposes fundamentals of the human cognitive linguistic dynamic process to models of cyber physical systems. Section 1.1 shows the scenario of autonomous systems citing, just to exemplify, an automated system with human intervention, a system that learns to react to the world around it, and advanced driver assistance systems. In this scenario, there are reports that the intelligent system did not understand what was assigned to it as a task, or it did not intervene correctly, which could lead to accidents. The origins of errors and algorithmic biases in these systems, called black box AI, are challenging for technicians and researchers. To uncomplicate the structure on which autonomous systems are based, we clarify, in Sect. 7.2, through Perelman's argumentative method, how the dynamic and universal process of human language and cognition takes place (Sect. 2.1) and how it can be applied to AI (Sect. 2.2). Understanding that language and cognition are a process and not a substance helps to shed light on the mysterious workings of the AI black box and helps to prevent its errors and failures. Advanced Driver Assistance Systems (ADAS) are described in more detail in Sect. 7.3, to resolve the causes of decisions coordinated by algorithms that provide, paradoxically, safety and risk. We show how this system merges physical and computational resources (Sect. 3.1), what is the framework behind the ADAS design (Sect. 3.2), and the logical and executive functions in autonomous driving systems related to what is at stake in the impenetrable box black AI (Sect. 3.3). Section 7.4 deals with how to rethink the interpretive activity of cyber-physical systems under a unifying context to ensure consistency in the system behavior and how to mitigate biases in the cognitive ability to interpret. In Sect. 4.1 we unravel the black box of logical and executive functions of autonomous driving systems. This is how we arrive at Sect. 7.5, in which we discuss how ADAS performs its learning through the integration of algorithmic core and context. This important integration process is what brings cognitive computing closer to the human cognitive linguistic dynamic process. Section 7.6 is devoted to the conclusion that integrating context and algorithmic sequence serves as an umbrella to encompass the diverse activities related to cyber systems, simplifying the complex application of different algorithms for different autonomous driving tasks. We show that the weakness of ADAS resides in not being able to align the task sequence with its environment and what is really at stake is *how* to pass instructions to the ADAS design, instead of *what* instructions to pass to ADAS.

7.1.1 Autonomous Systems Scenario: Some Mathematical Modeling Techniques for the Dynamics of Cyber-Physical Systems

Highly automated systems can cause harm to people and property due to their misuse or their design. Gillespie [4] suggests that the reliability of the automated system is achieved through intervention in the Autonomous Human-Machine Team (A-HMT-S), reallocating tasks and resources. The author states that this helps in approaching autonomy development problems for teams and systems with a human-machine interface. The difficulties encountered in teams of human beings-autonomous systems (A-HMT-S) are due to the frequent reconfiguration of systems, which are not always understandable or reproducible. To circumvent uncertainties in the interpretation of input information in artificial intelligence, the author suggests the use of a hierarchical architecture to improve the effectiveness of the design and development of the A-HMT-S through the use of specific tools machine learning, ML, the design decisions that ensure actions are taken based on authorization from the human team leader, and through a value adoption for tasks when setting priorities.

When it comes to automation of intelligent systems, there is a tendency, among scientists, to develop a robotic consciousness that learns to adapt to changes, although it is admitted that this subject is complex. [5] defines robotic consciousness as the machine's ability to recognize itself, imagine itself in the future and learn to react to the world around it. This was the goal of Kedar et al. [6, p. 7] when they created Spyndra, a quadruped robot, with an open-source machine learning platform to study machine self-awareness. The authors bet on the robot's self-simulation to predict the sensations of its actions. They compare a simulated gear and the actual gear to push the limits the machine presents in reshaping its own actions. The authors' hypothesis is that the system has self-awareness and record its own orientation and acceleration. Visual camera information is combined with deep learning networks for path planning. The experiment demonstrates that neither linear nor ridge regression accurately predicted the global measurement. The explanation found is that direction and orientation are related to yaw, and yaw is the least repeatable feature and difficult to predict. It was observed that the neural networks failed to make meaningful predictions, leading the authors to assume that the robot state depends on the robot's previous state [6, p. 8]. An extra perturbation was also identified in the simulated data due to the interference of the robot's contextual reality, since the simulation model assumed that the material is homogeneous [6, p. 9]. The authors promise to improve their machine learning model. They make available open-source control software and a data set for future researchers who want to develop a self-modeling platform through augmentation of feedback sensors and resources extracted from their simulation.

Autonomous systems deployed in cars, in turn, equipped with advanced driver assistance systems (ADAS) [3], perform the task of driving. It has been noted that while automation helps drivers, they must always be on the alert in case the computer does not know what to do or intervenes incorrectly. These risks are still not sufficiently recognized or managed. There are reports of situations where accidents occur because drivers cannot understand why the vehicle responds or does not respond in a specific way.

Artificial Intelligence, intelligent systems, machine learning, algorithms and neural networks have something in common and challenging in the development of a system that has its own conscience. When it comes to the correlations that the system or robot uses to identify the context and promote its adaptation to it, it is not enough to look at the superficial structures of cognition. No matter how many tasks and resources are reallocated, the resulting reconfiguration will not always be understandable or reproducible: neural networks end up failing to make predictions. We propose a look into the depths of cognition, at its origins, to teach cyber systems to intervene correctly to better manage risks.

The question is how to know which are the correlations that indicate a causal connection with the behavior of the system? This is an important basis for machine learning not to be vulnerable to human and algorithmic errors and biases. Errors and spurious correlations confuse the results obtained by the intelligent system. The challenge is still faced due to the complexity of neural network algorithms, called *black box* model. This expose people to danger by not knowing exactly how and why an algorithm arrived at a certain conclusion or decision. Much has been done to manage risk, eliminate errors, adopt best practice protocols, but we know that this is not enough. To better understand the reasons for this deadlock, we chose ADAS system [3] to discuss possible solutions so that it avoids errors and failures.

7.2 Disentangling the Cognitive Structure on Which Autonomous Systems Are Based: Perelman and Olbrechts-Tyteca's Methodology of Argumentation with an Appeal to Reality

Section 1.1 was dedicated to discussing the scenario of some autonomous systems, mentioning some of their flaws. This Section and this chapter in general aim to explain the foundations of cognition, whether human or machine. This is abstract knowledge because it addresses dynamic structures. For this reason, quantification, representation or performance techniques do not occupy a prominent place. The theoretical foundations of language and cognition, shared by humans and intelligent systems, make up much-needed knowledge for developers and technicians who design the algorithmic core of intelligent systems. It is these universal bases of language and cognition that construct information or that determine the relationship between the elements necessary for a system to carry out a certain task or decision. When seeking to build an AI tool that has the intuitiveness of human cognition, the elements exposed here are crucial.

Answering the complex question of what is at stake in the performance of self-driving cars requires pooling knowledge from multiple disciplines. AI imitates human behavior, and the use of neuroscience can help overcome some difficulties and find new alternatives to impasses. We unite neuroscience with the branch of autonomous AI systems to unravel the workings and weaknesses of ADAS. The increase in knowledge promoted by the exchange between human cognition and cognitive computing allows researchers and developers to optimize the self-experimentation of cyber-systems. This exchange takes place through a unique architecture: the cognitive linguistic dynamic process.

7.2.1 Human Cognitive Linguistic Process

There is still no concise and clear concept of what language/cognition is. Language is a system of conventional spoken or written symbols by which human beings communicate. This system groups or combines things or elements forming a complex or unitary whole, which, under a *dynamic*, involves value and gains the status of a *process*. We focus the content of this chapter on this dynamic face of language, understood as a *form* and not a *substance* [7]. We assume that, through Pereman and Olbrechts-Tyteca's [1] argument, human language and the language of intelligent systems are similar because they share the same and unique cognitive linguistic *process* (whether human or machine).

To establish the bridge that joins the human cognitive linguistic process to the cognitive process of intelligent systems, we take advantage of the approach of [1], to direct attention to *relationships*. Cognition and decision-making, as they are *processes* and have dynamic relationships between various elements, fit perfectly into their approach, which uses parameters of *value* and *context* with an appeal to reality. The qualitative knowledge of Neurolinguistics and Artificial Intelligence (Cognitive Computing) provided the meeting of a common element to the cognitive linguistic process that both incorporate. With the investigative focus on the similarity of *relationships* between these disciplines, we were able to identify a repetitive pattern in cognitive-linguistic functioning (Chap. 2 of [8]).

Recalling the examples in Sect. 7.1, we could observe that the reconfiguration of the tasks (their logical sequence) of the A-HMT-S project is not always comprehensible or reproducible. Spyndra, though self-aware, does not accurately predict the measurement linked to its own orientation. And, finally, the ADAS system, made clear the need for some element to manage events in cases where the computer does not know how to intervene correctly. The methodological approach of [1], by proposing attention to *relationships*, brought to this research on the cognitive linguistic process the opportunity to observe the existence of a standardized dynamic, present both in human cognition and in cognitive computing.

The argumentative method used in this chapter about what is at stake in the cognition of autonomous cars is not the analogy (which goes from the special to the generic), nor the hierarchy between elements. The focus of the method is the real observation of a repeated pattern in two branches of science (Neurolinguistics and Artificial Intelligence). This pattern plays the role of a bridge, which organizes, coordinates, unifies, and favors the exchange of information between disciplines and even between cognitive systems (whether human or machine). Autonomous car systems are designed to arrive at decision-making, which is, par excellence, the result of the cognitive linguistic process.

According to Neurolinguistics, the cognitive linguistic system of human beings encompasses the interconnection of neurons, glial cells, spinal cord, brainstem, cerebellum, and cerebrum [9]. This system somehow receives and processes electromagnetic stimuli such as light, mechanics such as waves, pressure, vibration, or touch; chemicals such as smell or taste; heat or cold [10, p. 26]. Its totality has not yet been reproduced in AI, which leads us to explore new avenues of investigation oriented towards the structural dynamics of the cognitive linguistic process, common to humans and AI. The immutable structural dynamics of cognitive linguistic behavior (Fig. 7.1) is put under the spotlight to show how it can be reproduced in its entirety in the behavior of intelligent systems.

In humans, stimuli enter the sensory organs and are taken to the central nervous system (brain at the center), where they are organized in a logical sequence so that they make sense. In AI equipped with a perceptual model or a multisensory combination design, the same process takes place. Environmental stimuli are captured by deep neural networks (vehicle location, pedestrian detection, traffic sign detection, etc.) and are taken to the algorithmic core (center) to be transformed into intelligible information for the AI, which may or may not be activated behavior to perform a task. (Figure created by the first author, art by Paulo Motta Monte Serrat. Icons retrieved from https://www.flaticon.com).

We clarify that we are analyzing cognitive linguistic dynamics (which is immutable), different from analyzes of specific models of autonomous vehicles,

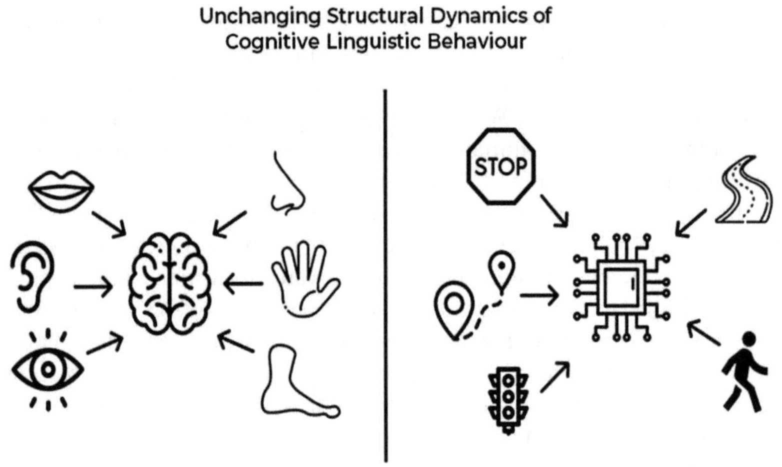

Fig. 7.1 Shows the unchanging structural dynamics of cognitive linguistic behavior in the individual (left side) and AI (right side)

which vary according to the tasks for which they were designed. In this Chapter we show that *all* of them are constituted by a uniform cognitive linguistic process, yet to be further explored. The analyzes of the dynamics of each of the autonomous systems can be done individually to elaborate ways of improvement in search of a model, or an organization of elements or even a network of connections that mimics human behavior.

Everything that is done to optimize an autonomous system needs to conform to the universal structure of the cognitive linguistic process present in AI and in humans. If the design of a given system (to perform a given task) meets that universal architecture, it will be successful. The positive result is achieved even though this system does not reproduce the completeness of human cognition with all its elements (neurons, glial cells, spinal cord, brainstem, cerebellum, and cerebrum). In other words, the autonomous vehicle model that conforms to the universality of the cognitive linguistic process acquires a universal coherence under the integration of the environment and algorithmic design in its core.

The universality of the cognitive linguistic process, described in the book *The natural language for Artificial Intelligence* [8], is represented by an algorithm that guarantees dynamism to language and cognition [8, 11], Chap. 10. This algorithm not only deals with events recorded in a chronological framework, but also with events located within an order and meaning provided by the context. We seek to make the machine learning operator aware that meaning and function come from a relationship between elements within the cognitive linguistic process. It is in this way that the intelligent system will be better adapted to its instrumentation.

7.2.2 AI Cognitive Linguistic Process

The reproduction of human language in intelligent systems depends on discerning the existence of two types of linguistic cognitive relationships: lower-range relationship and broader relationship (Fig. 7.2).

In Fig. 7.2, lower-range relationship represents a relationship that acts on specific elements of the core of the system (a sequence of tasks that will guide the behavior of the system under specific dependency relationships designated in the algorithmic core). Broader relationship represents the relationship between all the elements that make up the process that contribute to constituting the information, from the collection of stimuli or data to reaching cognition or concretized behavior. (Figure created by the first author, art by Paulo Motta Monte Serrat. Icons retrieved from https://www.flaticon.com).

A lower-range relationship (Fig. 7.2) acts on the specific elements of the core of the system (a sequence of tasks that will guide the behavior of the system). These are specific dependency relationships, which link the elements of the algorithm to each other so that the system fulfills the specific purpose for which it was intended. As a rule, these lower range relationships contain criteria that vary according to the model designed. A later element depends on an earlier element for its validity. There is also a broader relationship (Fig. 7.2) between the elements that make up the cognitive linguistic process. It is a hierarchical relationship among *all* the elements that contribute to constitute information, from the collection of stimuli/data to cognition or concretized behavior. This broader relationship creates and regulates the senses, making them effectively intelligible (that is, turning a stimulus into information understandable to humans). One should keep in mind that the cognitive linguistic process is fundamentally one and its concept must be considered unitarily by AI technicians. This universal faculty of the cognitive linguistic process oversees the

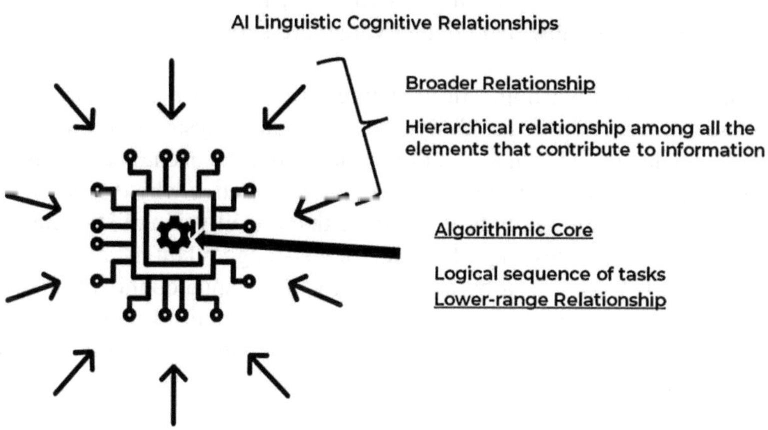

Fig. 7.2 Shows the AI's linguistic cognitive relationships: lower-range relationship and broader relationship

dynamic and universal flow that goes from the data/stimulus collection (from the context) to the cognitive center, where this stimulus/data is transformed into information or behavior. If the design of a system does not observe the universal feature of this cognitive linguistic flow, the purpose for which the autonomous system was designed may be jeopardized. Therefore, the specific criteria used for modeling each of the different autonomous systems should not be confused with the fundamental unit of the cognitive linguistic process embedded in all intelligent systems.

7.2.3 Recapping the Approach Discussed in This Section on Cognitive Structure

In this Section we explore a new avenue of investigation oriented towards the dynamic characteristic of the cognitive linguistic process. We emphasize that knowledge of how cognitive dynamics is carried out will help technicians to optimize intelligent systems. In the case of an autonomous system, it needs to comply with this dynamic, as it is a structure present both in the cognitive linguistic process of human beings and in the cognitive linguistic process of intelligent systems. It is, therefore, the universal structure of the cognitive linguistic process.

The concern with the dynamics of the construction of information or the algorithmic representation of the performance of a certain task by the intelligent system necessarily requires coherence in the integration of the environment and design in the algorithmic core. The concern with the representation of the dynamic process under which information is constructed for the system to perform a certain task necessarily requires planning coherence in the integration of the environment and design in the algorithmic core. This coherence in the algorithmic core guarantees the status of similarity with human cognition, with its dynamic sequences that imply relationships. The dynamic relationships carried out by the cognitive linguistic process involve parameters of value and context which are supported by the reality of the environment.

In short, the cognitive linguistic structure, common to humans and machines has an essential function in the design of autonomous systems: the bridge function, which organizes, coordinates, unifies and favors the exchange of information between human cognition and machine cognition, which is why it is so important. Furthermore, we highlight in Figs. 7.1 and 7.2 that cognitive linguistic relations can be discerned into two types: superficial and lower-range relations, and deep cognitive linguistic relations with a broader scope. Lower-range cognitive relations act on specific elements of the task sequence of a given system. Broad-ranging cognitive relationships have to do with a hierarchy of elements that build information or a sequence of tasks. This is the deep layer of cognition, shared by humans and machines.

7.3 Advanced Driver Assistance Systems (ADAS) Replacing Humans

Autonomous vehicles equipped with advanced driver assistance systems (ADAS) are intended to perform the task of driving, replacing human drivers [3]. ADAS has been scrutinized by the Dutch Safety Council, which has drafted the report "Who is in control? Road safety and automation in road traffic" [3], detailing safety issues in cases of shortages or problems in the safety of individuals. Driver assistance systems can be emergency braking, cruise control or even fundamentally changing the car's functions to take over driver tasks like steering, braking, and accelerating. When replacing the driver, ADAS makes decisions coordinated by algorithms providing, paradoxically, safety and risk. Although automation is destined to take the place of human drivers, they have not yet reached the stage where humans are superfluous. On the contrary, there is a need for a driver to be always alert in case the computer does not know what to decide or intervenes in the wrong way. What would be missing for automation to replace human drivers? This is what we discuss in the next section.

7.3.1 Merging Computational and Physical Resources in ADAS

The integration of cyber and physical elements in self-driving cars needs to be committed to responsible innovation [3]. We propose in this Sect. 7.3 that security should be considered from the design of the ADAS. To implement innovation in autonomous systems, an underlying structure of their design must be considered: the cognitive linguistic structure (as explained in Sect. 2.2). It should be considered the single cognitive linguistic framework for both humans and intelligent systems. This fundamental knowledge serves to inspire the formulation of algorithms and models of cyber-physical systems. It has been noticed that the interaction between humans and machines has mitigated the increase in the number of accidents. However, this is still not enough to manage and prevent them. ADAS is not yet mature to suppress the human driver [3]. The range of tasks required to operate the ADAS makes the driver less alert and generates conflicts that result in risks. There can still be cybersecurity risks when security system updates are not performed. This alters the functioning of the ADAS, without the driver being aware of it [3]. Our proposal goes beyond human-machine interaction, as it is related to the core of the intelligent system. We suggest understanding the cognitive linguistic process as a dynamic structure capable of mixing physical and computational resources with the ADAS algorithmic core.

7.3.2 Structure Behind the ADAS Design

Advanced driver assistance systems, ADAS, are equipped with tools whose objectives are defined by a sequence of tasks determined by algorithms (see lower-range relationships at Sect. 2.2). This is the algorithmic core of the autonomous system. Although determined to fulfill tasks, when it is linked to the individual's use, it becomes exposed to a wide range of contextual stimuli (see broader relationships at Sect. 2.2.) with which it will have to deal due to its deep learning algorithms. This occurs because the cognitive linguistic structure of the system has the same human cognitive linguistic structure, that is, it integrates two fronts: the contextual one, resulting from the collection of stimuli/data from the environment, and the logic one, which organizes these stimuli in a logical sequence, making them intelligible ones (turns stimuli into information) [8].

The fundamentals of human cognition as a dynamic process mixes the stimuli arising from the context with the logical sequence of the central cognitive system giving them a meaning. This fluid and changing composition of human cognition should inspire ADAS design so that it is able to adapt to different contexts while performing the main task for which it was designed. The mix of physical and computational resources in ADAS goes beyond the specific elements of its design. This can be observed when, at the time of an error or accident, the system's reaction to different contexts can be deficient and result in weaknesses in the performance of its final task. The fundamentals and principles of cognition serve as a guide to a more comprehensive imitation of human behavior, so that highly automated systems are successful when facing challenges. By bringing the two fronts of the cognitive linguistic process together at the core of autonomous systems, there will be less likelihood of damage to people and property and of misuse of system design.

The human cognitive linguistic structure to be imitated by ADAS must be guided not only by the algorithmic core (lower-range relationship, see Sect. 2.2), but also by receiving stimuli from its context (broader relationship, see Sect. 2.2). The union of these two fronts mimics human cognition encompassing environmental parameters, which makes the ADAS system dynamic, and its interpretive activity optimized. For this, the juxtaposition of both fronts is not enough. There is a need for an *organized combination* of tools that balance structural aspects of the algorithmic core with contextual aspects collected by the system. If ADAS only deals with behavior patterns determined by algorithms, the results of the system will not be satisfactory, since the data is static. When dealing with the competition between contextual stimuli and the sequence of tasks foreseen in the algorithmic core, the autonomous system works more intuitively, but this has not yet proved to be enough. Designers report that they cannot predict the results as it is a black box. The state of the art will be achieved when the unification of the fronts, contextual and logical, occurs in a hierarchically organized manner, ensuring sustainability in ADAS innovation, as the system starts to focus on *understanding* the contextual dynamics, reflecting results with fewer errors.

7.3.3 *Logical and Executive Functions in Autonomous Driving Systems: What is at Stake in the Impenetrable Black Box AI*

Autonomous systems integrate, as a rule, machine learning and deep learning algorithms for different tasks such as movement planning, vehicle localization, pedestrian detection, traffic sign detection, road sign detection, automated parking, vehicle cybersecurity and vehicle diagnostics failures [12]. The logical and executive functions of intelligent systems are linked to the activity of interpretation, that is, to the processing of semantically evaluable information. Monte-Serrat and Cattani [11] explain that information processing by AI should imitate the human dynamic cognitive linguistic process to result in the expected integration of the interpretive activity of the intelligent system.

The integration of algorithms to the foundations of the cognitive linguistic process allows computer scientists to optimize their cyber system. The overlapping of biological and intelligent systems reveals a universal hierarchical structure in charge of carrying out the interpretive activity [8, 11]. It is from this structure that we extract strategies that offer good instrumentation and guarantee safe performance for AI cognition. Describing in more detail, as a rule, the cyber system, to circumvent the situations of the environment and carry out its tasks, *interprets* data. The level of data interpretation by intelligent systems, despite the innovations in integrated detection and control, result in punctual solutions, reaching only specific applications.

There is a need to rethink the unifying context of cyber-physical systems about their interpretive activity. For now, what scientists have achieved is the use of open, flexible, and extensible architectures for cyber-physical systems; the use of principle-based compositions or integrations; activity in run-time operations to improve system performance and reliability. In short, what has been sought is that the sensitivity of the cyber-physical system to the context is *combined* with the ability to modify its behavior, *accommodating* variable configurations. However, the autonomous systems leave something to be desired, presenting defects and interpretive biases. And yet, for these systems to perform these *accommodations*, new approaches and human curation are needed to validate them. What has been noticed so far is that the accommodation of new fundamentals, methods and tools has been insufficient to mitigate errors in the interpretation of the autonomous system. In this chapter we take a step forward: Instead of *accommodation* we propose *integration*.

7.3.4 *Recapping the Fundamentals of Advanced Driver Assistance Systems (ADAS) Cognition*

This third Section shows what is missing for ADAS to imitate human cognition. We clarify that the construction of information or a sequence of tasks originates from the mix of stimuli arising from the context with the logical sequence of the central cognitive system. This is the fundamental structure of cognition, whose nature is a fluid, dynamic process, capable of adapting to different contexts. We show that the performance of ADAS when reacting to different contexts is still deficient and weak. What would ADAS be missing to reach the state of the art and imitate human behavior in the face of challenges? The computer does not know what to decide or how to intervene specifically because it does not faithfully reproduce this fundamental structure of the human cognitive linguistic process, articulating stimuli from the environment to the logical sequence. While ADAS juxtaposes tasks at its core, humans perform a hierarchically superior operation of combining stimuli or data in order to achieve balance in the operation that encompasses continuous changes over time.

7.4 Rethinking the Interpretive Activity of Cyber-Physical Systems Under a Unifying Context

Regarding deep neural networks, it has been claimed that the malfunction of intelligent systems is due to black box AI, which is related to the lack of knowledge of the algorithm's intended behavior. To deal with the complexity and mystique of the black box AI, it is necessary to understand that language and cognition form a structure that is related to the semantic dimension. Semantics comes not only from the linguistic system (logical functions of the system), but also from the context in which information is produced (such as movement planning, vehicle localization, pedestrian detection etc. [12]. Knowledge of the fundamental cognitive linguistic structure as a single process for humans and machines ensures consistency in system behavior and mitigates biases in system interpretation [8, 11]. The key to acceptable ADAS performance, therefore, lies in the dynamic aspect under which it interprets the information, making the system invariant to many input transformations and preventing it from misinterpreting the events to which it is exposed.

The interpretive activity of AI has focused on the use of multilayer neural networks designed to process, analyze, and interpret the vast amount of collected data. Cybertechniques expect intelligent systems to produce responses similar to human ones, but the results are subject to random interpretation and are often inconsistent with reality. To overcome this difficulty, Reinforcement Learning with Human Feedback (RLHF) techniques are used. Another interpretation technique that makes use of neural networks are knowledge graphs, but they also require exhaustive human curation for the system to interpret the relationships between entities in accordance with the real world. At the beginning of this Chapter, we cite the work of [4] who also suggests human intervention in what the author calls the Autonomous Human-Machine Team (A-HMT-S) to circumvent the defects presented by the intelligent system.

On the other hand, we have dynamic programming [13, 14] as an example of success in the mathematical optimization of cyber systems. It meets what we expose in this Chapter as the universal structure of language. Dynamic programming deals with the complexity of the context subdivided recursively [14], so that optimal solutions are found for the sub-elements of the context. This is a way to unify recursive relationships between the highest element value (broader relationship, see Sect. 2.2) with the smallest element values (lower-range relationship, see Sect. 2.2). Both encompass the universal structure of the dynamic cognitive linguistic process, promoting the union and simplification of decisions for autonomous systems through a recursive relation (also called Bellman's equation) [14].

7.4.1 Unraveling the Black Box of Logical and Executive Functions of Autonomous Driving Systems

Black box AI affects the interpretability of ADAS because this system makes use of deep neural networks. It has been observed that the critical stages of systems with autonomous driving are in the features related to the interpretive activity, such as, for example, in perception, information processing and modeling [15]. The inputs and operations of the algorithms are not visible to the user due to the complexity of the cyber-physical system. Impenetrability in the system stems from deep learning modeling that takes millions of collected data points as inputs and correlates that data to specific features to produce an output.

By constituting a cyber-physical system and dependent on interpretive activity, ADAS integrates the universal cognitive linguistic structure. The system makes use of the linguistic process on two fronts: via logical reasoning (as a sequence previously established by the algorithm designed by its technical developer) and via reception of stimuli (the repetition of the input of stimuli in the circuits of the neural networks, which, even being combined with reinforcement is still insufficient). This dual front of the ADAS cognitive linguistic process is self-directed and difficult to interpret by data scientists and users. Because it is not visualized or understood, the interpretive activity of the autonomous system is led to errors, from inconspicuous errors to errors that cause major problems, or even those that are impossible to repair. At a time before these problems, one could also identify AI bias (in the training data, for example) by the developers of the autonomous system, which could lead to potentially offensive results for those affected. How to act so that the self-directed activity of ADAS ceases to be a black box and interprets it in accordance with the human mind, to prevent problems and losses?

ADAS that adequately performs its tasks must have its universal linguistic cognitive structure organized according to a hierarchy of values. Values arising from context inputs (broader relationship, see Sect. 2.2) and values arising from the interpretive activity according to the algorithmic core model (lower range relationship, see Sect. 2.2), must come into play in a targeted manner, in order to organize the interpretive activity of the system before it accomplishes its ultimate goals (executive function). As the executive functions of ADAS are connected to deep neural networks responsible for collecting data from the environment, the collection of millions of data points may prevail over the interpretive activity of the algorithmic core, biasing it [11]. Although ADAS has a planned behavior (logical functions linked to the algorithmic core), if there is no organization of the cognitive linguistic activity of the system involving the broader and the lower-range relationships, it will not be adaptable to the changes that occur in the environment. On the other hand, if it is regulated and organized, the executive functions of the system will be flexible when errors and risks are detected.

The unification of the autonomous system (which is different from Reinforcement Learning from Human Feedback or human curation) is what will allow the monitoring of its decision making. The synchronized cognitive flexibility arising from the dynamic linguistic process (broader relationship unified with the lower-range relationship, see Sect. 2.2) allows it to adjust to unforeseen demands, overcoming sudden obstacles. Cognitive flexibility allows ADAS to face a variety of challenges, making the autonomous system more intuitive, which brings its mathematical modeling closer to the dynamic structure of human cognition. Both human cognition and AI cognition can *translate*, interpreting the real world, because they reflect the fundamental structure of the dynamic cognitive linguistic process, which is able to operate values to establish meanings, correlating logical pattern and contextual pattern [16].

ADAS modeling deals with *relationships* within a dynamic process that generates interpretation. Aware of this, it is assumed that the consistent interpretation of the events to which ADAS is exposed results from the *processing* of these relationships. There is imitation of the performance of human cognition to unify the operation of values (of the context) with the sequence of tasks (of the algorithmic core that determines the logical sequence of tasks to be executed). In this way, the supposed black box of autonomous systems has its functioning revealed by unifying mathematical relations (logic/previously categorized elements/frozen context) to non-mathematical relations (contextual/dynamic) [16]. The universal structure of the linguistic cognitive process makes it clear how the autonomous system makes use of the interpretive activity and how it can provide guidance consistently with the context to which the system is exposed. The cybersystems' developer needs to consider a hierarchy of values in the dynamic processing of the (interpretive) behavior of the system. The organization of this AI interpretive activity results in the valuation of categorized elements of the algorithmic core that are unified with the fluid values of the context of the environment to which the intelligent system is exposed. This hierarchy and unification optimize executive functions and makes the system more intuitive.

7.4.2 *Recapping About Dealing with the Complexity and Mystique of Black Box AI.*

In this Sect. 7.4, we show the need to understand that the cognitive linguistic structure, shared by humans and machines, is related to the semantic dimension. ADAS' semantic dimension deals with logical functions of the system and also with the dynamic context of the environment (movement planning, vehicle location, pedestrian detection etc.). The interpretative activity of ADAS covers dynamic aspects that cause input transformations, which can lead to erroneous interpretations of the environment. This cognitive functioning helps to unveil the AI black box. ADAS' core algorithmic collection of stimuli places them within a logical sequence of tasks. However, ADAS still does not present the superior operation of cognition, which hierarchically relates the elements it is dealing with. This lack of hierarchical dynamic organization of stimuli and data leads the intelligent system to present imperceptible errors and even errors that cause major problems. For ADAS to perform executive functions properly, mimicking human cognition, it must have its cognitive core organized according to a dynamic hierarchy of values. These values arise from context inputs (broader relation, see Sect. 2.2) and interpretive activity according to the central algorithmic model (lower range relation, see Sect. 2.2).

7.5 ADAS: Learning Coming Out of Integration Between Algorithmic Core and Context

Learning carried out by the autonomous driving system, when receiving stimuli from new interactions not foreseen in the algorithm, undergoes the reorganization of its neural circuits, similar to what happens in human learning. This process occurs according to the fundamentals of the human cognitive linguistic process [9]. Because it is a single structure, it overlaps in ADAS learning, which leads us to think that it is not regulated only by the algorithm (core), but also by complex aspects arising from the interaction of the system with the environment. How to perform the integration of both (algorithmic core and context) to reach the state of the art in intelligent systems?

The expected result of autonomous systems is that there is a structural and functional organization exactly as the result of the nervous system of individuals when reacting to contextual factors. Reports of failures pointed out in [3] show that autonomous systems still do not imitate human cognition satisfactorily. To resolve this impasse, we point out, as an example, Bellman's theory [17], which provides means to bring cognitive computation closer to the human cognitive linguistic dynamic process. Reproducing the biological mechanism of reorganizing neural circuits based on environmental stimuli in self-driving cars is not an easy task. For these systems to establish a memory and reorganize neural circuits to perform new tasks, the juxtaposition of different tools or mechanisms is not enough. It is necessary to consider another aspect of human cognition, that, in addition to the interaction with the environment, there is a specific chronological pattern to be imitated.

The unification of the two features of human cognition (logical and contextual) provides the model for autonomous driving systems to reach the state of the art. On the one hand, we have the logical pattern, understood as logical reasoning 'if P then Q' [18, 19], [2], which can be represented by the logical sequence of tasks described in the algorithmic core. And on the other hand, we have the chronological pattern arising from the facts of the environment, whose stimuli are received by the sensory organs—visual, auditory, olfactory, tactile, gustatory or proprioceptive [9]. The latter can be represented by tools that enable the autonomous vehicle to perceive its surroundings [12].

The *unification* or *synchronization* of these two features, proposed in this chapter, imitates the role of the human central nervous system, when it starts to make a unique and specific direction that we call processing of semantically evaluable information. This is how human neuroplasticity increases flow between neural circuits. Pre-existing synapses are activated to increase the efficiency of information exchange. Something that has been learned must be reactivated. Learning that mimics human cognition and optimizes the ADAS autonomous driving system involves recording, storing "relevant" knowledge, skills, and attitudes for the system to perform its tasks well. The hierarchical dynamic processing (synchronized chronology) between the system's interaction with the world and the learning process clarifies what is at stake in the performance of the cognitive-linguistic functions of the ADAS. Although there is much to be explored in the unknown field of autonomous vehicle cognition, it is assumed that the recursive processing of the main cognitive-linguistic functions, by imitating human cognition, serves as a guide for ADAS to receive, process, store and use information. Reports of errors and defects verified so far [3] show that learning by the intelligent system requires a reorganization of neuronal connections that are stimulated by external information. Our proposal is that the new organizational pattern of ADAS learning be the unification, as Bellman [17] teaches, of the algorithmic core of the system with its neural circuits that react to the environment.

7.5.1 The ADAS Learning Process: Principles that Organize the 'Way of Doing'

What is at stake in the behavioral ability of the intelligent system to carry out the tasks that have been assigned to it is the *process* of learning. This reflection, when carried over to ADAS learning, takes us beyond the dependence on its deep neural networks, and leads us to consider the stimuli that have their origin in the environment that surrounds the autonomous driving vehicle. The deep neural algorithms (core of the intelligent system) are responsible for only a part of the cognitive process of ADAS. The other part of its cognition is based on experiences in the environment, which interfere, often in ways not foreseen by its designers, in the activity of deep neural networks, resulting in AI black box biases.

Knowledge of how human learning works, which, upon receiving stimuli, reorganizes its neural circuits [9] makes it clear that ADAS will be defective if its learning is regulated only by the algorithmic core of logical sequence. The state of the art will be found when the ADAS algorithmic core is chronologically integrated with complex aspects arising from environment stimuli. How to perform this integration? The *way of doing* physical-cybernetic systems is as important as the tool used in them. This chapter brings a warning to researchers and developers of intelligent systems that expressing mathematical propositions in logical sequences is not enough. The system needs to understand its context to respond appropriately with a behavior. The algorithmic core does not accurately describe the environment. There is a need for the system to be powered by another type of information to receive stimuli from real events. How can the (logical) algorithmic core integrate these real events into their contextual structure, whose order of meaning cannot be summarized as a mere logical sequence of tasks? Taking these questions into account and Carl Sagan's assertion that science is not just a body of knowledge, but also a way of thinking [20], we list some principles that organize the *way of doing* the design of an intelligent autonomous driving system:

1. the cognitive linguistic system of AI must be understood not as a substance, but as a form, that is, a dynamic process;
2. the cognitive linguistic process of cyber-physical systems must have its components inspired by the human cognitive linguistic process, which has two fronts: a contextual one and a logical one;
3. The contextual front must align the design of the autonomous system's cognition in different spatial and temporal scales to respond to dynamic events;
4. the logical front must configure the sequence of tasks that may or may not result in decision-making;
5. All the above organizing principles make up the interpretive activity of cyber-physical systems. They must, therefore, be designed in a unifying way, like an umbrella, to ensure consistency in the behavior of autonomous driving systems, integrating context stimuli into the algorithmic sequence.

7.5.2 Recapping About Learning Accomplished by ADAS

Section 7.5 highlights that the learning carried out by the autonomous driving system has not yet reached the way in which the universal structure of cognition merges logical sequence of tasks with the dynamics of stimuli received from the environment. ADAS, when receiving stimuli from new unforeseen interactions, does not organize its neural circuits satisfactorily, which prevents it from reaching the state of the art of imitating human cognition.

ADAS learning, in addition to being regulated by the task sequence core, needs to reach complex aspects arising from the system's interaction with the environment.

The juxtaposition of different tools is still not enough. A memory capable of organizing neural circuits to perform new tasks under a specific chronological pattern to be imitated is necessary. It is the synchronization between the logical sequence 'if P then Q' to the chronological pattern resulting from the facts of the environment that will avoid errors and defects of ADAS. Our proposal is that ADAS learning synchronization is carried out in the form of the unification of the system's algorithmic core with its neural circuits that react to the environment. Our contribution, therefore, lies in suggesting *how* to make physical-cybernetic systems. For this reason, this Chapter does not focus on intelligent systems tools. We do not bring tools, but rather a body of knowledge to overcome the difficulties of integrating the algorithmic core of autonomous systems with real events in their contextual structure. Within this purpose, we have brought in this section some principles that organize the way of designing an intelligent autonomous driving system.

7.6 Conclusion

The integration between context and algorithmic sequence developed by the cognitive linguistic dynamic process serves as an umbrella to encompass the various activities related to cybersystems. We cite Richard Bellman's solution process as an example for cybernetic projects involving dynamic programming, i.e., to find the best decisions in a problem-solving process, one must seek one solution after another, nesting smaller decision problems within of major decisions [17]. Contrasting Bellman's solution adopted by us, we can observe that the application of different algorithms for different autonomous driving tasks is a complex task. [12] claim that the complexity of autonomous vehicles implies the use of more than a single algorithm, since the vehicle's activity provides information from different perspectives. They suggest for faster execution the tree model as a learning model, for motion planning, they suggest the dynamic model to reduce the planner execution time; Reinforcement Learning (RL) for speed control; for pedestrian detection, they propose an algorithm that combines a five-layer convolution neural network and a classifier; for lane recognition, a steerable fusion sensor capable of remaining unchanged on structured and unstructured roads.

We seek, in understanding the basic functioning of the human cognitive linguistic process, a way to simplify this task. We show that the *interaction* of the human being with the world is essential for the *development and learning processes*. This interaction deserves to be highlighted in the development of autonomous cars, no matter how diverse the tools used are. What is at stake in the intelligence of autonomous cars is not just the tool used, but *how* it works, *how* the human-machine-environment interaction is carried out. The expected result of autonomous systems is that there is a structural and functional organization similar to that of the nervous system of individuals that can be altered by contextual factors. We suggest that this integration be done recursively according to Bellman's theory [17], synchronizing the algorithmic core to the collection of stimuli from the context in which ADAS is operating.

The credibility of ADAS will increase as it develops its capacity for self-regulation within the proposed hierarchy, optimizing its capacity for self-guidance. This skill includes developing strategies, seeking information on its own, solving problems, making decisions, and making choices - regardless of human support. This learning is slow and requires constant adjustments. The interpretation performed by ADAS goes beyond the information [21] in its algorithm, as there is cognitive overload generated by the contexts to which it is exposed, and this also shapes its cognition. ADAS needs protection for a proper interpretation of the context combined with concentration on the sequence of tasks given by the algorithm. In this way, the autonomous system is not reduced to identifying something, but to *thinking* about something. In other words, it is not about *what* to learn, but *how* to behave considering different perspectives [22].

The fragility of ADAS lies in its cognitive activity in not being able to distinguish information, in verifying the sequence of tasks with its environment and in aligning its decision-making with the context in which it finds itself. The fundamentals of the linguistic cognitive structure described in this chapter says less in terms of performance or quantification techniques and more in terms of cognitive process. Thinking about the *how* rather than the *what* legitimizes universality and makes the cognitive-linguistic process a less obscure notion. Where is the characteristic of universality of the cognitive linguistic process capable of unraveling the AI black box? In the structure, in the dynamic process carried out both by the cognitive faculties of individuals and by the cognition of artificial intelligence. When ADAS does not reach this universality, it does not acquire the necessary cognitive legitimacy to keep up to date, which leaves it susceptible to weaknesses.

What is really at stake is *how* to pass instructions to the ADAS design, rather than *what* instructions to pass to ADAS. The perspective of the universality of ADAS cognition, therefore, does not lie in the statistical data it collects, nor in the combination of different algorithms, but in its ability to *properly* process different contextual situations. The universal structure of the cognitive linguistic process reveals the way in which human cognition processes information. Inspiring the design of cyber systems in this universal structure means finding solutions to the security issues that exist in the cyber-physical systems of autonomous driving vehicles. In addition to mentioning dynamic programming [13, 14] as an example of success in the mathematical optimization of cybernetic systems, we disclose that the challenges in implementing the approach proposed in this Chapter by applying real-time cognition to cybernetic systems represent the new directions of our research. It is moving towards publishing new studies that teach intelligent systems not only to identify something, but also to think about something. The universality of the cognitive-linguistic process is leading the way for us to resort to new mathematical techniques that, as far as we know, have not yet been related to language and cognition. These new techniques will convey the embryonic aspect of cognition, preventing the researcher or technician from getting lost in the complex aspects of the superficial layers of the cognitive linguistic process. In this new approach, aspects of memory and representation that organize neural circuits to perform new tasks are being considered. We believe that this new point of view will be able to meet the real chronological pattern of the human cognitive linguistic process. In this way, it will be possible to design cyber systems that are able to synchronize learning in order to unify their algorithmic core with neural circuits that react to the environment.

References

1. Perelman, C., Olbrechts-Tyteca, L.: The New Rhetoric: Treatise on Argumentation. Wilkninson, J. (transl). University of Notre Dame Press 1973
2. Monte-Serrat, D., Cattani, C.: The natural language for artificial intelligence. Elsevier-Academic Press, 233p (2021)
3. Board, D.S.: Who is in control? Road safety and automation in road traffic (2019). https://www.onderzoeksraad.nl/en/page/4729/who-is-in-control-road-safety-and-automation-in-road-traffic. Accessed 28 Jan 2023. Distributed to GRVA as informal document GRVA-05-48 5th GRVA, 10–14 February 2020, agenda item 3
4. Gillespie, T.: Building trust and responsibility into autonomous human-machine teams. Front. Phys. **10**, 942245 (2022)
5. Lipson, H.: Cosa accadra allumanitá se si dovesse creare la 'conscienza robotica'? In Dagospia (2023). https://www.dagospia.com/rubrica-29/cronache/cosa-accadra-39-all-39-umanita-39-se-si-dovesse-creare-338944.htm. Accessed 28 Nov 2023
6. Kedar, O., Capper, C., Chen, Y. S., Chen, Z., Di, J., Elzora, Y., Lipson, H.: Spyndra 1.0: An Open-Source Proprioceptive Robot for Studies in Machine Self-Awareness (2022). https://www.creativemachineslab.com/uploads/6/9/3/4/69340277/spyndra_summer_report_v3.pdf. Accessed 27 Jan 2022
7. Saussure, F.: Cours de linguistique Générale, 3rd edn. In: Bally, C., Sechehaye, A. (Eds.). Payot, Paris (1916)
8. Monte-Serrat, D., Cattani, C.: Connecting different levels of language reality. In: The Natural Language for Artificial Intelligence, pp. 7–15. Elsevier-Academic Press (2021)
9. Copstead, L.E., Banasik, J.: Pathophysiology, 5th edn. Elsevier Inc (2013)
10. Amaral, A.L., Guerra, L.: Neuroscience and education: looking out for the future of learning. Translation Mirela C. C. Ramacciotti. Brasília: SESI/DN, 270p (2022). https://static.portaldaindustria.com.br/media/filer_public/7c/15/7c153322-d2e7-44e3-86b1-aeaecfe8f894/neuroscience_and_learning_pdf_interativo.pdf. Accessed 20 Jan 2023
11. Monte-Serrat, D., Cattani, C.: Interpretability in neural networks towards universal consistency. Int. J. Cogn. Comput. Eng. **2**, 30–39 (2021)
12. Bachute, M.R., Subhedar, J.M.: Autonomous driving architectures: insights of machine learning and deep learning algorithms. Mach. Learn. Appl. **6**, 100164 (2021)
13. Cormen, T.H., Leiserson, C.E., Rivest, R.L., Stein, C.: Introduction to Algorithms, 2nd edn, pp. 344. MIT Press & McGraw–Hill, ISBN 0-262-03293-7 (2001)
14. Dixit, A.K.: Optimization in Economic Theory, 2nd edn. Oxford University Press, p. 164. ISBN 0-19-877211-4 (1990)
15. Gruyer, D., Magnier, V., Hamdi, K., Claussmann, L., Orfila, O., Rakotonirainy, A.: Perception, information processing and modeling: critical stages for autonomous driving applications. Annu. Rev. Control. **44**(2017), 323–341 (2017). https://doi.org/10.1016/j.arcontrol.2017.09.012
16. Monte-Serrat, D.: Operating language value structures in the intelligent systems. Adv. Math. Models Appl. **6**(1), 31–44 (2021)
17. Dreyfus, S.: Richard Bellman on the birth of dynamic programming. Oper Res Informs **50**(1), 48–51 (2002). ISSN 1526-5463
18. Monte-Serrat, D.: Neurolinguistics, language, and time: Investigating the verbal art in its amplitude. Int. J. Percept. Publ. Health, IJPPH **1**(3) (2017)

19. Monte-Serrat, D., Belgacem, F.: Subject and time movement in the virtual reality. Int. J. Res. Methodol. Soc. Sci. **3**(3), 19 (2017)
20. Sagan, C.: The Demon-Haunted World: Science as a Candle in the Dark. Ballantine Books (2011)
21. Cormen, E., Inc.: Language. Library of Congress, USA (1986)
22. Monte-Serrat, D., Cattani, C.: Applicability of emotion to intelligent systems. In: Information Sciences Letters, vol. 11, pp. 1121–1129. Natural Sciences Publishing, New York (2022)

Chapter 8
Intelligent Under Sampling Based Ensemble Techniques for Cyber-Physical Systems in Smart Cities

Dukka Karun Kumar Reddy, B. Kameswara Rao, and Tarik A. Rashid

Abstract Cyber-Physical Systems (CPSs) represent the next evolution of engineered systems that seamlessly blend computational and physical processes. The rise of technologies has brought about a heightened focus on security, making it a noteworthy concern. An intelligent ML-based CPS plays a pivotal role in analysing network activity within the CPS by leveraging historical data. This enhances intelligent decision-making to safeguard against potential threats from malicious hackers. The inherent uncertainties in the physical environment, CPS increasingly depend on ML algorithms capable of acquiring and leveraging knowledge from historical data to enhance intelligent decision-making. Due to limitations in resources and the complexity of algorithms, conventional ML-based CPSs face challenges when employed for operational detection in the critical infrastructures of smart cities. A lightweight intelligent CPS that is optimal, inexpensive, and can minimise the loss function is required. The widespread adoption of high-resolution sensors results in the presence of datasets with high dimensions and class imbalance in numerous CPS. Under-sampling-based ensemble algorithms ensures a better-equipped process to handle the challenges associated with imbalanced data distributions. The under-sampling-based ensemble technique solves class imbalance by lowering the majority class and establishing a balanced training set. This strategy improves minority class performance while reducing bias towards the majority class. The experimental findings validate the effectiveness of the proposed strategy in bolstering the security of the CPS environment. An assessment conducted on the MSCA benchmark IDS dataset

---

D. K. K. Reddy (✉)
Department of Computer Science Engineering, Vignan's Institute of Engineering for Women (Autonomus), Visakhapatnam, Andhra Pradesh 530046, India
e-mail: karun.redy@gmail.com

B. K. Rao
Department of Computer Science and Engineering, GITAM (Deemed to be University) Visakhapatnam Campus, Visakapatnam, Andhra Pradesh 530045, India
e-mail: kamesh3410@gmail.com

T. A. Rashid
Erbil, Kurdistan Region, Iraq
e-mail: tarik.ahmed@ukh.edu.krd

J. Nayak et al. (eds.), *Machine Learning for Cyber Physical System: Advances and Challenges*, Intelligent Systems Reference Library 60,
https://doi.org/10.1007/978-3-031-54038-7_8

affirms the promise of this approach. Moreover, the suggested method surpasses conventional accuracy metrics, striking a favourable balance between efficacy and efficiency.

Keywords Cyber physical systems · Under sampling techniques · Ensemble learning · Machine learning · Anomaly detection · Multi-step cyber-attack (MSCA)

Abbreviations

AUC	Area under the ROC Curve
BCC	Balance Cascade Classifier
BN	Bayesian Network
BRFC	Balanced Random Forest Classifier
CI	Critical Infrastructure
CNN	Convolutional Neural Networks
CPS	Cyber Physical Systems
DBN	Deep Belief Network
DL	Deep Learning
DR	Detection Rate
EBNN	Extremely Boosted Neural Network
EEC	Easy Ensemble Classifier
FPR	False Positive Rate
ICS	Intelligent Control Systems
ICT	Information and Communication Technology
ID	Intrusion Detection
IDS	Intrusion Detection System
IML	Intelligent Machine Learning
IoT	Internet of Things
k-NN	k-Nearest Neighbour
LSTM	Long Short-Term Memory
ML	Machine Learning
MSCA	Multi-Step Cyber-Attack
NN	Neural Network
PSO	Particle Swarm Optimization
RNN	Recurrent Neural Network
RUSBC	Random Under Sampling Boost Classifier
SBS	Sensor-Based Systems
SPEC	Self-Paced Ensemble Classifier
TPR	True Positive Rate
UBC	Under Bagging Classifier
WUP	World Urbanization Prospect

8.1 Introduction

At present, more than 50% of the global population lives in cities, and this trend is projected to continue as urban areas grow in both population and size. As per the UN WUPs, it is projected that by 2050, approximately 66% of the global population will reside in cities [1]. To address the escalating complexity of modern urban landscapes, several projects have been initiated to amalgamate advanced technological solutions, thereby elevating the sophistication of urban design and management. Prominent examples of these intelligent urban solutions include the implementation of ICT technologies in areas such as enhanced power grids for reduced energy loss, progressive transportation systems along with connected vehicle innovations to boost city mobility, and optimized infrastructures aimed at diminishing hazards and bolstering operational effectiveness [2, 3]. The development of novel information and communication technologies, such as the Cloud Computing, CPS, Big Data, and IoT has made these advancements possible. The growing interest in incorporating the concept of CPS into the realm of smart cities, has garnered increasing attention recently. CPS represent the fusion of ICT with physical infrastructure and systems, empowering cities to meet the growing demand for greater sustainability, efficiency, and improved quality of life for their inhabitants, thereby advancing their smartness [4]. This concept of smartness is closely tied to awareness, which involves the capability to identify, perceive, or be aware of objects, events, or physical arrangements. The significant advancements in sensor and wireless technologies have led to the capability to accurately monitor and capture physical phenomena in the environment. This data can then be preprocessed using embedded devices and seamlessly transmitted wirelessly to networked applications capable of performing sophisticated data analysis and processing [5]. CPS have deeply integrated CIs with human life. So, it becomes imperative to prioritize the security considerations of these systems. Model-based design and analysis, including the use of attack and countermeasure models, offer significant potential in tackling the security challenges associated with CPS [6]. IML-based CPSs play a crucial role in the development and sustainability of smart cities. These systems seamlessly integrate physical infrastructure with advanced ML capabilities, enabling cities to enhance efficiency, safety, and quality of life for their residents. IML-CPS facilitates real-time monitoring and data-driven decision-making, enabling city authorities to optimize traffic management, energy consumption, waste disposal, and emergency response systems. Moreover, these systems can predict and mitigate potential issues, contributing to more resilient and sustainable urban environments. By harnessing the power of ML and data analytics, IML-CPS empowers smart cities to not only address current challenges but also anticipate and adapt to future urban complexities, making them more liveable, sustainable, and responsive to the needs of their citizens.

The objective of the chapter is

i. To develop an anomaly-based detection system tailored for CPS environments characterized by resource constraints, capable of assessing the categorization of network traffic into normal or anomalous events.

ii. Under-sampling-based ensemble techniques robust the ID by reducing the class imbalance.
iii. The efficacy of the proposed approach is evaluated using the MSCA standard IDS dataset. The results confirm that the suggested method outperforms default ML parameters in terms of both accuracy and efficiency. The experimental outcomes of the research substantiate this assertion.

The following sections delineate the remaining segments as: Sect. 8.2 gives a brief introduction of CPS structure and its workflow. Section 8.3 illustrates the limitations of feature selection and hyperparameter tuning for anomaly detection in CPS. The Under-sampling-based ensemble techniques used in the experiments are outlined in Sect. 8.4. In Sect. 8.5, a compare of prior surveys that focused on IDS results of the MSCA dataset is studied. Section 8.6 provides a concise overview of the experimental setup and descriptions of the dataset. The analysis and discussion of the results are presented in Sect. 8.7, while Sect. 8.8 serves as the conclusion of the paper.

8.2 Cyber Physical System

A CPS is a system that intricately combines both physical and computational elements to function cohesively. CPS potentially consists of ICS (Cyber System) and SBS (Physical System) [7]. SBS, as demonstrated through technologies like wireless sensor networks and intelligent building management systems, utilize a network of distributed sensors to gather data about the environment and system operations. This information is then sent to a centralized system for analysis and processing. CPS act as a conduit linking the tangible, physical world with the digital domain, a place where data undergoes storage, processing, and transformation. CPS, which amalgamate computing, communication, and control functionalities, have emerged as a pioneering frontier in the advancement of physical device systems. CPS is characterized as an interconnected assembly of loosely integrated distributed cyber systems and physical systems, managed, and regulated through user-defined semantic rules. The network serves as the conduit bridging the cyber and physical domains, creating a sprawling, heterogeneous, real-time distributed system [8]. A CPS comprises four fundamental components: physical elements, a sensing network, control node (computing device), and a communication network. Figure 8.1 illustrates the CPS system model. The physical components represent the systems of interest that require monitoring and safeguarding. The sensing network consists of interconnected sensors distributed to observe the physical environment. As an integral component of the CPS, the sensing network actively engages in a closed-loop process encompassing sensing, computing, decision-making, and execution [6, 7]. These sensor-generated data are then transmitted to control node for processing and analysis. Computational intelligence methods are applied to make informed decisions and control actuators, ultimately influencing the behaviour of the physical components. The control nodes

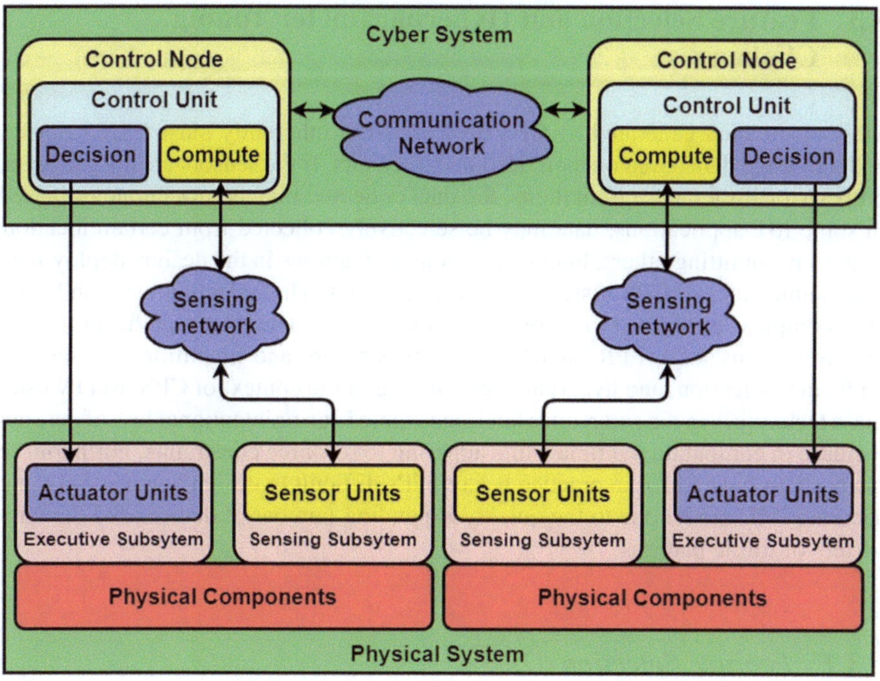

Fig. 8.1 The general structure of CPS

are interconnected through a communication network, facilitating efficient coordination to execute essential computational tasks, particularly those involving spatial data.

The integration of intelligence into CPS empowers them to perform intricate tasks within dynamic environments and even in unforeseen circumstances. In dealing with the inherent uncertainty of the physical world, ML offers statistical solutions that consistently strive for optimal decision-making. The diverse spectrum of ML techniques enables the identification of patterns within the gathered sensor data, facilitating functions like anomaly detection for ensuring system safety, behaviour recognition for comprehending the surrounding environment, and prediction for system optimization and planning

8.3 Feature Selection and Hyperparameter Tuning Challenges

Most of the data generated by CPS devices is not inherently biased. CPS devices collect a huge data based on their design and sensors. If the sensors are not calibrated properly or if they have limitations, the data collected may be inaccurate or biased. In some IoT applications, data may be selectively collected from certain locations or devices, omitting others, human decisions and actions in the design, deployment, and maintenance of CPS systems can introduce bias. This selection bias can lead to an incomplete or skewed view of the overall system. Due to a significant number of false alarms, high FPR, and low DR, researchers and practitioners often rely on feature selection, and hyperparameter tuning in the context of CPS. But by using these techniques in the smart cites landscape there is an unintentional loss of data and increase in computational time while adhering to resource constraints. Furthermore, many CPSs have failed in practice because it's difficult to design a quick, light, and accurate IML model due to the quickly expanding number of devices and the large variety of traffic patterns.

8.3.1 Feature Selection

Feature selection for CPS faces several limitations. Firstly, the multidimensional nature of CPS data often involves a high volume of features, making it challenging to identify the most relevant ones efficiently. Additionally, CPS data can exhibit dynamic and nonlinear relationships, and feature selection methods may struggle to capture complex patterns adequately. Furthermore, some CPS applications demand real-time processing, limiting the time available for exhaustive feature selection procedures. Data quality issues, including noise and missing values, can also hinder the accuracy of feature selection outcomes. Lastly, the diversity of CPS domains, from healthcare to industrial automation, poses unique challenges, as feature selection techniques may need to be tailored to specific application contexts, making it crucial to consider these limitations when implementing feature selection strategies for CPS.

8.3.2 Hyperparameter Tuning

Hyperparameter tuning, while a valuable technique in ML and artificial intelligence, presents notable limitations when applied to CPS. First, CPS often involve real-time or safety-critical operations, where computational overhead and latency introduced by hyperparameter optimization can be impractical. Second, CPS may have resource-constrained environments, making it challenging to execute computationally intensive tuning algorithms [9]. Additionally, the dynamic and complex nature of CPS behaviour makes it difficult to define a static set of hyperparameters that can adequately adapt to changing conditions. Furthermore, tuning hyperparameters in CPS may not guarantee optimal performance across all scenarios, as they often operate in highly diverse and unpredictable environments. Lastly, validating the effectiveness of tuned hyperparameters in CPS may require extensive testing, which can be time-consuming and expensive, potentially undermining the benefits of optimization. Therefore, while hyperparameter tuning can enhance CPS performance, careful consideration of its limitations and trade-offs is essential to ensure safe and efficient deployment in real-world applications.

Applying feature selection to biased data in CPS presents challenges primarily because feature selection techniques typically assume that the data is unbiased and that features are selected based on their ability to contribute valuable information to the modelling or analysis process. When data is biased, it means that certain aspects or groups within the data are disproportionately represented, which can lead to skewed feature selection results. While hyperparameter tuning can optimize ML models for improved performance, it primarily focuses on adjusting parameters related to model complexity, learning rates, and regularization. Bias in CPS data often arises from systematic errors, skewed sampling, or structural issues within the data, which are not directly resolved by hyperparameter tuning.

Sampling techniques play a crucial role in addressing class imbalance issues in CPS datasets. While feature selection and hyperparameter tuning are essential components of building effective ML models for CPS applications, sampling techniques often take precedence when dealing with imbalanced data. Under-sampling is often considered a better strategy than over-sampling in the context of addressing class imbalance in datasets. Under-sampling involves reducing the number of instances from the majority class, thus bringing the class distribution closer to balance. This approach is preferred when there is a significant amount of data available for the majority class, and the minority class contains valuable, albeit limited, information. In CPS attacks, where data is often scarce and expensive to collect, under-sampling can help preserve critical instances of the majority class while still addressing the imbalance issue. It ensures that the model is not overwhelmed by the majority class, which might dilute the detection capacity for the minority class, making it more effective in identifying rare and potentially harmful cyber-physical attacks in CPS scenarios.

8.4 Proposed Methodology

Sampling techniques can be highly valuable in the context of CPS for protecting CI. CPS involves the integration of physical processes with digital systems, and protecting these systems is paramount in safeguarding CI. Sampling allows for the efficient collection of data from various sensors and components within the CPS network. By strategically selecting data points to monitor and analyse, sampling reduces the computational burden and network traffic while still providing insights into system behaviour and anomalies. This approach aids in early detection of cyber threats, such as intrusions or malfunctions, by focusing on critical data points and enabling rapid response to potential security breaches. Furthermore, sampling can help optimize resource allocation and prioritize security measures, ensuring that the most critical aspects of the infrastructure are continuously monitored and protected, ultimately enhancing the resilience and security of CI in the realm of CPS. Using under-sampling technique in conjunction with ensemble learning can be highly beneficial for enhancing the security of CPS safeguarding CI. CPS environments generate vast amounts of data, and analysing all of it in real-time can be challenging. Sampling allows us to efficiently select a representative subset of this data for analysis. Ensemble learning, on the other hand, leverages multiple ML models to improve accuracy and robustness. By combining these two approaches, CPS can effectively identify and respond to security threats and anomalies in real-time. Ensemble models can integrate diverse sources of information from various sensors and devices within the infrastructure, while sampling ensures that the models receive manageable and relevant data streams [10]. This combination enhances threat detection, reduces false positives, and allows for more efficient resource allocation, ultimately bolstering the resilience and security of CI in the face of cyber threats [11].

Ensemble methods that incorporate under-sampling are advanced strategies designed to tackle the challenge of class imbalance in IML tasks. These methods aim to counteract the bias in models arising from imbalanced datasets. Common approaches include modifying the training data's distribution through resampling or adjusting the weights of different classes. The effectiveness of ensemble learning techniques, when combined with under-sampling, lies in their ability to integrate outcomes from multiple classifiers. This integration often leads to a reduction in variance, a common issue in methods that rely on resampling or reweighting. By training multiple models on sampled subsets, these ensembles capture diverse patterns and combine predictions to make accurate decisions. The objective is to mitigate bias towards the minority class and optimise accuracy and effectiveness, especially in classifying anomaly detection tasks.

8.4.1 Under-Sampling Ensemble Techniques

Ensemble techniques that utilize under-sampling are a group of ML strategies. They tackle the issue of class imbalance within datasets by reducing the size of the majority class or by varying the training data. These techniques aim to improve the performance of predictive models when dealing with imbalanced datasets. The Algorithm 1 gives a brief generalized working representation of the under-sampling technique. These methods generate instances for the majority class or under sample existing instances, creating a more balanced dataset for training. By doing so, under-sampling ensemble techniques help prevent the model from being biased toward the majority class, resulting in better generalization and improved classification accuracy, especially when dealing with rare or underrepresented classes.

When compared to current imbalance learning techniques, SPE demonstrates notable efficacy, especially on datasets characterized by large-scale, noise-ridden, and heavily imbalanced conditions. The BCC works well when you want to improve the classification performance of the minority class, which is often the case in real-world scenarios. It sequentially trains multiple classifiers, focusing on the hardest-to-classify minority instances, and iteratively builds a balanced dataset. BRFC algorithm combines the power of RFs with the capability to balance class weights, resulting in improved performance by assigning more importance to the minority class without entirely neglecting the majority class. The EEC addresses this issue by creating multiple balanced subsets from the majority class and combining them with the entire minority class. By repeatedly training classifiers on these balanced subsets, EEC helps improve the model's ability to identify and classify instances from the minority class, making it a valuable choice for tackling imbalanced classification problems where ensuring a proper balance between precision and recall is crucial. RUS Boost Classifier is particularly effective when you want to strike a balance between addressing class imbalance and maintaining computational efficiency in scenarios with limited computational resources. UBC addresses the trade-off between bias and variance in prediction models by reducing the model's variance. It helps mitigate overfitting in models. These approaches help in achieving better classification results for the minority class while maintaining high accuracy for the majority class, making it a valuable tool when handling imbalanced data.

Table 8.1 displays the under-sampling ensemble technique classifiers considered for the proposed study, along with their algorithmic representations, excluding UBC, as it is similar to RUSBC but includes balancing the training set. The Algorithm 2–6 gives a detailed working representation of the above-mentioned under-sampling technique.

Table 8.1 Under-sampling-based ensembles

References	Model	Description	Advantages	Disadvantages
[12]	SPEC	Combines under-sampling with self-paced learning, gradually selecting informative majority class instances to create balanced subsets	Robustness to noisy data, improved convergence, enhanced generalization, and adaptability to data distribution	Sensitivity to hyperparameters, risk of underfitting, and increased complexity
[13]	BCC	An iterative ensemble approach that starts with an under-sampled dataset and iteratively increases the minority class size by adding misclassified majority class instances	Reduced training time, and improved classification performance	Sensitivity to noise, hyperparameter tuning, and potential overfitting
[14]	BRFC	Combines random forests with under-sampling, creating balanced subsets by randomly under-sampling the majority class	Reduces bias, unbiased model evaluation, and feature importance	Computational complexity, loss of information, and parameter tuning
[13]	EEC	Combines under-sampling with boosting, under-sampling the majority class to create balanced subsets and focusing on misclassified instances using boosting	Improved minority class detection, and ensemble robustness	Potential information loss, sensitivity to sampling variability, and limited applicability
[15]	RUSBC	Combines random under-sampling with boosting, under-sampling the majority class and applying boosting to assign higher weights to misclassified instances	Improved generalization, and reduces computation time	Sensitivity to under-sampling ratio, and loss of diversity
[14]	UBC	Combines under-sampling with bagging, creating diverse subsets by randomly under-sampling the majority class and aggregating the predictions of base classifiers	Reduces variance, and handles high-dimensional data	Sensitivity to noise limited improvement with strong base learners, and resource intensive

Algorithm 1: Under Sampling

Input:
- Data: Original dataset
- Class A: Instances from the minority class
- Class B: Instances from the majority class
- IR: Class imbalance ratio
- UR: Desired undersampling ratio

Output:
- Balanced Dataset

Step 1: $\text{IR} = \dfrac{(\text{Number of Class B instances})}{(\text{Number of Class A instances})}$

Step 2: $\text{UF} = \dfrac{\text{IR}}{\text{UR}}$

Step 3: *if* UR < 1:
    Number of Instances to Select = floor(UF * Number of Class B instances)
    Randomly Select Instances from Class B
    Balanced Dataset = Concatenate (Class A, Selected Instances from Class B)
else:
    Balanced Dataset = Concatenate (Class A, Class B)

Step 4: return Balanced Dataset

Algorithm 2 : SPEC

Input	D Training set N majority in D P minority in D f Base classifier n number of base classifiers H Hardness function k number of bins i initialized to zero
Output	Final ensemble $F(x) = \frac{1}{n}\sum_{m=1}^{n} f_m(x)$
step 1	Train classifier f_0 using random under sample majority subsets N_0' and P, where $\lvert N_0' \rvert = \lvert P \rvert$
	repeat
step 2	i = i + 1
step 3	Ensemble $F_i(x) = \frac{1}{i}\sum_{j=0}^{i-1} f_j(x)$
step 4	Cut majority set into k bins w. r. t. $H(x, y, F_i)$: $B_1, B_2, \ldots, B_k$
step 5	Average hardness contribution in l^{th} bin: $h_l = \sum_{s \in B_l} \frac{H(x_s, y_s, F_i)}{\lvert B_l \rvert}, \forall l = 1, \ldots k$
step 6	Update self paced factor $\alpha = \tan(\frac{i\pi}{2n})$
step 7	Unnormalized sampling weight of lth bin: $P_l = \frac{1}{h_l + \alpha}, \forall l = 1, \ldots k$
step 8	Under sample from l^{th} bin with $\frac{P_l}{\sum_m P_m} \lvert P \rvert$ samples
step 9	Train f_i using newly under − sampled subset
	until i = n

Algorithm 3 : BCC

Input D Training set
N majority in D
P minority in D , where $|P| < |N|$
s_i Number of iterations to train AdaBoost ensemble H_i
T Number of subsets from N
f False poistive rate(FPR)
i initialized to zero

Output Final ensemble $H(x) = \frac{1}{n}\sum_{i=1}^{T} \sum_{j=1}^{s_i} \alpha_{i,j}\, h_{i,j}(x) - \sum_{i=1}^{T} \Theta_i$

repeat
step 1 i = i + 1
step 2 Randomly sample a subset N_i from N, $|N_i| = |P|$
step 3 Learn H_i using P and N_i.H_i is an AdaBoost ensemble with s_i weal class and corrsponding weights $\alpha_{i,j}$. The ensemble threshold is Θ_i i.e.,

$$H_i(x) = \sum_{j=1}^{s_i} \alpha_{i,j}\, h_{i,j}(x) - \Theta_i$$

step 4 Adjust Θ_i, such that H_i FPR is f
step 5 Remove from N all examples that are correctly classifed by H_i
until i = T

Algorithm 4 : BRFC

step 1 In each cycle of the random forest process, select a bootstrap sample fro the smaller class. Then, choose an equivalent number of cases from the larger class, using replacement.

step 2 Develop an unpruned classification tree to its full extent using the data. This tree should be constructed using the CART methodology, with one key variation: At every decision point, rather than examining all variable for the best division, limit the search to a randomly chosen subset of m_{t} variables.

step 3 Execute the steps 1 & 2 repeatedly as many times as necessary. Compile the outcomes from the collective ensemble and derive the final decision based on this aggregation.

Algorithm 5 : EEC

Input D Training set
P minority in D
N majority in D, where $|P| < |N|$
T Number of subsets from N
s_i Number of iterations to train AdaBoost ensemble H_i
i initialized to zero

Output Final ensemble $H(x) = \sum_{i=1}^{T} \sum_{j=1}^{s_i} \alpha_{i,j}\, h_{i,j}(x) - \sum_{i=1}^{T} \theta_i$

repeat

step 1 i = i + 1
step 2 Randomly sample a subset N_i from N, $|N_i| = |P|$
step 3 Learn H_i using P and N_i. H_i is an AdaBoost ensemble with s_i weal class and corrsponding weights $\alpha_{i,j}$. The ensemble threshold is θ_i i.e.,

$$H_i(x) = \sum_{j=1}^{s_i} \alpha_{i,j}\, h_{i,j}(x) - \theta_i$$

until i = T

Algorithm 6 : RUSBoost

Input S Training set
X Feature space
Y Class labels
h_t WeakLearner
x_i point in X
y_i point in Y
t initialized to zero

Output Final ensemble $H(x) = \underset{y \in Y}{\operatorname{argmax}} \sum_{t=1}^{T} h_t(x, y)\, log \frac{1}{\alpha_t}$

step 1 Initialize $D(i) = \frac{1}{m}$ for all i

repeat

step 2 t = t + 1
step 3 Create temporary training dataset S_t' with distribution D_t' using random under sampling.
step 4 Call WeakLearn providing it with examples S_t' and their weights D_t'.
step 5 Hypothesis $h_t : X \times Y \rightarrow [0,1]$
step 6 Compute the loss $\epsilon_t = \sum_{(i,y):y_i \neq y} D_t(i)(1 - h_t(x_i, y_i) + h_t(x_i, y))$
step 7 Compute the weight update parameter: $\epsilon_t = \frac{\epsilon_t}{1 - \epsilon_t}$
step 8 Update $D_t : D_{t+1}(i) = D_t(i)\alpha_t^{\frac{1}{2}(1 + h_t(x_i, y_i) - h_t(x_i, y: y \neq y_t))}$
step 9 Normalize D_{t+1}: $Z_t = \sum_i D_{t+1}(i)$, $D_{t+1}(i) = \frac{D_{t+1}(i)}{Z_t}$

until t = T

8.5 Related Works

The objective is to enhance the performance of the underrepresented class while maintaining or improving the performance of the overrepresented class. To enhance the efficacy and accuracy of a ML algorithm in the IDS, the research aims to tackle the issue of class imbalance in ML algorithms concerning the MSCA dataset.

In a study by Jamal et al. [16], an IDS that utilized DL techniques like CNN and DBN has been proposed. The aim was to improve the performance of the IDS while reducing training and response times. The researchers evaluated the effectiveness of their framework by conducting experiments on the MSCAD dataset. The results showed that their proposed approach achieved exceptional performance, with an accuracy rate of 99.6% without using any balancing strategies, 97.6% accuracy rate with the use of SMOTE, and 98.1% accuracy rate with the combination of SMOTETomek for the dataset they investigated.

An advanced neural network approach is employed to evaluate its performance in predicting MSCA [17]. The accuracy achieved with different algorithms is reported as 94.09% for Quest model, 97.29% for BN, and 99.09% for NN. Evaluation of the MSCA dataset demonstrates the proposed EBNN attain high accuracy of 99.72% in predicting MSCA. However, limitations are noted in addressing the class imbalance, particularly for Web_Crawling and HTTP_DDoS attacks with low-density counts. These precise predictions are crucial for effective real-time cyber-attack management.

An attention-based RNN model for detecting MSCA in networks is proposed in [18]. The model incorporates a LSTM unit with an Attention layer. Feature selection is performed using the PSO metaheuristic, resulting in a 72.73% reduction in the dataset, improved computational efficiency, and reduced time consumption, with an accuracy of 99.83% and a DR increase of over 1%. However, it is vital to note that the model has limitations in effectively handling low-density count data of ICMP_Flood and Web_Crwling attacks, which means it does not fully address the class imbalance problem.

Alheeti et al. [19] proposes an intelligent IDS that leverages the k-NN algorithm to differentiate between authentic and tampered data. The system's performance is evaluated using the MSCAD to identify new attack types. Experimental results show that the k-NN based approach improves detection performance, increasing accuracy to 82.59% while minimizing false alarms.

8.6 Experiment Setup and Datasets Descriptions

This section provides a concise overview of both the system environment and the dataset employed in the study. The procedures for collecting the dataset and conducting experiments are outlined here, encompassing the materials and methods

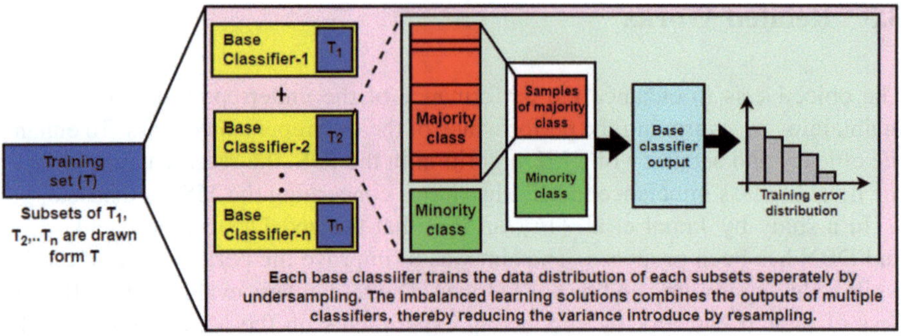

Fig. 8.2 General representation of under-sampling ensemble technique

integral to the comprehensive framework used to validate the experiment's performance metrics and outcomes. The key processes within this paradigm involve data collection and observation. Throughout this phase, the acquired dataset undergoes close monitoring to identify various types of information. The dataset is preprocessed, there was no need for additional pre-processing. The data feature vectors for the training and testing sets are partitioned in an 80:20 ratio, with 103,039 instances utilized for training and 25,760 instances for testing. The learning process utilizes the training data to develop a final model. In this study, sampling is applied to the majority class (non-intrusion instances) to address class distribution imbalance. The proposed work's diagrammatic representation is illustrated in Fig. 8.2.

8.6.1 System Environment

The testing platform utilized was the Google Colab Notebook. The imbens.ensemble framework is open-source and designed to harness the capabilities of ensemble learning for tackling the challenge of class imbalance.

8.6.2 Dataset Description

As the MSCA environment experiences rapid growth in tandem with the increasing prevalence of networks and applications, there is a rising need for a dependable IDS to safeguard networks and devices. To effectively address the unique features of emerging threats, particularly in the context of MSCA, the availability of a current and dependable dataset becomes imperative for robust IDS implementation. This research introduces a novel benchmark MSCA dataset for analysing cyberattacks, encompassing two distinct attack scenarios [20]. The primary setup focuses on password cracking attacks, while the next setup centers on volume-based DDoS attacks.

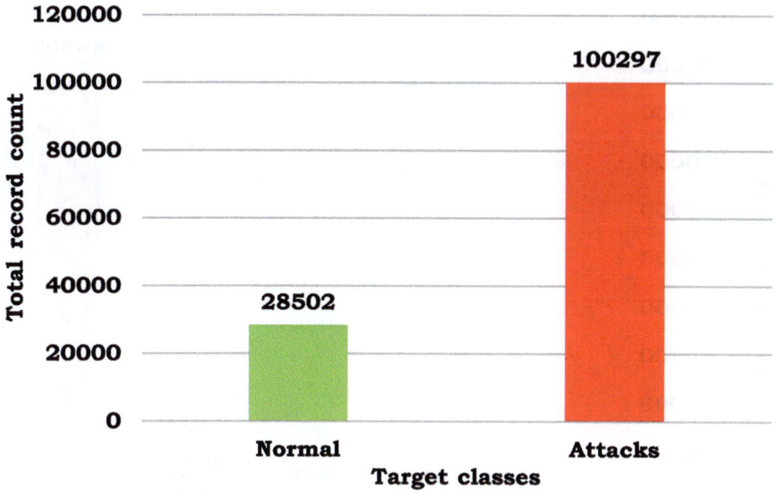

Fig. 8.3 Label distribution of MSCA dataset

The dataset has been meticulously annotated, comprising six PCAP-processed files and 77 network feature files acquired through Wireshark analysis. It is organized into normal and anomalous network traffic categories, and the distribution of the MSCA dataset is illustrated in Figs. 8.3, 8.4 and 8.5.

8.7 Results and Discussion

The experimental results indicate that the under-sampling classifiers SPEC, UBC, and BCC accurately detect network anomalies. Figures 8.6, 8.7, 8.8, 8.9, 8.10 and 8.11 shows the under-sampling classifiers training distribution with respect to the estimators. For a fair study of the proposed work all the models were seeded with $n\_estimators = 100$. Precision, recall, and F1-score metrics were used to assess the performance of each algorithm. Tables 8.2, 8.3, 8.4, 8.5, 8.6 and 8.7 present the evaluation metrics TPR, FPR, precision, recall, AUC, F1-score, error rate, and accuracy are used for under sampling ensemble techniques. In the experiment conducted on the MSCA dataset to address class imbalance problems, SPEC achieved the highest accuracy among all six classifiers. SPEC consistently achieved an average accuracy of approximately 0.9613 in all cases, indicating its outstanding classification correctness. UBC and BCC also exhibited promising results with slightly lower accuracy, demonstrating commendable predictive correctness. Despite the significantly lower density count of cases related to ICMP_Flood, Web_Crwling, and HTTP_DDoS compared to Port_Scan and Brute_Force, all the classifiers achieved decent accuracy. UBC, BRFC, RUSBC, and BCC showed accuracy in the range of 0.97 to 0.99. However, the EEC classifier exhibited lower accuracy of 0.88. It appears

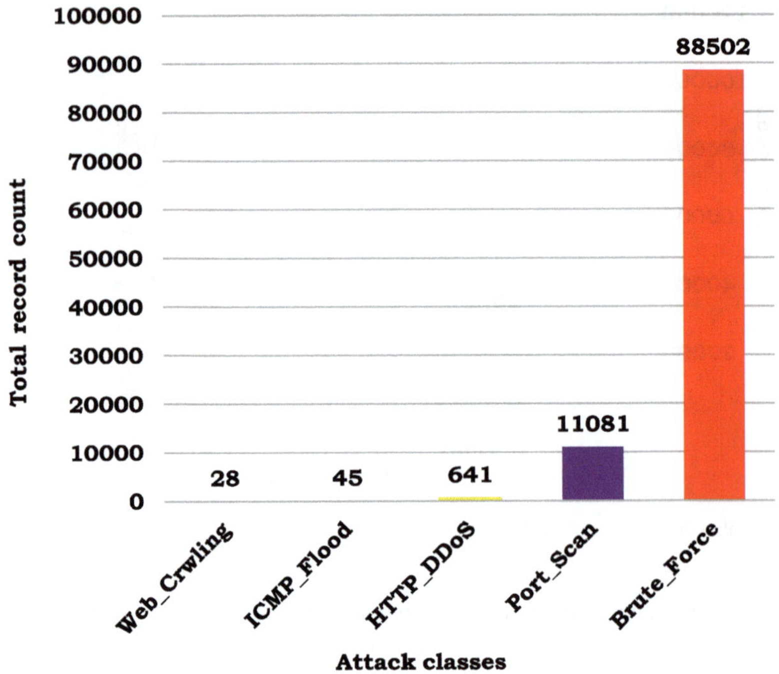

Fig. 8.4 Attacks distribution of MSCA dataset

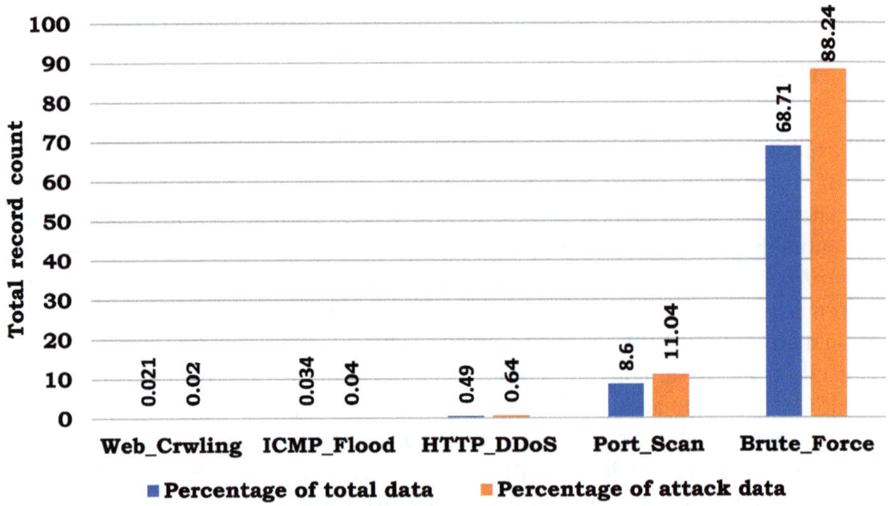

Fig. 8.5 Attack distribution with total data and attack data

that the EEC classifier struggled to effectively address class imbalance using under-sampling techniques, suggesting the need for further research using over-sampling in the case of the EEC classifier. The precision, recall, F1-score, and accuracy values were the lowest for cases associated to ICMP_Flood, and Web_Crwling anomalies due to their small number of instances. Nonetheless, the precision and recall metrics exhibited consistently high values across different anomalies, particularly notable in the case of Brute_Force and Normal. The evaluation metrics of the proposed work are visually depicted in Figs. 8.12 and 8.13. Table 8.8 illustrates the weighted average of the under-sampling-based ensembles techniques. Table 8.9 shows a comparison study of various researchers work on MSCA dataset. It is worth noting that relying solely on a single rule to detect intrusions based on typical traffic patterns often leads to false positive results. Anomaly-based CPS models consider any traffic deviating from the normal pattern as abnormal. The utilization of under-sampling techniques helps address this issue. While under sampling can be effective in balancing imbalanced data, there are some challenges to consider when deploying it in real-time applications. As the data is constantly changing in real-time applications, it may be difficult to maintain a balanced dataset. It is important to carefully monitor and adjust the sampling technique to ensure accurate results.

#Estimators	Class Distribution	Data: train Metric acc	balanced_acc	weighted_f1
1	{'Brute_Force': 23, 'HTTP_DDoS': 23, 'ICMP_Flood': 23, 'Normal': 23, 'Port_Scan': 23, 'Web_Crwling': 23}	0.857	0.871	0.886
10	{'Brute_Force': 23, 'HTTP_DDoS': 23, 'ICMP_Flood': 23, 'Normal': 23, 'Port_Scan': 23, 'Web_Crwling': 23}	0.943	0.944	0.957
20	{'Brute_Force': 23, 'HTTP_DDoS': 23, 'ICMP_Flood': 23, 'Normal': 23, 'Port_Scan': 23, 'Web_Crwling': 23}	0.952	0.951	0.961
30	{'Brute_Force': 23, 'HTTP_DDoS': 23, 'ICMP_Flood': 23, 'Normal': 23, 'Port_Scan': 23, 'Web_Crwling': 23}	0.955	0.955	0.964
40	{'Brute_Force': 23, 'HTTP_DDoS': 23, 'ICMP_Flood': 23, 'Normal': 23, 'Port_Scan': 23, 'Web_Crwling': 23}	0.953	0.953	0.963
50	{'Brute_Force': 23, 'HTTP_DDoS': 23, 'ICMP_Flood': 23, 'Normal': 23, 'Port_Scan': 23, 'Web_Crwling': 23}	0.955	0.955	0.964
60	{'Brute_Force': 23, 'HTTP_DDoS': 23, 'ICMP_Flood': 23, 'Normal': 23, 'Port_Scan': 23, 'Web_Crwling': 23}	0.956	0.956	0.965
70	{'Brute_Force': 23, 'HTTP_DDoS': 23, 'ICMP_Flood': 23, 'Normal': 23, 'Port_Scan': 23, 'Web_Crwling': 23}	0.957	0.957	0.966
80	{'Brute_Force': 23, 'HTTP_DDoS': 23, 'ICMP_Flood': 23, 'Normal': 23, 'Port_Scan': 23, 'Web_Crwling': 23}	0.958	0.957	0.966
90	{'Brute_Force': 23, 'HTTP_DDoS': 23, 'ICMP_Flood': 23, 'Normal': 23, 'Port_Scan': 23, 'Web_Crwling': 23}	0.960	0.960	0.969
100	{'Brute_Force': 23, 'HTTP_DDoS': 23, 'ICMP_Flood': 23, 'Normal': 23, 'Port_Scan': 23, 'Web_Crwling': 23}	0.962	0.961	0.971
final	{'Brute_Force': 23, 'HTTP_DDoS': 23, 'ICMP_Flood': 23, 'Normal': 23, 'Port_Scan': 23, 'Web_Crwling': 23}	0.962	0.961	0.971

Fig. 8.6 SPEC training distribution with metrics

#Estimators	Class Distribution	Data: train Metric acc	balanced_acc	weighted_f1
1	{'Brute_Force': 23, 'HTTP_DDoS': 23, 'ICMP_Flood': 23, 'Normal': 23, 'Port_Scan': 23, 'Web_Crwling': 23}	0.857	0.871	0.886
10	{'Brute_Force': 23, 'HTTP_DDoS': 23, 'ICMP_Flood': 23, 'Normal': 23, 'Port_Scan': 23, 'Web_Crwling': 23}	0.948	0.944	0.959
20	{'Brute_Force': 23, 'HTTP_DDoS': 23, 'ICMP_Flood': 23, 'Normal': 23, 'Port_Scan': 23, 'Web_Crwling': 23}	0.969	0.957	0.978
30	{'Brute_Force': 23, 'HTTP_DDoS': 23, 'ICMP_Flood': 23, 'Normal': 23, 'Port_Scan': 23, 'Web_Crwling': 23}	0.944	0.941	0.964
40	{'Brute_Force': 23, 'HTTP_DDoS': 23, 'ICMP_Flood': 23, 'Normal': 23, 'Port_Scan': 23, 'Web_Crwling': 23}	0.949	0.940	0.964
50	{'Brute_Force': 23, 'HTTP_DDoS': 23, 'ICMP_Flood': 23, 'Normal': 23, 'Port_Scan': 23, 'Web_Crwling': 23}	0.956	0.946	0.969
60	{'Brute_Force': 23, 'HTTP_DDoS': 23, 'ICMP_Flood': 23, 'Normal': 23, 'Port_Scan': 23, 'Web_Crwling': 23}	0.960	0.951	0.972
70	{'Brute_Force': 23, 'HTTP_DDoS': 23, 'ICMP_Flood': 23, 'Normal': 23, 'Port_Scan': 23, 'Web_Crwling': 23}	0.939	0.936	0.959
80	{'Brute_Force': 23, 'HTTP_DDoS': 23, 'ICMP_Flood': 23, 'Normal': 23, 'Port_Scan': 23, 'Web_Crwling': 23}	0.934	0.933	0.956
90	{'Brute_Force': 23, 'HTTP_DDoS': 23, 'ICMP_Flood': 23, 'Normal': 23, 'Port_Scan': 23, 'Web_Crwling': 23}	0.922	0.924	0.945
100	{'Brute_Force': 23, 'HTTP_DDoS': 23, 'ICMP_Flood': 23, 'Normal': 23, 'Port_Scan': 23, 'Web_Crwling': 23}	0.919	0.923	0.943
final	{'Brute_Force': 23, 'HTTP_DDoS': 23, 'ICMP_Flood': 23, 'Normal': 23, 'Port_Scan': 23, 'Web_Crwling': 23}	0.919	0.923	0.943

Fig. 8.7 BCC training distribution with metrics

#Estimators	Data: train Metric acc	balanced_acc	weighted_f1
100	0.941	0.930	0.954

Fig. 8.8 BRFC training distribution with metrics

#Estimators	Data: train Metric acc	balanced_acc	weighted_f1
100	0.821	0.808	0.872

Fig. 8.9 EEC training distribution with metrics

#Estimators	Class Distribution	Data: train Metric acc	balanced_acc	weighted_f1
1	{'Brute_Force': 23, 'HTTP_DDoS': 23, 'ICMP_Flood': 23, 'Normal': 23, 'Port_Scan': 23, 'Web_Crwling': 23}	0.682	0.330	0.668
10	{'Brute_Force': 23, 'HTTP_DDoS': 23, 'ICMP_Flood': 23, 'Normal': 23, 'Port_Scan': 23, 'Web_Crwling': 23}	0.538	0.334	0.580
20	{'Brute_Force': 23, 'HTTP_DDoS': 23, 'ICMP_Flood': 23, 'Normal': 23, 'Port_Scan': 23, 'Web_Crwling': 23}	0.561	0.562	0.696
30	{'Brute_Force': 23, 'HTTP_DDoS': 23, 'ICMP_Flood': 23, 'Normal': 23, 'Port_Scan': 23, 'Web_Crwling': 23}	0.506	0.541	0.591
40	{'Brute_Force': 23, 'HTTP_DDoS': 23, 'ICMP_Flood': 23, 'Normal': 23, 'Port_Scan': 23, 'Web_Crwling': 23}	0.541	0.366	0.625
50	{'Brute_Force': 23, 'HTTP_DDoS': 23, 'ICMP_Flood': 23, 'Normal': 23, 'Port_Scan': 23, 'Web_Crwling': 23}	0.524	0.457	0.613
60	{'Brute_Force': 23, 'HTTP_DDoS': 23, 'ICMP_Flood': 23, 'Normal': 23, 'Port_Scan': 23, 'Web_Crwling': 23}	0.525	0.534	0.618
70	{'Brute_Force': 23, 'HTTP_DDoS': 23, 'ICMP_Flood': 23, 'Normal': 23, 'Port_Scan': 23, 'Web_Crwling': 23}	0.363	0.404	0.484
80	{'Brute_Force': 23, 'HTTP_DDoS': 23, 'ICMP_Flood': 23, 'Normal': 23, 'Port_Scan': 23, 'Web_Crwling': 23}	0.545	0.498	0.644
90	{'Brute_Force': 23, 'HTTP_DDoS': 23, 'ICMP_Flood': 23, 'Normal': 23, 'Port_Scan': 23, 'Web_Crwling': 23}	0.696	0.559	0.742
100	{'Brute_Force': 23, 'HTTP_DDoS': 23, 'ICMP_Flood': 23, 'Normal': 23, 'Port_Scan': 23, 'Web_Crwling': 23}	0.553	0.525	0.666
final	{'Brute_Force': 23, 'HTTP_DDoS': 23, 'ICMP_Flood': 23, 'Normal': 23, 'Port_Scan': 23, 'Web_Crwling': 23}	0.553	0.525	0.666

Fig. 8.10 RUSBC training distribution with metrics

Fig. 8.11 UBC training distribution with metrics

#Estimators	Data: train Metric acc	balanced_acc	weighted_f1
100	0.950	0.944	0.959

Table 8.2 SPEC evaluation metrics

Evaluation metrics	ICMP_Flood	HTTP_DdoS	Brute_Force	Web_ Crwling	Port_Scan	Normal
TPR	0.8	0.97	0.99	0.6	0.97	0.83
FPR	0.01	0	0.03	0	0	0
F1-score	0.05	0.7	0.99	0.025	0.95	0.9
Precision	0.028	0.55	0.98	0.012	0.94	0.99
Error rate	0.01	0.003	0.01	0.009	0.007	0.03
AUC	0.89	0.98	0.98	0.79	0.98	0.91
Individual accuracy	0.98	0.99	0.98	0.99	0.99	0.96
Accuracy	9613%					

Table 8.3 BCC evaluation metrics

Evaluation metrics	ICMP_Flood	HTTP_DdoS	Brute_Force	Web_ Crwling	Port_Scan	Normal
TPR	0.7	0.96	0.99	0.6	0.96	0.64
FPR	0.01	0	0.03	0.05	0	0
F1-score	0.04	0.72	0.99	0	0.96	0.78
Precision	0.02	0.57	0.98	0	0.96	0.99
Error rate	0.01	0	0.01	0.05	0	0.07
AUC	0.84	0.98	0.98	0.77	0.98	0.82
Individual accuracy	0.98	0.99	0.98	0.94	0.99	0.92
Accuracy	92%					

Table 8.4 BRFC evaluation metrics

Evaluation metrics	ICMP_Flood	HTTP_DdoS	Brute_Force	Web_ Crwling	Port_Scan	Normal
TPR	0.8	0.88	0.99	0.6	0.91	0.75
FPR	0.01	0	0.05	0.01	0	0
F1-score	0.03	0.49	0.98	0.01	0.93	0.86
Precision	0.01	0.34	0.97	0	0.95	0.99
Error rate	0.01	0	0.01	0.01	0.01	0.05
AUC	0.89	0.94	0.97	0.97	0.95	0.87
Individual accuracy	0.98	0.99	0.98	0.98	0.98	0.94
Accuracy	93.9%					

Table 8.5 EEC evaluation metrics

Evaluation metrics	ICMP_Flood	HTTP_DdoS	Brute_Force	Web_ Crwling	Port_Scan	Normal
TPR	0.7	0.91	0.88	0.8	0.9	0.62
FPR	0	0	0.01	0.11	0	0.06
F1-score	0.77	0.44	0.93	0.002	0.94	0.67
Precision	0.87	0.29	0.99	0.001	0.99	0.73
Error rate	0	0.01	0.08	0.11	0	0.13
AUC	0.84	0.95	0.93	0.84	0.95	0.77
Individual accuracy	0.99	0.98	0.91	0.88	0.99	0.86
Accuracy	82.6%					

Table 8.6 RUSBC evaluation metrics

Evaluation metrics	ICMP_Flood	HTTP_DdoS	Brute_Force	Web_ Crwling	Port_Scan	Normal
TPR	0	0.96	0.54	0.2	0.67	0.53
FPR	0.006	0.31	0.04	0.02	0.04	0.067
F1-score	0	0.027	0.69	0.002	0.63	0.6
Precision	0	0.014	0.96	0.001	0.6	0.68
Error rate	0.007	0.3	0.32	0.02	0.06	0.15
AUC	0.49	0.82	0.75	0.58	0.81	0.73
Individual accuracy	0.99	0.69	0.67	0.97	0.93	0.84
Accuracy	55.4%					

Table 8.7 UBC evaluation metrics

Evaluation metrics	ICMP_Flood	HTTP_DdoS	Brute_Force	Web_Crwling	Port_ Scan	Normal
TPR	0.7	0.89	0.99	0.6	0.9	0.82
FPR	0.007	0	0.05	0.012	0.003	0.003
F1-score	0.069	0.47	0.98	0.018	0.93	0.89
Precision	0.03	0.31	0.97	0.009	0.95	0.98
Error rate	0.007	0.009	0.018	0.012	0.01	0.041
AUC	0.84	0.94	0.97	0.79	0.95	0.9
Individual accuracy	0.99	0.99	0.98	0.98	0.98	0.95
Accuracy	95.04%					

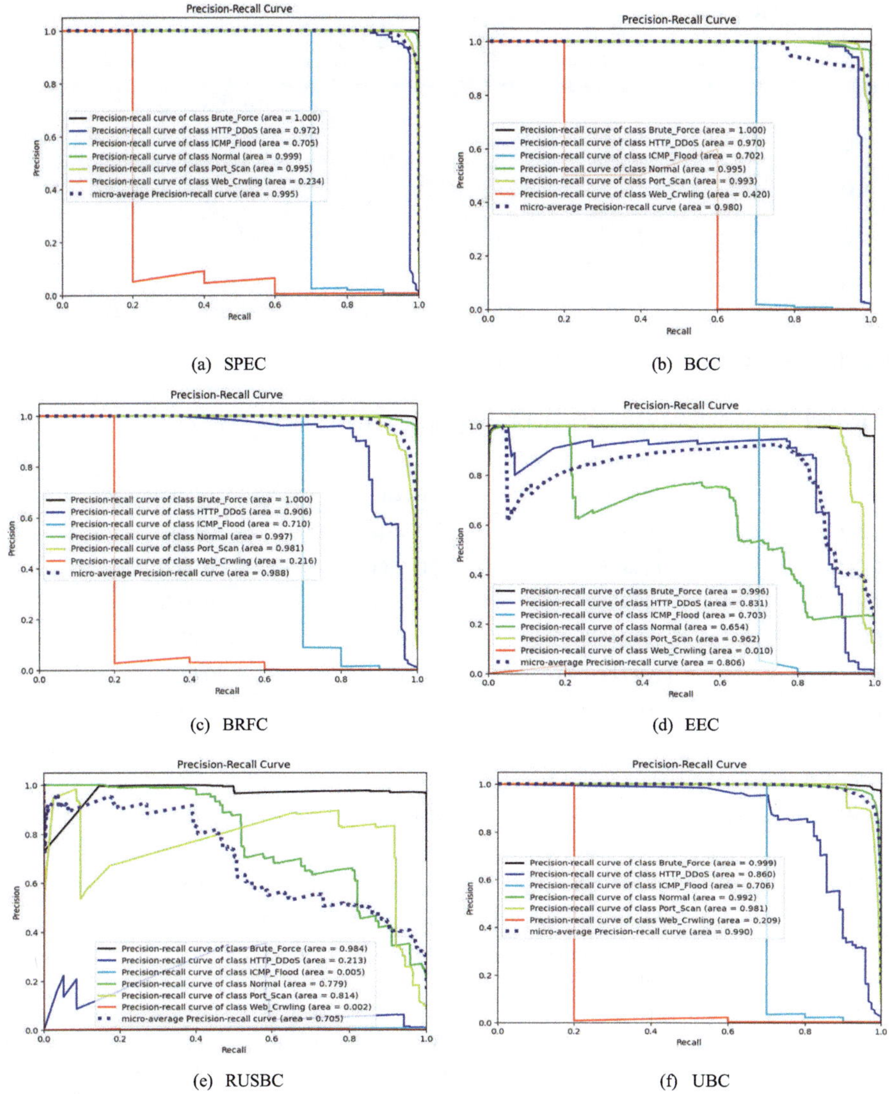

(a) SPEC (b) BCC

(c) BRFC (d) EEC

(e) RUSBC (f) UBC

Fig. 8.12 Precision-recall curve of under-sampling-based ensembles techniques

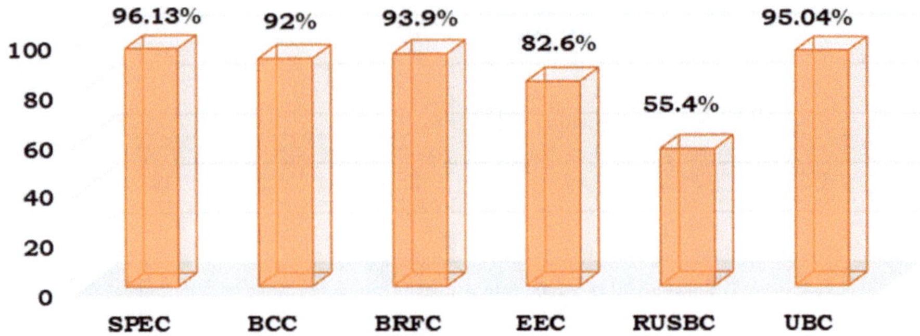

Fig. 8.13 Accuracies of the proposed models

Table 8.8 Weighted average of the under-sampling-based ensembles techniques

Model	Evaluation metrics		
	Precision	Recall	F1-score
SPEC	0.9823	0.9613	0.9696
BCC	0.9828	0.9200	0.9428
BRFC	0.9751	0.9396	0.9530
EEC	0.9345	0.8262	0.8759
RUSBC	0.8711	0.5541	0.6670
UBC	0.9744	0.9504	0.9599

Table 8.9 Comparison of various researchers work

Various researchers work	Model	Metrics	Proposed work (SPEC)
[16]	CNN-DBN	F1-score = 80%	F1-score = 96.96%
	SMOTE + CNN-DBN	F1-score = 97.5%	
	SMOTETomex + CNN-DBN	F1-score = 97.9%	
[17]	EBNN	Individual accuracies (Web_Crwling = 0%, HTTP_DdoS = 94.1%, Port_Scan = 95.8%)	Individual accuracies (Web_Crwling = 99%, HTTP_DDoS = 99%, Port_Scan = 99%, Web_Crwling = 99%)
[18]	Attention-based RNN with PSO	TPR (ICMP Flood = 70%, Web_Crwling = 20.01%)	ICMP Flood = 80%, Web_Crwling = 60%
[19]	k-NN	Accuracy = 82.59%	Recall = 96.13%

8.8 Conclusion

The research introduced a potential approach to enhance the security of CPS in smart city environments. This was achieved by addressing the class-imbalance issue encountered by machine learning algorithms, employing under-sampling ensemble techniques. Refining the data through under-sampling offers advantages over developing complex ML models. It helps address the class imbalance problem and improves accuracy without the drawbacks associated with complex model development. Experimental findings demonstrate that under-sampling classifiers SPEC, UBC, and BCC exhibit exceptional accuracy in detecting network anomalies. SPEC surpasses other classifiers, achieving an average accuracy of 96.13%. Despite the class imbalance, all classifiers demonstrate strong performance, with high precision and recall for most anomalies. These results highlight the significance of under-sampling techniques in anomaly detection. Relying solely on a single rule for intrusion detection based on traffic patterns can yield false positives, making under-sampling preferable over complex ML models. The main advantage is that it can help improve the accuracy of predictive models by giving equal weight to all classes. However, it may result in unintentional loss of data from the majority class, sometimes leading to biased results. To further improve the predictive models, a combination of different sampling techniques like reweighting-based ensembles, and compatible ensembles, create a more balanced dataset.

References

1. Ghaemi, A.A.: A cyber-physical system approach to smart city development. In: 2017 IEEE International Conference on Smart Grid and Smart Cities (ICSGSC), IEEE, pp. 257–262. https://doi.org/10.1109/ICSGSC.2017.8038587
2. Wang, C. et al.: Dynamic Road Lane management study: A Smart City Application To cite this version: HAL Id: hal-01259796 A Smart City Application (2019)
3. Reddy, D.K.K., Behera, H.S., Naik, B.: An intelligent security framework for cyber-physical systems in smart city. In: Big Data Analytics and Intelligent Techniques for Smart Cities, vol. 10, no. 16, pp. 167–186. CRC Press, Boca Raton (2021). https://doi.org/10.1201/978100318 7356-9
4. Nam, T., Pardo, T.A.: Conceptualizing smart city with dimensions of technology, people, and institutions. In: Proceedings of the 12th Annual International Digital Government Research Conference: Digital Government Innovation in Challenging Times, pp. 282–291. ACM, New York, NY, USA (2011). https://doi.org/10.1145/2037556.2037602
5. Neirotti, P., De Marco, A., Cagliano, A.C., Mangano, G., Scorrano, F.: Current Trends in Smart City Initiatives: Some Stylised Facts, vol. 38 (2014). https://doi.org/10.1016/j.cities.2013.12.010
6. Sallhammar, K., Helvik, B.E., Knapskog, S.J.: Incorporating attacker behavior in stochastic models of security (2005)
7. Nayak, J., Kumar, P.S., Reddy, D.K.K., Naik, B., Pelusi, D.: An intelligent security framework for cyber-physical systems in smart city. In: Big Data Analytics and Intelligent Techniques for Smart Cities, pp. 167–186. Wiley, Boca Raton (2021)
8. Tang, B. (2016). Toward Intelligent Cyber-Physical Systems: Algorithms, Architectures, and Applications (2016)

9. Reddy, D.K.K., Nayak, J., Behera, H.S.: A hybrid semi-supervised learning with nature-inspired optimization for intrusion detection system in IoT environment. In: Lecture Notes in Networks and Systems, vol. 480 LNNS, pp. 580–591 (2022). https://doi.org/10.1007/978-981-19-3089-8_55
10. Reddy, D.K.K., Behera, H.S.: CatBoosting Approach for Anomaly Detection in IoT-Based Smart Home Environment, pp. 753–764 (2022). https://doi.org/10.1007/978-981-16-9447-9_56
11. Reddy, D.K.K., Behera, H.S., Pratyusha, G.M.S., Karri, R.: Ensemble Bagging Approach for IoT Sensor Based Anomaly Detection, pp. 647–665 (2021). https://doi.org/10.1007/978-981-15-8439-8_52
12. Liu, Z., et al.: Self-paced Ensemble for Highly Imbalanced Massive Data Classification. In: 2020 IEEE 36th International Conference on Data Engineering (ICDE), IEEE, Apr. 2020, pp. 841–852. https://doi.org/10.1109/ICDE48307.2020.00078
13. Liu, X.Y., Wu, J., Zhou, Z.H.: Exploratory undersampling for class-imbalance learning. IEEE Trans. Syst. Man Cybern. B Cybern.Cybern. B Cybern. **39**(2), 539–550 (2009). https://doi.org/10.1109/TSMCB.2008.2007853
14. Chen, C., Liaw, A.: Using Random Forest to Learn Imbalanced Data
15. Seiffert, C., Khoshgoftaar, T.M., Van Hulse, J., Napolitano, A.: RUSBoost: a hybrid approach to alleviating class imbalance. IEEE Trans. Syst. Man Cybern. Part A Syst. Humans **40**(1), 185–197 (2010). https://doi.org/10.1109/TSMCA.2009.2029559
16. Jamal, M.H., et al.: Multi-step attack detection in industrial networks using a hybrid deep learning architecture. Math. Biosci. Eng.Biosci. Eng. **20**(8), 13824–13848 (2023). https://doi.org/10.3934/mbe.2023615
17. Dalal, S., et al.: Extremely boosted neural network for more accurate multi-stage Cyber-attack prediction in cloud computing environment. J. Cloud Comput. **12**(1) (2023). https://doi.org/10.1186/s13677-022-00356-9
18. Udas, P.B., Roy, K.S., Karim, M.E., Azmat Ullah, S.M.: Attention-based RNN architecture for detecting multi-step cyber-attack using PSO metaheuristic. In: 3rd International Conference on Electrical, Computer and Communication Engineering, ECCE 2023 (2023). https://doi.org/10.1109/ECCE57851.2023.10101590
19. Alheeti, K.M.A., Alzahrani, A., Jasim, O.H., Al-Dosary, D., Ahmed, H.M., Al-Ani, M.S.: Intelligent detection system for multi-step cyber-attack based on machine learning. In: Proceedings—International Conference on Developments in eSystems Engineering, DeSE, vol. 2023-Janua, pp. 510–514 (2023). https://doi.org/10.1109/DeSE58274.2023.10100226
20. Almseidin, M., Al-Sawwa, J., Alkasassbeh, M.: Generating a benchmark cyber multi-step attacks dataset for intrusion detection. J. Intell. Fuzzy Syst. **43**(3), 3679–3694 (2022). https://doi.org/10.3233/JIFS-213247

Chapter 9
Application of Deep Learning in Medical Cyber-Physical Systems

H. Swapnarekha and Yugandhar Manchala

Abstract The integration of IoT devices to healthcare sector has enabled remote monitoring of patient data and delivery of suitable diagnostics whenever required. Because of the rapid advancement in embedded software and network connectivity, Cyber physical systems (CPS) have been widely used in the medical industry to provide top-notch patient care in a variety of clinical scenarios because of the quick advancements in embedded software and network connectivity. Due to the heterogeneity of the medical devices used in these systems, there is a requirement for providing efficient security solutions for these intricate environments. Any alteration to the data could have an effect on the patient's care, which may lead to accidental deaths in an emergency. Deep learning has the potential to offer an efficient solution for intrusion detection because of the high dimensionality and conspicuous dynamicity of the data involved in such systems. Therefore, in this study, a deep learning-assisted Attack Detection Framework has been suggested for safely transferring healthcare data in medical cyber physical systems. Additionally, the efficacy of the suggested framework in comparison to various cutting-edge machine and ensemble learning techniques has been assessed on healthcare dataset consisting of sixteen thousand records of normal and attack data and the experimental findings indicate that the suggested framework offers promising outcomes when compared with the state-of-the-art machine learning and ensemble learning approaches.

Keywords Cyber physical system · Machine learning · Healthcare sector · Deep neural network · Cyber-attacks · Deep learning

H. Swapnarekha (✉)
Department of Information Technology, Aditya Institute of Technology and Management, Tekkali 532201, India
e-mail: swapnarekha23@gmail.com

Y. Manchala
Department of Information Technology, Vardhaman College of Engineering, (Autonomous), Hyderabad, Telangana 501218, India

J. Nayak et al. (eds.), *Machine Learning for Cyber Physical System: Advances and Challenges*, Intelligent Systems Reference Library 60,
https://doi.org/10.1007/978-3-031-54038-7_9

9.1 Introduction

Due to the expeditious advancement of technology in control theory and network communications, CPSs have emerged as a critical research area among researchers from both industry and academia. Cyber physical system is a device that is controlled and supervised tightly by computer-based algorithms. In cyber physical systems, the amalgamation of networking, computing and physical devices provide continuous connectivity between services of cyber and physical systems [1]. The basic interpretation of CPS is to carry out critical tasks by incorporating intelligence in everyday activities of real-world applications such as smart grids [2], smart cities [3], 5G cellular networks [4], healthcare systems [5], sustainable developments [6], robotics systems [7], and so on.

Over the last few years, the rapid transformation in the healthcare sector is providing a vast scope for research due to the advancement of the computing technologies in the medical field in order to provide quality services to people. Health always plays a pivotal role in the society's advancement. The lack of healthcare physicians in several countries causes a drop in the quality of medical services and enhancement in healthcare cost. Therefore, for the sustainment of health, Healthcare systems are established to meet the demands of the needy people by the medical experts in the associated fields. In recent years, healthcare systems have adapted low power and low-cost sensors for supporting variety of medical applications such as efficient monitoring of remote patients, early diagnosis of disease and medical emergencies. The primary concern that arises from the transmission of data through various communication media and low-cost sensors in the healthcare sector is the interpretation of enormous amount of data in real time. Therefore, there is a need for developing Medical cyber-physical systems (MCPs) that combines the features of cyber world with dynamics of real world for efficient monitoring and processing of patient's information and for making autonomous decisions without the involvement of physicians and caregivers [8, 9]. The primary concern in designing Medical cyber-physical system is security as patient data is confidential from ethical and legitimate point of view. The heterogeneity nature of medical devices and recent advancement in wireless and mobile technologies introduced enormous attacks and vulnerabilities in the MCPS which can lead to unauthorized access to patient's personal details. Sometimes, these attacks can also cause false diagnosis and improper treatment which may result in loss of human life. Thus, in order to protect patient information and deliver high-quality services, precise access control must be implemented on patient data. [10].

In recent years, protection of sensitive data has become an important research area among researchers because of the increased number of cyber-attacks. Over the past few decades, machine learning (ML) techniques have been used efficiently deployed in various application domains such as natural language processing [11], classification of images [12], speech recognition [13], detection of malware [14] and so on because of their capability in analyzing and addressing the complex issues. As machine learning approaches have shown significant performance in distinct application areas over traditional based algorithms, these approaches have also been used in the detection of attacks and vulnerabilities in CPS related to medical sector. For the efficient detection of cyber-attacks in healthcare system, an IWMCPS (improved wireless medical cyber-physical system) framework that makes use of ML technique has been presented by Alzahrani et al. [15]. The planning and monitoring of resources, computational and safety core and communication and monitoring core are the three key components of the suggested framework. Additionally, real patient data and security attack data were used to assess the suggested framework. The empirical findings show that the recommended framework achieved a higher detection accuracy rate of 92% with less computational expense. A novel framework based on RFE (Recursive Feature elimination) and MLP (multi layer perceptron) has been developed by Kilincer et al. [16] for the identification of cyber-attacks in healthcare sector. The optimal features have been selected by the RFE approach using kernel function of LR (Logistic regression) and XGBRegressor. To enhance the performance of the suggested approach, tenfold cross-validation approach and hyperparameter optimization algorithm have been used to adjust the parameters of the MLP. Then the model has been validated on various standard datasets related to IoMT cybersecurity and the results reveal that the suggested approach has attained an accuracy of 99.99% with ECU-IoHT dataset, 99.94% with ICU dataset, 98.12% with ToN-IoT dataset and 96.2% with WUSTL-EHMS dataset respectively. A unique approach known as MCAD (machine learning-based cyberattack detector) for cyberattack detection in healthcare system has been developed by Halman and Mohammed JF Alenazi [17]. In order to obtain normal and abnormal traffic, the developed approach makes use of L3 (three-layer learning) switch application by deploying it on Ryu controller. The suggested model has been validated, and results of the experiments indicate that the MCAD model performed better, achieving an F1-score of 98.82% on abnormal data and 99.98% on normal data, respectively. Moreover, the MCAD model ability has also been measured various network key parameter indicators and results show that throughput of the MCAD model has been enhanced by 609%, delay reduced by 77% and jitter reduced by 23% respectively. Though several approaches of ML have been used in the classification of cyber-attacks, they are not competent of providing unique feature descriptor because of their drawbacks in model complexity.

Nowadays, deep learning approaches has resulted in major advancement in distinct application domains over the standard machine learning approaches because of their improved learning capabilities in solving problems of the real-world applications, high level feature extraction, and discovery of hidden patterns. For the identification of unknown attacks, an advanced intrusion detection system that makes use of deep neural network has been proposed by Maithem Mohammed and Ghadaa A. Al-Sultany [18]. The suggested model has been evaluated on KDD cup 99 dataset and the outcomes shows that model attained superior performance with detection rate of 99.98%. Cil et al. [19] have suggested a framework that makes use of deep neural network model for the DDoS attack identification from network traffic. Further, the authors have conducted experiments on CICDDoS2019 dataset using deep neural network model. The experimental results shows that the suggested approach detects and classify network attacks with an accuracy of 99.99% and 94.75% respectively. A deep neural network (DNN) model consisting of one input layer, three hidden layers and one output layer has been suggested by Tang et al. [20] for accomplishing flow-based anomaly detection. The NSL-KDD dataset was used to validate the DNN model, and the results show that this ML technique is superior to other ones at accurately detecting zero-day assaults. Li et al. [21] have proposed an enhanced DNN model known as HashTran-DNN for the classification of Android malware. In order to preserve locality features, the input samples are transformed using hash function. To enhance the he performance of the system, denoising task has been carried by HashTran-DNN by utilizing auto encoder that attains locality information in potential space. From the empirical outcomes, it is observed that HashTran-DNN can detect four distinct attacks more effectively when compared with the standard DNN. For efficient and reliable online monitoring of AGVs (automated guided vehicles) against cyber-attacks, an integrated IoT framework that makes use of DNN with ReLu was suggested by Elsisi et al. [22]. The developed framework along with distinct deep learning and machine learning approaches namely 1D-CNN (one dimensional convolutional neural network), SVM, decision tree, XGBoost and random forest were trained and validated on real AGV dataset and various types of cyber-attacks such as pulse attack, ramp attack, sinusoidal attack and random attack. From the empirical findings, it is clear that the suggested integrated IoT framework attained better detection accuracy of 96.77% when compared with other standard deep learning and machine learning approaches.

Presently, deep neural networks are the basis for many contemporary artificial intelligence applications because of their superior performance in various application domains over the traditional machine learning approaches. The DNN model is capable of learning series of hidden patterns hierarchically as it comprises of set of stacked layers. Moreover, DNNs offer superior performance over various machine learning approaches as they are capable of extracting high-level features with fewer parameters. Keeping in view all these aspects, a deep neural network approach has been developed in this study for the classification of cyber-attacks in medical cyber physical systems. The following are the major contributions of this study.

1. An intelligent security framework based on deep neural network has been developed for the detection of cyber-attacks in healthcare sector.
2. The suggested framework has been validated using WUSTL -EHMS 2020 dataset that consist of network traffic indicators collected from the patient's biometric data.
3. Further, the performance of the suggested framework along with various traditional machine learning and ensemble learning approaches has been validated using various performance metrics to show the efficacy of the suggested approach.

The chapter remaining sections are structured as follows. Section 9.2 outlines the study of the literature on machine learning techniques for detecting cyberattacks in the healthcare industry, as well as their shortcomings. Methodology of the proposed approach has been represented in Sect. 9.3. The environmental setup and dataset description has been described in Sect. 9.4 and evaluation metrics and comparative analysis of proposed DNN model along with other considered model has been described in Sect. 9.5. Finally, the conclusion and future scope of work has been represented in Sect. 9.6.

9.2 Related Study

The recent advances in the field of machine learning have attracted several researchers to carry out their research work in the detection of attacks in medical cyber physical systems. This section describes some of the recent research endeavors undertaken for the detection of cyber attacks in the MCPS.

To protect patient data in healthcare networks, AlZubi et al. [5] have presented the CML-AD framework (cognitive machine learning attack detection). The patient-centric design-based plan that has been suggested minimises the local load resulting from the numerical outcomes while simultaneously guaranteeing the security of patient data in MCPs. Further, the empirical outcomes also indicate that the suggested approach has attained 96.5%, 98.2%, 97.8% of prediction ratio, accuracy ratio and efficiency ratio respectively when compared with other existing approaches.

Schneble et al. [23] have proposed a unique paradigm based on ML technique for intrusion detection in healthcare cyber physical system. To reduce the computation and communication associated with solutions based on conventional machine learning approaches, authors have explored the conception of federated learning in the suggested framework. Then the suggested framework has been evaluated on real-time patient dataset for determining the security attacks and the empirical outcomes indicates that the suggested framework not only detects security attacks with an accuracy of 99% but also minimizes the communication overhead.

A novel real-time healthcare system based on ensemble classifier for detection of cyber-attacks has been suggested by Kumar and Bharathi [24]. Initially, the authors have utilized greedy routing technique for the creation and placement of sensor node and an agglomerative mean shift maximization clustering approach for the normalization and grouping of transmitted data. A feature extraction process that makes use of multi-heuristic cyber ant optimization approach is used for the extraction of abnormal features from health data. Then, the suggested framework makes use of XGboost classifier for the detection of security attacks. Ultimately, the findings of the experiment demonstrate that the suggested framework performs better in identifying cyberattacks within the healthcare system.

An inventive security framework based on machine learning approach has been developed by Sundas et al. [25], for the identification of harmful attacks in smart healthcare system. The suggested system observes and compares the vitals of various devices connected to the smart health system in order to differentiate the normal activity from abnormal activity. Moreover, the framework utilizes distinct machine learning approaches such as Random Forest, Artificial Neural Network, K-nearest neighbor and decision tree for the identification of harmful attacks in healthcare systems. Further, the suggested framework has been trained on twelve harmless occurrences collected from eight distinct smart medical devices and the empirical results indicate that suggested frameworks is reliable with success rate and F1-score of 91% and 90% respectively.

Tauqeer et al. [26] have developed a unique method for the identification of cyber-attacks in an IoMT environment that combines three machine learning techniques: SVM, Random Forest, and GBoost. The network and biometric feature-rich WUSTL EHMS 2020 dataset has been used to assess the proposed methodology. To improve the system's performance, preprocessing methods including feature selection and cleaning were first performed to the dataset. With an accuracy of 95.85%, 96.9%, and 96.5%, respectively, the suggested techniques GBoost, SVM and random forest achieved greater performance, according to the empirical results. Table 9.1 lists the numerous studies that have been done on applying machine learning techniques to identify cyberattacks in the healthcare system.

9.3 Proposed Approach

This section describes about the mathematical background and structure of the proposed Deep Neural network model.

9.3.1 Mathematical Background

Deep neural networks are formed by the combination of feedforward neural networks that does not contain the feedback connection. The three significant layers such as input, hidden and output layers are the basic components of the feedforward neural network. The architectural layout of the deep neural network is illustrated in Fig. 9.1.

The preprocessed data is fed into the network through the input layer. The amount of input features that the network receives is equal to the number of neurons in the input layer. Equation (9.1) illustrates how the input layer with "N" input features is represented.

$$X = [x_1, x_2, \ldots, x_N] \tag{9.1}$$

DNN's can have more than one hidden layer. Each of the hidden layers contains units with weights that are used for performing activation processes of the units obtained from the previous layer. The mathematical expression described in Eq. (9.2) represents the mapping function of the neuron in the hidden layer.

$$h(x) = f\left(x^t w + b\right) \tag{9.2}$$

Table 9.1 Various works on the detection of cyber attacks using ML approaches

Author and year	Objective	Dataset	Approach	Results	Observations	References
Gupta et al. (2022)	Intrusion detection in IoMT network	Wustl-ehms-2020	Random Forest with Grid Search	Accuracy = 94.23%, F1 score = 93.8%	The dataset considers only two types of attacks such as data alteration and data spoofing	[27]
Kumar et al. (2021)	For detection of cyber attacks in IoMT network	ToN-IoT dataset	An ensemble approach that makes use of decision tree, random forest and naïve bayes at first level and XGBoost at next level	Attained an accuracy of 96.35%	False Alarm rate is very high	[28]
Zachos et al. (2021)	To detect malicious attacks in IoMT network	TON_IoT	Naïve Bayes, random forest, decision tree, linear regression, SVM and KNN	Decision tree, KNN and random forest performed better when compared with other approaches	Not considered computational overhead on gateway and sensors	[29]
Hady et al. (2020)	For detection of Man-In-The-Middle attacks (MITM) in healthcare system	Real time dataset consisting of 16,000 records of normal and MITM attack packets	Random Forest, ANN, SVM and KNN	ANN attained better performance with AUC score = 92.98%	Dataset is imbalance	[30]
Saba (2020)	For detection of intrusion in smart city hospitals	KDDCup-99 dataset	Bagged Decision Tree, random forest, extra trees, AdaBoost, Stochastic Gradient Boost, SVM, Logistic, CART	Bagged Decision tree obtained better performance with an accuracy of 93.2%	Not considered the security of smart city hospitals	[31]

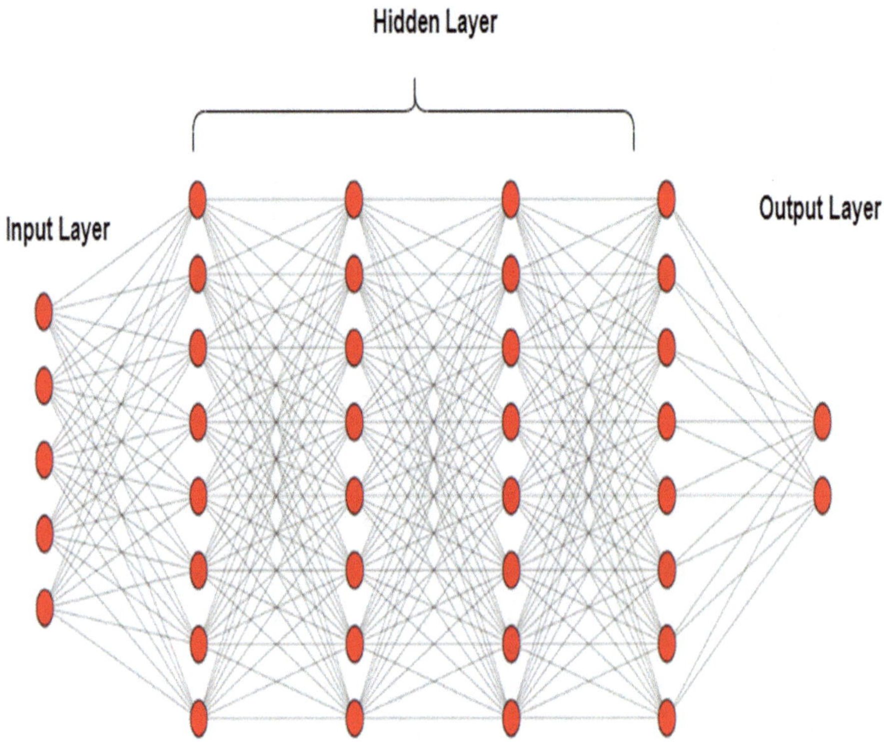

Fig. 9.1 General architectural layout of deep neural network

In the above Eq. (9.1), $h, f, x, w \text{ } and \text{ } b$ are used for representing hidden layer, activation function, input vector, weight vector and bias. Generally, sigmoid, rectified linear unit and hyperbolic tangent function are the typical activation functions used in neural network. As ReLu activation function minimizes vanishing gradient descent problem, it offers better results despite of non-linearity and non-differentiability at zero value when compared with other activation function. Therefore, in the proposed architecture, ReLu is the activation function used at the hidden for obtaining smooth approximation as shown in Eq. (9.3).

$$ReLu(x) = \max(0, x) \tag{9.3}$$

The sigmoid activation function, as demonstrated in Eq. (9.4), is used at the output layer to assign an estimated label to the input data that flows through the network.

$$sigmoid(x) = \frac{1}{e^{-x} + 1} \tag{9.4}$$

The inputs from hidden layer are processed at the output layer through the activation function and produces the outputs of the deep neural network which is represented as shown in Eq. (9.5).

$$sigmoid(X)_j = \frac{e^{X_j}}{\sum_{k=1}^{k} e^{X_k}} \tag{9.5}$$

where vector of inputs transferred to the output layer is represented as 'X' and the number of output units is represented by 'j' and $j = 1, 2, \ldots, k$.

The network training with huge dataset is carried out with the above-mentioned DNN setup using the inputs at the input layer to produce respective class output. Further, the weight of each input neuron is iteratively modified in order to reduce the errors occurred at the training phase.

9.3.2 *Optimization Using Adam Optimizer*

The hyperparameter that impacts the training of the deep neural network is learning rate. Hence, there is need to adopt efficient neural network architecture and parameters to curtail the errors occurred during the training phase. These hyperparameters have direct impact on the performance of the network architecture. In this study, Adam optimizer has been chosen which optimized the hyperparameter using first and second moment estimates of the gradients [32]. The primitive functionality of the Adam optimizer has been depicted in Fig. 9.2.

In the above Fig. 9.2, $f(\theta)$, α, β_1, β_2, θ_t, λ represents the objective function, step size, exponential decay rates, convergence parameter and tolerance parameter respectively. The equations for updating and calculating time step, gradient descent, first and second moment estimates, unbiased first and second moment estimates and objective function parameters of the Adam optimizer are represented in Eq. (9.6) to Eq. (9.12)

$$t \leftarrow t + 1 \tag{9.6}$$

$$g_t \rightarrow \nabla_\theta f_t(\theta_t - 1) \tag{9.7}$$

$$m_t \leftarrow \beta_1 \cdot m_{t-1} + (1 - \beta_1) \cdot g_t \tag{9.8}$$

$$v_t \leftarrow \beta_2 \cdot v_{t-1} + (1 - \beta_2) \cdot g_t^2 \tag{9.9}$$

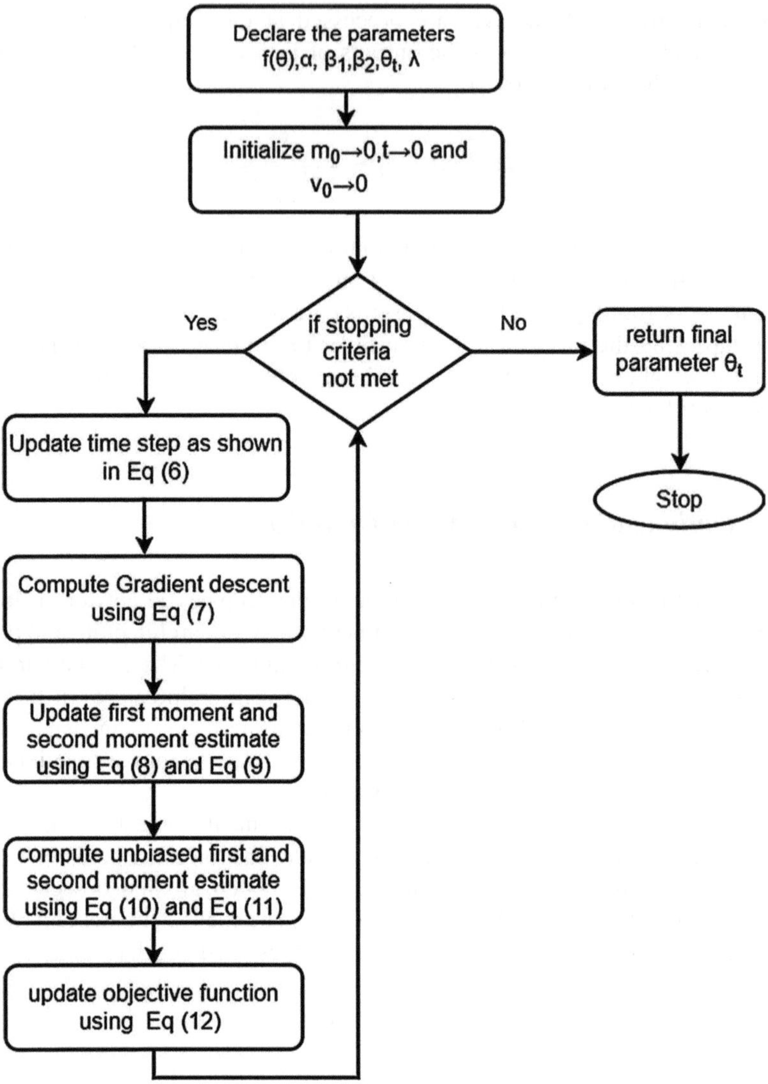

Fig. 9.2 Basic working of Adam optimizer

$$\widehat{m_t} \leftarrow m_t/\left(1-\beta_1^t\right) \tag{9.10}$$

$$\widehat{v_t} \leftarrow v_t/\left(1-\beta_2^t\right) \tag{9.11}$$

$$\theta_t \leftarrow \theta_{t-1} - \alpha \cdot \widehat{m_t}/\sqrt{\left(\widehat{v_t}\right)} + \lambda \tag{9.12}$$

9.4 Description of Dataset and Environmental Setup

This section covers the dataset and environmental setup that were utilized in the experiments utilizing the recommended methodology as well as different machine learning and ensemble learning techniques.

9.4.1 *Environmental Setup*

The suggested approach as well as other machine learning and ensemble learning techniques have been simulated using the following system requirements. The environmental setup consists of HP Pavilion × 360 desktop with Windows 10 64-bit operating system, Intel (R) Core (TM) i7-10510U CPU with a capacity of 2.30 GHz processor and 16 GB RAM. Further, the experiments are carried using Python software. For better analysis of the data, it makes use of Python libraries such as Pandas, Numpy and Imblearn. The visualization of data has been done using matplotlib framework. Additionally, it applies ensemble learning and machine learning techniques by utilizing the sklearn and Mlxtend libraries. Figure 9.3 shows the general framework of the suggested methodology.

9.4.2 *Dataset Description*

In this work, the WUSTL-EHMS-2020 dataset has been used to train and evaluate the suggested DNN strategy in conjunction with other machine learning and ensemble learning techniques. The dataset includes biometric information about patients as well as network flow indicators that were gathered from the real-time enhanced health monitoring system testbed medical sensors, network, gateway and control unit with visualization constitutes the basic components of the EHMS testbed. The data collected from medical sensors connected to patient's body is transferred to the gateway. The gateway then transfers the data to server through router or gateway for visualization purpose. Both the network traffic data and sensor data generated in the testbed is utilized for the detection of threats. In addition, an attack dataset was produced by injecting three attacks in the dataset such as spoofing attack, man-in-the-middle attack and data injection. The ARGUS (Audit Record Generation and Utilization System) tool was used to gather both network traffic and biometric data of patient in the form of csv file [33]. The dataset comprises 16,318 samples in total, of which 14,272 are samples pertaining to regular network records and 2046 samples are samples of network attacks. A total of 44 features were included in the dataset: 35 of these had to do with network traffic, 8 had to do with the biometric data of the patients, and 1 was used as a label feature. The parameters such as temperature, heart rate, pulse rate, systolic blood pressure, diastolic blood pressure, respiration

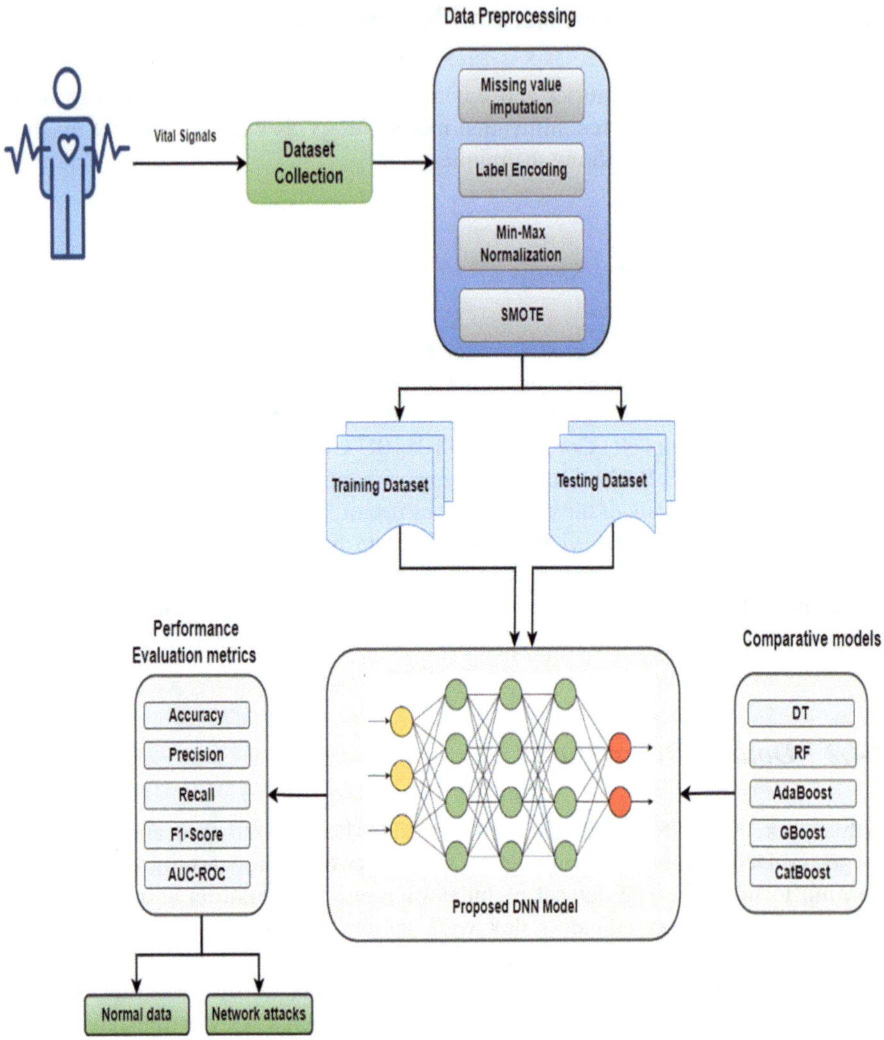

Fig. 9.3 Overall framework of the proposed approach

rate, ECG ST segment data and peripheral oxygen saturation are related to biometric data and the remaining forty-three features are related to network traffic data. The entire dataset is categorized into two distinct classes namely attack data represented with “0” and normal data represented with “1”.

9.4.3 Data Preprocessing

As the application of preprocessing approaches results in enhancing the performance of the system, various preprocessing techniques have been applied on the WUSTL-EHMS-2020 dataset. Initially, missing value imputation has been applied to reduce the missing values. Then label encoding is applied on the target column to convert categorical data into numerical data. The WUSTL-EHMS-2020 dataset considered in the study consists of both continuous and discrete values. Due to the combination of continuous and discrete values, features in the dataset consisting of varying values. To resolve this problem, min-max normalization has been applied which yields more flexibility in the design of neural network models. Moreover, this approach does not insert any bias in the system as it preserves all relationships in the data precisely. Therefore, the application of min-max normalization approach has resulted in the normalization of features by scaling them to suitable range of values for the classifier [34]. As the dataset considered in the present study is imbalance with 2046 network attack samples and 14,272 normal samples, synthetic minority oversampling technique (SMOTE) has been applied. In order to alleviate the imbalance, SMOTE was used to generate synthetic samples from the minority class. Following the application of preprocessing techniques, data is partitioned into an 80:20 ratio, with 80% of the data used for training the model and 20% utilized for validating the model.

9.5 Analysis of Empirical Findings

The several assessment metrics that were utilized to validate the model are shown in this section. Additionally, this section offers an examination of the outcomes produced with the suggested DNN strategy in addition to other ML techniques such as Decision tree (DT), Random Forest (RF) and ensemble learning approaches such as Adaptive Boost (AdaBoost), Gradient Boost (GBoost) and Categorical Boost (CatBoost).

9.5.1 Metrics Used in Validation of Model

In this study, various evaluation measures such as accuracy, F1-score, precision, recall and AUC-ROC results are employed in conjunction with other conventional ML and ensemble learning techniques to validate the effectiveness of the proposed DNN strategy. The mathematical equations used for representing accuracy, F1-score, precision and recall are displayed in Eqs. (9.13)–(9.16).

$$Accuracy = \frac{True\ Positive + True\ Negative}{Total\ no.\ of\ samples} \tag{9.13}$$

Table 9.2 Parameter values used in proposed DNN approach

Number of hidden layers	4
No. of neurons in hidden layer	128
Activation function in hidden layer	ReLu
Learning rate in optimizer	0.01
Optimizer	Adam
Output layer activation function	Sigmoid

$$F1 - score = \frac{2 * (recall * precision)}{recall + Precision} \tag{9.14}$$

$$Precision = \frac{True\ Positive}{Total\ predcited\ Postive} \tag{9.15}$$

$$Recall = \frac{True\ Positive}{Total\ actual\ Positive} \tag{9.16}$$

In Eq. (9.15), Total predictive positive is total no. of True positive + False Positive samples, whereas in Eq. (9.16) Total actual positive is total no. of True Positive + False Negative samples.

9.5.2 Comparative Assessment of Findings

This section compares the proposed DNN model's performance against that of existing conventional machine learning and ensemble learning algorithms such as DT, RF, AdaBoost, GBoost, and CatBoost. Tables 9.2 and 9.3 show the parameters used in training the suggested DNN technique as well as other approaches.

Furthermore, K-fold cross validation has been used to validate the suggested DNN model in conjunction with further ML and ensemble techniques. K-1 folds are used for model training in K-fold validation, while the remaining data is used for model

Table 9.3 Parameter values used in other ML and Ensemble approaches

Model name	Parameters
RF	*max_features = 'sqrt', criterion = 'gini', n_estimators = 50*
DT	*min_samples_leaf = 1,criterion = 'gini', min_samples_split = 2, splitter = 'best'*
AdaBoost	*learning_rate = 1.0, algorithm = 'SAMME.R', n_estimators = 50*
CatBoost	*n_estimators = 50*
GBoost	*min_samples_split = 2, subsample = 1.0, max_depth = 3, loss = 'log_losz', n_ estimators = 50,* *learning_rate = 0.1, min_samples_leaf = 1, criterion = 'friedman_mse'*

testing. In a similar manner, the procedure is repeated K times, with the end result serving as the cross-validation result. The WUSTL-EHMS-2020 dataset is split into tenfolds for this investigation. Table 9.4 shows the outcomes of the DNN model's tenfold cross-validation as well as other alternative methods. Table 9.4 shows that the suggested DNN model outperformed previous approaches in terms of cross validation accuracy.

Table 9.5 depicts a comparison of several evaluation metrics such as precision, recall, F1-score, AUC-ROC, and accuracy utilised in assessing the suggested DNN and other established techniques. From the Table 9.5, it is noticed that proposed DNN model surpassed other considered approaches with a precision of 0.9999, recall of 1.0, F1-score of 1.0, AUC-ROC of 0.9999 and accuracy of 100% respectively. From all the models, AdaBoost model obtained lowest accuracy of 98.38%. The other models DT, RF, GBoost, CatBoost obtained an accuracy of 99.42%, 99.35%, 99.07%, 99.98% respectively. Table 9.5 further demonstrates that, in comparison to machine learning approaches, ensemble approaches performed better. This is mainly because of combination of multiple models in ensemble approaches which results in minimizing the variance and bias of the model. Moreover, the proposed DNN model attained superior performance over ensemble approaches because of their capability in optimizing features while extracting.

The confusion matrix of the suggested DNN model and other models under consideration is shown in Fig. 9.4a–f. From the Fig. 9.4a, it is observed that in decision tree out of 4276 network attack samples, 4227 samples were correctly classified as network attack samples. 49 Samples of normal data were incorrectly classified as network attack samples. All 4288 samples of normal data are correctly classified. It is observed from Fig. 9.4b, random forest correctly classified 4223 and 4286 samples of network attack and normal data. 53 samples of normal data and 2 samples of network data were incorrectly classified in random forest model. From the confusion matrix of AdaBoost in Fig. 9.4c it is observed that all samples of network attack data are

Table 9.4 Results of tenfold cross validation

No. of stratified folds (SFd)	DT	RF	AdaBoost	GBoost	CatBoost	Proposed DNN
SFd1	99.42	99.35	98.38	99.07	99.98	100.0
SFd2	99.45	99.38	98.31	99.12	99.93	100.0
SFd3	99.40	99.32	98.45	99.05	99.96	99.99
SFd4	99.31	99.24	98.33	99.07	99.99	100.0
SFd5	99.42	99.55	98.32	99.08	99.98	99.99
SFd6	99.43	99.32	98.38	99.06	99.92	100.0
SFd7	99.45	99.35	98.27	99.18	99.98	100.0
SFd8	99.42	99.38	98.40	99.10	99.93	99.99
SFd9	99.40	99.34	98.34	99.07	99.96	100.0
SFd10	99.42	99.32	98.32	99.12	99.93	99.99

Table 9.5 Evaluation metric of the suggested DNN and other considered models

Classification model	Precision	Recall	F1 score	ROC-AUC	Accuracy (%)
DT	0.9887	1.0	0.9943	0.9942	99.42
RF	0.9877	0.9995	0.9936	0.9935	99.35
AdaBoost	1.0	0.9678	0.9836	0.9839	98.38
GBoost	1.0	0.9815	0.9907	0.9907	99.07
CatBoost	0.9997	1.0	0.9998	0.9998	99.98
Proposed DNN	0.9999	1.0	1.0	0.9999	100.0

correctly classified. Out of 4288 normal samples, only 4150 samples were correctly classified as normal data samples. 138 samples of network attack data are incorrectly classified as normal data samples. From Fig. 9.4d, It is noticed that GBoost was able to classify correctly all samples of network attack data. Whereas only 4209 samples of normal data out of 4288 samples were correctly classified and 79 samples of network attack data were incorrectly classified as normal data samples. From Fig. 9.4e, it is noticed that 4275 samples of network attack data and 4288 samples of normal data are correctly classified. Only one sample from network attack data was incorrectly classified as normal data sample. Finally, the confusion matrix of proposed DNN model in Fig. 9.4f represents that all 4276 and 4288 samples of network attack data and normal samples were correctly classified as network attack and normal data samples.

The AUC-ROC curve results for the DT, RF, AdaBoost, GBoost, CatBoost, and suggested DNN model are shown in Figs. 9.5, 9.6, 9.7, 9.8, 9.9 and 9.10. Figure 9.10 shows that, in comparison to other traditional methods, the suggested DNN model achieved an AUC-ROC of 1.00 for both network attack and normal data class labels. Additionally, the suggested DNN approach's macro- and micro-average ROC curve values were both identical to 1.00, indicating that every occurrence was correctly identified. This suggests that, in comparison to other approaches, the recommended method can examine every case in the data.

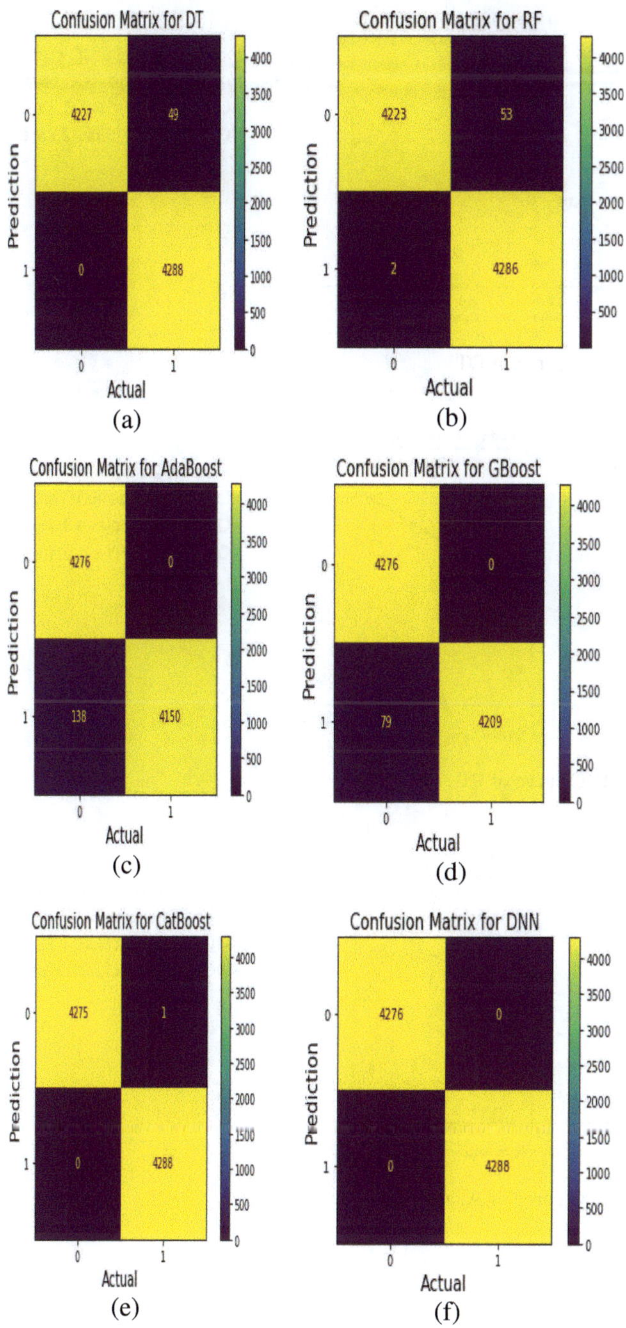

Fig. 9.4 Confusion matrix of **a** DT, **b** RF, **c** AdaBoost, **d** GBoost, **e** CatBoost, **f** proposed DNN

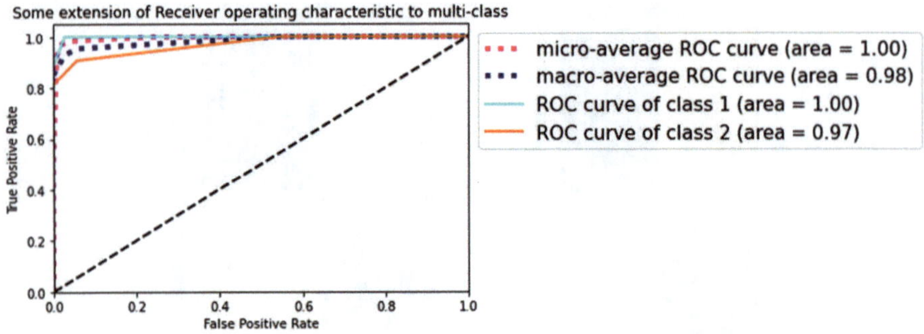

Fig. 9.5 AUC-ROC curve of DT

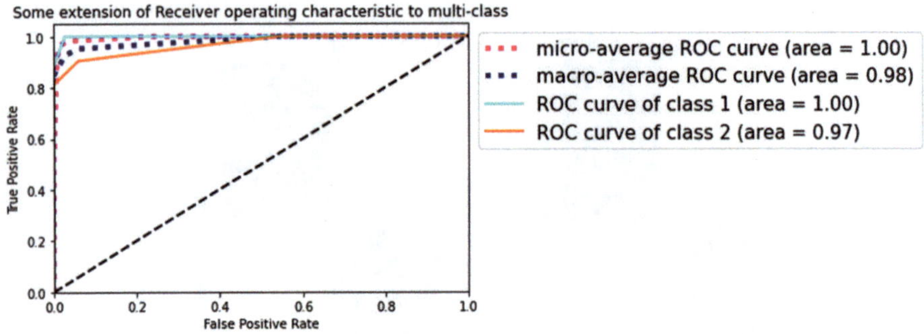

Fig. 9.6 AUC-ROC curve of RF

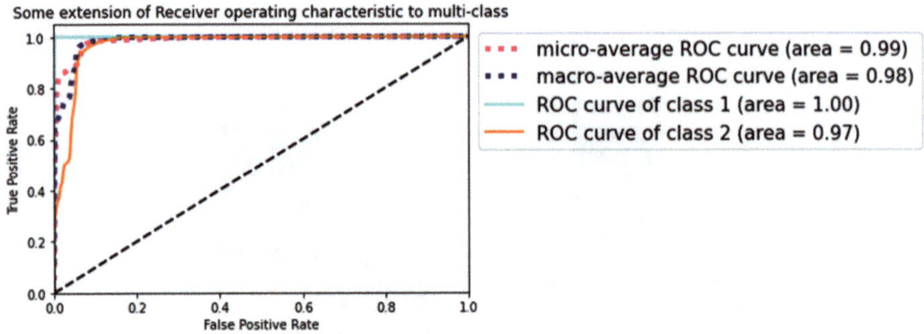

Fig. 9.7 AUC-ROC curve of Adaboost

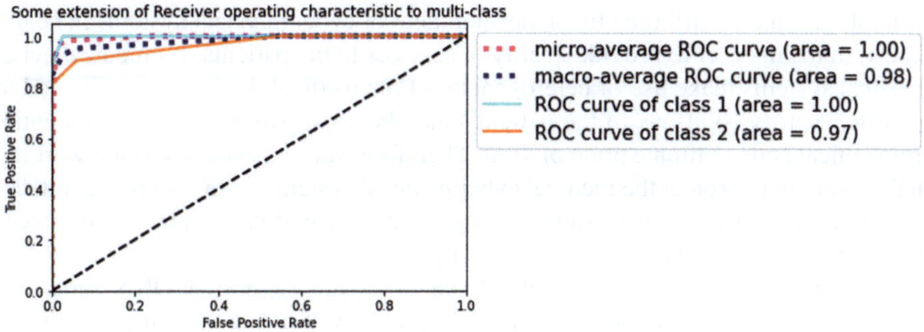

Fig. 9.8 AUC-ROC curve of GBoost

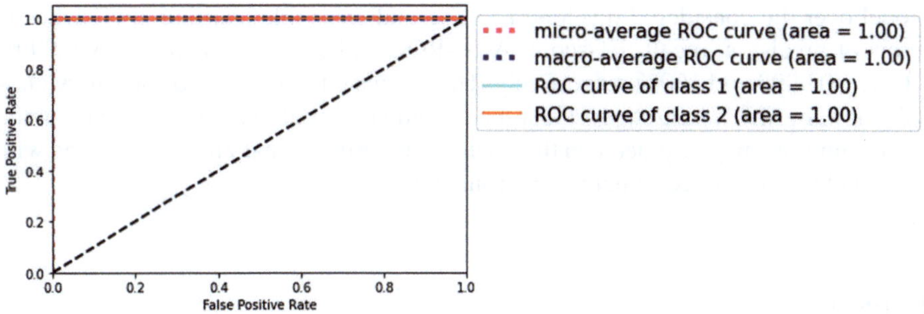

Fig. 9.9 AUC-ROC curve of CatBoost

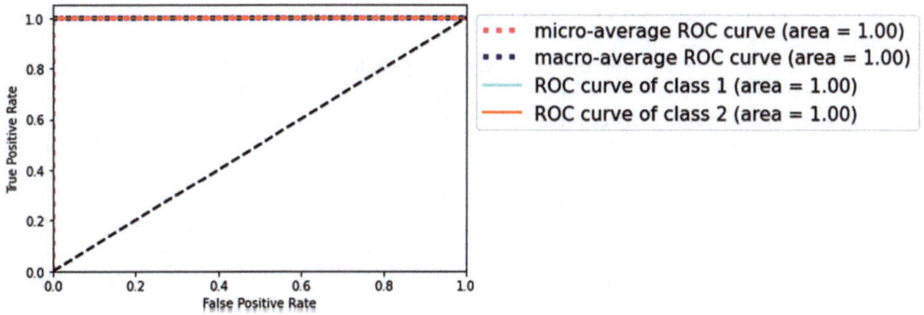

Fig. 9.10 AUC-ROC curve of proposed DNN model

9.6 Conclusion

From the past few decades, the cost of healthcare services has been tremendously increased due to the advancement in technology as well as growing population all over the world. In addition, the rapid advancement in IoT technology has lead to the monitoring and diagnosing of patients from remote locations. The integration of IoT technologies has contributed to the development of cyber physical systems in health care systems in order to provide quality of services to the patients. As medical cyber physical systems make use of heterogeneity of the medical devices, there is need to provide security solutions to the patient's data because patient data is confidential from ethical and legitimate point of view. Therefore, various researchers are working in this domain to protect the medical cyber physical systems from unforeseen attack. This study proposes deep learning assisted attack detection framework for safely transferring patient's data in medical cyber physical system.

This research has shown the effectiveness of the suggested DNN model in detecting cyberattacks in the healthcare system, as well as the outcomes of several machine learning and ensemble learning techniques, including DT, RF, AdaBoost, GBoost, and CatBoost approaches. The experiment has been carried with WUSTL-EHMS-2020 dataset and the results demonstrate the superiority of the proposed DNN model over the considered machine learning and ensemble learning approaches in terms of precision, recall, F1-score, AUC-ROC and accuracy with values 0.9999, 1.0, 1.0, 0.9999 and 100% respectively. The suggested work is used for binary classification. Further research work can be extended on multiclass categorization and more emphasis may be placed on the security and privacy challenges associated with several cloud/fog-based dynamic environments.

References

1. Murguia, C., van de Wouw, N., Ruths, J.: Reachable sets of hidden cps sensor attacks: analysis and synthesis tools. In: IFAC-PapersOnLine 50.1, pp. 2088–2094 (2017)
2. Jha, A.V., et al.: Smart grid cyber-physical systems: Communication technologies, standards and challenges. Wirel. Netw. **27**, 2595–2613 (2021)
3. Habibzadeh, H., et al.: A survey on cybersecurity, data privacy, and policy issues in cyber-physical system deployments in smart cities. Sustain. Cities Soc. **50**, 101660 (2019)
4. Atat, R., et al.: Enabling cyber-physical communication in 5G cellular networks: challenges, spatial spectrum sensing, and cyber-security. IET Cyber Phys. Syst. Theory Appl. **2**(1), 49–54 (2017)
5. AlZubi, A.A., Al-Maitah, M., Alarifi, A.: Cyber-attack detection in healthcare using cyber-physical system and machine learning techniques. Soft Comput. **25**(18), 12319–12332 (2021)
6. Ahmed, A.A., Nazzal, M.A., Darras, B.M.: Cyber-physical systems as an enabler of circular economy to achieve sustainable development goals: a comprehensive review. Int. J. Precis. Eng. Manuf. Green Technol. 1–21 (2021)
7. Rajawat, A.S., et al.: Cyber physical system fraud analysis by mobile robot. Machine Learning for Robotics Applications, pp. 47–61 (2021)
8. Haque, S.A., Aziz, S.M., Rahman, M.: Review of cyber-physical system in healthcare. Int. J. Distr. Sensor Netw. **10**(4), 217415 (2014)
9. Dey, N., et al.: Medical cyber-physical systems: a survey. J. Med. Syst. **42**, 1–13 (2018)
10. Sliwa, J.: Assessing complex evolving cyber-physical systems (case study: Smart medical devices). Int. J. High Perform. Comput. Netw. **13**(3), 294–303 (2019)
11. Nagarhalli, T.P., Vaze, v., Rana, n.k.: Impact of machine learning in natural language processing: a review. In: 2021 Third International Conference on Intelligent Communication Technologies and Virtual Mobile Networks (ICICV). IEEE (2021)
12. Nahid, A.Al, Kong, Y.: Involvement of machine learning for breast cancer image classification: a survey. Comput. Math. Meth. Med. (2017)

13. Vashisht, V., Pandey, A.K., Yadav, S.P.: Speech recognition using machine learning. IEIE Trans. Smart Process. Comput. **10**(3), 233–239 (2021)
14. Singh, J., Singh, J.: A survey on machine learning-based malware detection in executable files. J. Syst. Architect. **112**, 101861 (2021)
15. Alzahrani, A., et al.: Improved wireless medical cyber-physical system (IWMCPS) based on machine learning. Healthcare **11**(3). MDPI (2023)
16. Kilincer, I.F., et al.: Automated detection of cybersecurity attacks in healthcare systems with recursive feature elimination and multilayer perceptron optimization. Biocybernet. Biomed. Eng. **43**(1), 30–41 (2023)
17. Halman, L.M., Alenazi, M.J.F.: MCAD: a machine learning based cyberattacks detector in software-defined networking (SDN) for healthcare systems. IEEE Access (2023)
18. Maithem, M., Al-Sultany, G.A.: Network intrusion detection system using deep neural networks. J. Phys. Conf. Ser. **1804**(1). IOP Publishing (2021)
19. Cil, A.E., Yildiz, K., Buldu, A.: Detection of DDoS attacks with feed forward based deep neural network model. Expert Syst. Appl. **169**, 114520 (2021)
20. Tang, T.A., et al.: Deep learning approach for network intrusion detection in software defined networking. In: 2016 International Conference on Wireless Networks and Mobile Communications (WINCOM). IEEE (2016)
21. Li, D., et al.: Hashtran-dnn: a framework for enhancing robustness of deep neural networks against adversarial malware samples (2018). arXiv:1809.06498
22. Elsisi, M., Tran, M.-Q.: Development of an IoT architecture based on a deep neural network against cyber attacks for automated guided vehicles. Sensors **21**(24), 8467 (2021)
23. Schneble, W., Thamilarasu, G.: Attack detection using federated learning in medical cyber-physical systems. In: Proceedings of 28th International Conference on Computing Communication Networks (ICCCN), vol. 29 (2019)
24. Kumar, C.N.S.V.: A real time health care cyber attack detection using ensemble classifier. Comput. Electr. Eng. **101**, 108043 (2022)
25. Sundas, A., et al.: HealthGuard: an intelligent healthcare system security framework based on machine learning. Sustainability **14**(19), 11934 (2022)
26. Tauqeer, H., et al.: Cyberattacks detection in IoMT using machine learning techniques. J. Comput. Biomed. Informatics **4**(01), 13–20 (2022)
27. Gupta, K., et al.: A tree classifier based network intrusion detection model for Internet of Medical Things. Comput. Electr. Eng. **102**, 108158 (2022)
28. Kumar, P., Gupta, G.P., Tripathi, R.: An ensemble learning and fog-cloud architecture-driven cyber-attack detection framework for IoMT networks. Comput. Commun. **166**, 110–124 (2021)
29. Zachos, G., et al.: An anomaly-based intrusion detection system for internet of medical things networks. Electronics **10**(21), 2562 (2021)
30. Hady, A.A., et al.: Intrusion detection system for healthcare systems using medical and network data: a comparison study. IEEE Access **8**, 106576–106584 (2020)
31. Saba, T.: Intrusion detection in smart city hospitals using ensemble classifiers. In: 2020 13th International Conference on Developments in eSystems Engineering (DeSE). IEEE (2020)
32. Yazan, E., Fatih Talu, M.: Comparison of the stochastic gradient descent based optimization techniques. In: 2017 International Artificial Intelligence and Data Processing Symposium (IDAP). IEEE (2017)
33. Argus. https://openargus.org. Accessed 14 Nov 2023
34. Priddy, K.L., Keller, P.E.: Artificial Neural Networks: An Introduction, vol. 68. SPIE Press (2005). https://doi.org/10.1117/3.633187

Chapter 10
Risk Assessment and Security of Industrial Internet of Things Network Using Advance Machine Learning

Geetanjali Bhoi, Rajat Kumar Sahu, Etuari Oram, and Noor Zaman Jhanjhi

Abstract Securing IIoT networks is crucial for maintaining seamless operations, safeguarding sensitive industrial data, and averting safety risks. It helps managing financial exposure, protects intellectual property, and ensures compliance with regulations. Due to interconnected nature of IIoT devices, the looming threat of cyber incidents that could disrupt industries and supply chains. Machine learning is crucial for securing IIoT networks through tasks such as anomaly detection, predictive analytics, and adaptive threat response. By analyzing extensive datasets, it identifies patterns, detects deviations from normal behavior, and proactively addresses potential security threats, thereby fortifying the resilience and efficacy of IIoT network defenses. In this study, an optimized Gradient Boosting Decision Tree based model has been trained on a IIOT data to identify anomalies pattern and normal behavior. The trained model is tested and found efficient as compare to many machine learning model.

Keywords IIOT · Anomaly detection · Gradient boosting decision tree · Gravitational search algorithm

G. Bhoi (✉) · R. K. Sahu · E. Oram
Department of Computer Application, Veer Surendra Sai University of Technology, Burla, Odisha 768018, India
e-mail: gbhoi_phdca@vssut.in

R. K. Sahu
e-mail: rksahuphdca23@vssut.ac.in

E. Oram
e-mail: eoram_mca@vssut.ac.in

N. Z. Jhanjhi
School of Computer Science, Taylor's University, Subang Jaya, Malaysia
e-mail: noorzaman.jhanjhi@taylors.edu.my

J. Nayak et al. (eds.), *Machine Learning for Cyber Physical System: Advances and Challenges*, Intelligent Systems Reference Library 60,
https://doi.org/10.1007/978-3-031-54038-7_10

10.1 Introduction

The Industrial Internet of Things (IIoT) is often described as a revolution that is fundamentally changing how business is conducted. However, it is actually a progression that began more than 15 years ago with technology and features created by forward-thinking automation vendors. The full potential of the IIoT may not be realized for another 15 years as global standards continue to develop. The changes to the industry during this time will be significant, but the good news is that machine builders can now maximize their returns by combining new IIoT technologies with their current investments in people, end-users, and technology. As part of the Internet of Things (IoT), the IIoT is also known as Industry 4.0. According to current estimates, industrial IoT will continue to rise exponentially. As we approach a world with more than 75 billion connected devices by 2025, about a third will be used in manufacturing-related industrial applications. By connecting industrial machines and devices, manufacturing and industrial processes can be improved using the IIoT. Data analytics can be achieved by monitoring, collecting, exchanging, and analyzing large amounts of data using IIoT applications. In turn, companies will be able to make more informed, data-driven decisions in their business operations. While IoT and IIoT share similar basic principles, their purposes differ. IoT is about connecting physical objects to the internet, such as smart devices, vehicles, home appliances, and more. Agriculture, transportation, manufacturing, gas and oil, and other businesses are using the IIoT to connect devices and machines. Among the IIoT devices in this network are sensors, controllers, industrial control systems, and other connected devices used for monitoring productivity and assessing machine performance. The combination of edge computing and actionable insights from analytics allows machines to do autonomous or semi-autonomous activities without the need for human intervention at a speed that is unimaginably faster than humans.

Today, industry is experiencing a number of technology trends driven by the IIoT. When this technology gains momentum, a whole new industry will be created. As a result, industries worldwide will be able to benefit from a data-driven, digital world in the future. The widespread embrace of the IIoT is anticipated to surge considerably with the expanding count of interconnected devices. A major goal of the IIoT is to provide real-time information about processes and efficiency through the connection of devices and machines. IIoT devices connected to sensors collect and store a large amount of data. A business can then make data-driven decisions with the help of this data, which is then transformed into actionable insights.

Industrial IoT includes sensor-driven computing, data analytics, and intelligent machine applications with the goals of scalability, efficiency, and interoperability. The integration of this technology allows for automation of critical infrastructure, which increases business efficiency [1]. Even with improvements in productivity, there are still issues that need to be resolved, chief among them being the critical security of industrial infrastructure and its elements. Cyberattacks on vital industries and infrastructure are becoming more frequent, which presents a serious risk and can result in large losses. As such, it is critical to learn from these events and recognize that industries are becoming easy targets for cybercriminals. It becomes imperative that IIoT security issues be resolved. The confluence of information technology (IT) with operational technology (OT) is a common definition of IIoT. Whereas OT deals with the plant network where production takes place, IT handles the enterprise network. To avoid security breaches in IIoT infrastructure, these two components have different security requirements that need to be carefully taken into account.

A common paradigm in the field of information technology security is the client-server model, in which protocols such as Transmission Control Protocol (TCP), Internet Protocol (IP), Hypertext Transfer Protocol (HTTP), or User Datagram Protocol (UDP) are used to facilitate communication between these entities. Successful attacks in this domain typically result in financial or reputational damage, with safety threats being a rare occurrence [2]. On the contrary, OT systems were initially designed to ensure the safe and reliable operation of industrial processes. In contrast to IT systems, security considerations were not initially integrated into the conception of OT components and subsystems. To counter this, security measures for OT include isolating OT networks and implementing physical security measures.

However, these security controls exhibit unreliability due to inherent loopholes that can be exploited for attacks. While isolating OT networks may serve to thwart external network-based attacks, it proves inadequate in preventing threats originating within the network itself. In an isolated network, the deployment of malware becomes a potent strategy for compromising the system. Consequently, there is a pressing need to delve into the examination of potential attacks at different levels of the IIoT architecture.

Cyber attackers engage in the theft or destruction of information within computers and network structures, fuelling the landscape of cyberwarfare. Various attack types, such as data theft, botnet, Man-in-the-Middle (MitM), network scanning, port-sweep, port-scan attacks, and address-sweep contribute to the vulnerability of systems. The diverse array of IoT devices introduces the risk of unsecured connections, both Machine-to-People (M2P) and Machine-to-Machine (M2M), providing hackers with easy access to crucial information. This not only infringes upon privacy and network space usage but also leads to operational disruptions and financial losses [3]. Importantly, there have been cyberattacks against industrial IoT systems, as evidenced by the well-known attack on the Ukrainian power grid in 2015. In one instance, nearly 230,000 subscribers' electricity was interrupted due to cybercriminals gaining remote access to the control unit [4]. Similarly, an attack that targeted a Taiwanese chip factory with an IIoT network in 2018 caused damages estimated to be at USD 170 million [5] in losses of over USD 170 million [6]. Even with the inevitable trend toward greater reliance on automation and digitalization, industrial enterprises are actively looking for improved ways to fortify their IIoT networks. The financial consequences are significant: IIoT firms who do not put in place appropriate mitigation techniques against cyber-attacks on their networks could end up spending up to USD 90 trillion by 2030 [7].

An Intrusion Detection System (IDS) is essential for protecting the privacy and information integrity of transmitted data and for strengthening the security of the IIoT network. Its main goal is to automatically identify, record, address, and stop any malevolent or intrusive activity that could compromise the security of an IIoT network [8]. An intrusion detection system's efficacy is determined by how well it can identify attacks with a high degree of precision and a low false-positive rate [3]. Additionally, an IDS should excel in identifying the initiation of probing activities by hackers, a vital step in establishing a secure IIoT environment [9].

Smart data generation is a new factor in a synthetic net of outcomes IIoT. Utmost diligence is to try and automate the system of developing and producing products. Using Machine Learning (ML) in Industry 4.0 is a crucial factor to take advantage of the IIoT. Due to the added scale of stationed outstation IoTs in synthetic, the IIoT becomes miscellaneous, distinctive, and stoutly changeable. An IIoT device typically consists of information processing technology, a phase of wise manipulation, and a community conversation generator. ML uses cunning techniques to virtually attach the real global. Many businesses use machine learning methods and algorithms to save operating and product costs. The use of ML to identify and detect malicious activity within target networks has gained more attention in recent times. This technology is particularly appealing for next-generation IoT networks due to its ability to strike a balance between performance efficiency and computational costs. Researchers have made significant strides in developing advanced IDS methods leveraging ML techniques, resulting in promising outcomes in terms of attack detection performance [10]. However, a significant challenge linked to current IDS datasets lies in their considerable size, encompassing both the quantity of network traces and the dimensions of the feature space. Additionally, an uneven distribution of sample numbers for each sort of assault plagues the IDS dataset. This imbalance has posed a barrier for previous ML or deep learning (DL) models, hindering their ability to attain high performance in detecting specific attack types. In this chapter, an EL based model is designed to detect anomalies in IIoT network. Here, gradient boosted decision tree is used with its optimized hyperparameters using gravitational search algorithm. The remaining contents are organized in to following sections: literature survey presented in Sects. 10.2 and 10.3 presents methodology used in the proposed model followed by Result analysis and Conclusion in Sects. 10.4 and 10.5 respectively.

10.2 Literature Survey

Gao et al. study [11] delves into noncoherent maximum likelihood detection in large-scale SIMO systems for industrial IoT communication. Their proposed scheme focuses on optimizing power consumption and reducing latency, resulting in an energy-efficient and low-latency method for massive SIMO using noncoherent ML detection. Through simulations, the authors prove that their proposal surpasses existing methods in terms of energy efficiency and latency. This study is a crucial resource for IoT communication professionals and researchers, particularly those operating in industrial environments. Zolanvari et al. [12] explore how imbalanced datasets impact the security of industrial IoT systems that utilize machine learning algorithms. Their research analyses the effects of class imbalance on classification accuracy and highlights the significance of balanced datasets in achieving precise outcomes. The authors present an oversampling method, Synthetic Minority Oversampling Technique (SMOTE), as a solution for addressing class imbalance. They demonstrate through experimentation that SMOTE can improve the classification accuracy of minority classes, thus enhancing the overall security of industrial IoT systems. The study offers a valuable resource for professionals and researchers aiming to employ machine learning for securing industrial IoT systems. Zolanvari et al. [13] introduce a novel ML approach to examine network vulnerabilities in industrial IoT systems. Their multi-stage framework incorporates anomaly detection, vulnerability assessment, and risk analysis to pinpoint security threats in IoT networks. Through experimentation with a network traffic dataset, the authors illustrate the high effectiveness of their approach in identifying vulnerabilities and categorizing them according to their level of risk. The article provides a critical resource for researchers and professionals working on IoT security, especially those concentrating on the analysis of network vulnerabilities in industrial environments. In their work, Latif et al. [14] suggest a novel approach for detecting attacks in the IIoT using a lightweight random neural network. The proposed scheme employs an unsupervised learning algorithm that utilizes temporal correlations in data to detect anomalies caused by malicious activities. The system is assessed using a readily accessible dataset, revealing enhanced performance in terms of both detection computational complexity and accuracy when compared to existing methods. The study provides important findings for those working in the field of IIoT security and suggests a promising approach for detecting attacks in these systems.

In their publication, Mudassir et al. [15] introduce a new method for identifying botnet attacks in IIoT systems by utilizing multilayer deep learning strategies. The authors emphasize the importance of their research in tackling security risks associated with IIoT systems and showcase the efficiency of their approach via experimental studies. In their research article, Qolomany et al. [16] introduce a novel strategy to improve Federated Learning (FL) in IIoT and Smart City services. The authors integrate Particle Swarm Optimization (PSO) into their proposed method to optimize model accuracy and minimize communication overhead. By conducting empirical evaluations and comparing their approach with conventional FL and non-FL techniques, the researchers demonstrate the effectiveness of their approach. The study underscores the potential of their approach in resolving FL-related challenges in IIoT and Smart City applications. In their paper, Ksentini et al. [17] introduce an innovative Fog-enabled IIoT network slicing model that leverages Multi-Objective Optimization (MOO) and ML techniques. The proposed approach considers both Quality of Service (QoS) requirements and resource limitations while slicing IIoT networks. By combining MOO with ML, the authors optimize the network slicing process and enhance performance. Experimental evaluations of the method on real-world scenarios demonstrate its superiority over traditional techniques. The study underscores the potential of their approach to improve IIoT network slicing, boost network efficiency, and provide insights for system designers and developers. The research presents a significant contribution to the field of IIoT network slicing. In their research, Marino and his team [18] propose a distributed system for detecting faults in IIoT using machine learning techniques. The system uses data-driven models generated from sensor readings to achieve scalable fault detection quality. The team demonstrated the effectiveness of their approach in detecting faults in an industrial pump system, achieving high detection rates while minimizing false positives. The study emphasizes the potential of machine learning-based fault diagnosis systems for the industrial IoT industry.

In their study, Taheri and colleagues [19] present a federated malware detection architecture, FED-IIoT, designed for Industrial IoT (IIoT) systems. The proposed architecture operates on a collaborative model that permits sharing and processing of data across multiple IIoT networks while guaranteeing data privacy and security. The authors assess the effectiveness of their approach using real-world datasets, showcasing its superiority over existing centralized and distributed detection approaches in terms of accuracy and detection rates. The research emphasizes the significance of a robust and secure federated approach for malware detection in IIoT systems. In their work, Yazdinejad et al. [20] put forward an ensemble deep learning model designed to detect cyber threats within the IIoT framework. The authors emphasized that IIoT systems are highly vulnerable to malicious attacks, with potentially catastrophic consequences. In tackling this challenge, the suggested model employs a blend of a CNN (Convolutional Neural Network), a RNN (Recurrent Neural Network), and a LSTM (Long Short-Term Memory) network to detect and highlight suspicious activities. The model underwent evaluation using a publicly accessible dataset, demonstrating superior performance compared to conventional machine learning methods in both accuracy and speed. Le et al. [21] explored the application of the XGBoost algorithm to enhance the accuracy of IDS in the context of IIoT, specifically in scenarios involving imbalanced multiclass classification. The authors argued that detecting cyber-attacks on IIoT systems is crucial for maintaining sustainability and avoiding environmental damage. The XGBoost model was evaluated on a publicly available dataset, and it outperformed other ML techniques in terms of F1-score and overall accuracy. Mohy-Eddine et al. [22] presented an intrusion detection model for IIoT systems that leverages ensemble learning techniques. The authors highlighted the need for robust threat detection mechanisms to protect IIoT systems from cyber-attacks. Their proposed model combines multiple machine learning algorithms to detect unusual patterns of behaviour. The model was tested on a real-world IIoT dataset, and its performance surpassed that of other machine learning approaches in terms of detecting malicious activity while minimizing false-positive alerts. Rashid et al. [23] introduced an innovative approach to enhance intrusion detection in IIoT networks by employing federated learning. The authors argued that conventional intrusion detection techniques are inadequate for identifying sophisticated cyber-attacks on IIoT systems. Their proposed model takes advantage of the distributed learning capabilities of edge devices to enhance detection accuracy while safeguarding data privacy. The model's efficacy was evaluated with a publicly accessible dataset, and it demonstrated superior performance compared to current state-of-the-art methods in terms of accuracy. Rafiq and colleagues [24] introduced an innovative technique to thwart evasion attacks by malicious actors in IIoT networks. The authors highlighted the inadequacy of conventional intrusion detection systems for detecting such attacks. Their proposed approach utilizes a dynamic and adaptive strategy to detect and respond to potential threats. The effectiveness of the approach was assessed using an openly accessible IIoT dataset, with the results showcasing superior detection accuracy when compared to current methods.

In this chapter, a model based on EL is crafted for the detection of anomalies in IIoT networks. The model employs a gradient-boosted decision tree with optimized hyperparameters achieved through the gravitational search algorithm.

10.3 Methodology

In this section, the detail implementation of Gradient Boosting Decision Tree (GBDT) based model for malicious access detection and its hyperparameter optimization using gravitational search algorithm (GSA) is discussed. Further, the detail on working procedure of GBDT and GSA have been presented.

10.3.1 Gradient Boosting Decision Tree

EL is one of the ML approaches in which multiple models (homogeneous/heterogeneous) are combined to do a specific machine learning task such as classification, regression etc. It also refers to creating a strong model out of combing multiple weak models (Fig. 10.1). Gradient boosting decision tree (GBDT) [25] comes under family of boosting algorithm where each weak model (usually a single decision tree) learns from the error of previous weak model. Here, a model is added over another model iteratively to create a strong model. The basic steps involved in GBDT are as follows:

i. Construction of initial model (often a single decision tree) from the training data which is refers as a weak learner.
ii. Making prediction by using initial model on the training data.
iii. Identifying variations between predicted values and actual targets, referred to as residuals, through the utilization of a differentiable loss function. Calculate the residuals by subtracting the predicted values from the true values. This

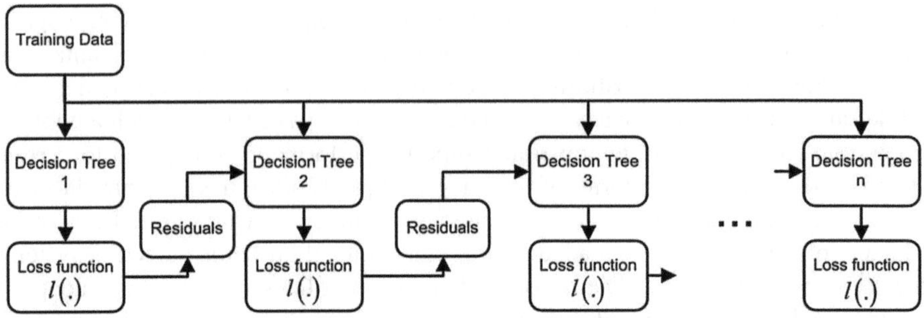

Fig. 10.1 Gradient boosting decision tree working schema

adjustment is accomplished using the negative gradient of the employed loss function.

iv. The adjusted model is used for learning rate adjusted predictions and model is updated.
v. Repetition of step 1 to 4 by adding model sequentially and incrementally over previous model. Repetitions are continued for the selected number of estimators (models).
vi. The final prediction is made by using sum of prediction of all added models (weak learners).
vii. Regularization techniques are used to control the depth of the decision tree using selected subsample of data which helps to avoid overfitting.
viii. Finally, the ensemble gradient boosted weak models are used for making final prediction on the testing data.

10.3.2 Gravitational Search Algorithm

The Gravitational Search Algorithm (GSA) is a metaheuristic optimization method developed by Rashedi et al. [26]. The GSA is inspired by the fundamental principles of gravitational forces and motion. In this algorithm, candidate solutions are represented as masses, and the gravitational forces between these masses are used to update their positions iteratively. The algorithm aims to find optimal solutions by mimicking the gravitational interactions that occur in celestial bodies. In the proposed methodology, the algorithm conceptualizes agents as objects, evaluating their performance based on their masses. The interaction among these objects is governed by gravitational forces, leading to a collective motion where all objects gravitate toward those with heavier masses. This gravitational force serves as a direct means of communication, facilitating cooperation among masses. Notably, masses representing superior solutions move at a slower pace than lighter ones, ensuring a focus on exploiting promising solutions. Within the GSA, each mass, or agent, is characterized by four attributes: passive gravitational mass, active gravitational mass, position, and inertial mass. The position signifies a potential solution to the problem, and the calculation of gravitational and inertial masses is achieved through a fitness function. Essentially, each mass encapsulates a solution, directing the algorithm in the iterative process of refining gravitational and inertial masses. Over time, the expectation is that masses will converge toward the heaviest mass, representing an optimal solution within the search space. The fundamental steps of the GSA, can be outlined as follows:

i. Randomly generate a set of initial solutions, where each solution is represented as a mass in the search space.
ii. Evaluate each solution' fitness to determine its mass based on a fitness function. Calculate the gravitational acceleration acting on each mass using the fitness values.
iii. Utilize the gravitational force between masses to update their positions in the search space. Adjust the positions of masses according to their masses and the gravitational forces.
iv. Implement boundary checking to ensure that the updated positions of masses remain within the defined search space.
v. Recalculate the masses of solutions based on their updated positions and recompute the gravitational accelerations.
vi. Iteratively perform the steps of updating positions, checking boundaries, and calculating mass-acceleration until a specified stopping criterion is met, whether it be reaching a maximum number of iterations or attaining a satisfactory solution.
vii. The algorithm outputs the best solution found during the iterations as the optimized solution to the given problem.

10.3.3 Proposed Method

In this work, GBDT based model for malicious access detection and its hyperparameter optimization using gravitational search algorithm has been designed (Algorithm 1 to Algorithm 3). The used model GBDT (for malicious attack detection) is affected by various hyperparameters such as learning rate, maximum depth, number of estimators, and bin sub-sample size. In this section, we have used GSA for finding optimal hyperparameters combination that produce better prediction performance of GBDT. In this study, following hyperparameters are considered: maximum depth ($\rho 1$), learning rate ($\rho 2$), number of estimators ($\rho 3$), and bin sub-sample size ($\rho 4$). The GSA (Algorithm 1) starts with initial population (θ) has n number of hyperparameter set $\theta = \{\theta_1, \theta_2 ... \theta_n\}$ drawn from hyperparameters space with following ranges: $\rho 1_i \in (1, 16)$, $\rho 2_i \in (0, 1)$, $\rho 3_i \in (1, 31)$, and $\rho 4_i \in (0, 1)$. The goal is to explore optimal hyperparameter set $\theta_i{}^* = \{\rho 1_i, \rho 2_i, \rho 3_i, \rho 4_i\}$ in the search space.

Algorithm 1: GSA for finding optimal hyperparameter set for GBDT

Begin

Set the GSA parameters: β, Gravitational constant g, g_0

K=1

While (1)

Calculate fitness of each θ_i in θ:

For each θ_i in θ

$f_{\theta_i} \leftarrow GBDT(\theta_i, P)$ (Algorithm 2)

$f = f \cup f_{\theta_i}$

Find out θ_{best} (θ_i with best fitness) and θ_{worst} (θ_i with worst fitness)

Compute mass M_{θ_i} of each θ_i in θ:

For each θ_i in θ, Calculate the mass m_{θ_i} as in Eq 1 and M_{θ_i} as in Eq.2.

Calculate the force F_{θ_i,θ_j} acts on mass M_{θ_i} from M_{θ_j}:

For $i = 1$ to n

For $j = 1$ to n

Calculate the force F_{θ_i,θ_j} using Eq.3.

For each in θ_i in θ, Calculate the total force F_{θ_i} (Eq.4) acts on mass M_{θ_i}

For each in θ_i in θ, Find acceleration a_{θ_i} from mass M_{θ_i} and force F_{θ_i} using Eq.5 and Eq.4.

Find next velocity and net position of each θ_i:

For $i = 1$ to n

$V_{\theta_i}{}^{next} = rand(0,1) \times V_{\theta_i} + a_{\theta_i}$

$\theta_i{}^{next} = \theta_i + V_{\theta_i}{}^{next}$

If (K==Max OR improvement in fitness if less than a threshold value)

Exit from While

Else

Update: $V = V^{next}, \theta = \theta^{next}$

K = K + 1

Return best θ_i from θ

The proposed approach has been presented by using Algorithm 1, Algorithm 2, Algorithm 3. The steps of GSA for finding optimal hyperparameter set for GBDT while training with the used dataset are presented in Algorithm 1. Algorithm 2 represent the steps for calculating fitness of i th hyperparameter θ_i on the given dataset P. Algorithm 2 mix use of Algorithm 3 to construct a decision tree regressor which is bounded by maximum depth ($\rho 1$).

$$m_{\theta_i} = \frac{f_{\theta_i} - f_{\theta_{worst}}}{f_{\theta_{best}} - f_{\theta_{worst}}} \tag{10.1}$$

Algorithm-2: $GBDT(\theta_i, P)$

INPUTS: $\theta_i = \{\rho 1_i, \rho 2_i, \rho 3_i, \rho 4_i\}$: i^{th} hyperparameter set

$P = (P_i, t_i)_{i=1}^{N}$: Obesity data including physical condition, eating habits, and obesity types. $T = (t_i)_{i=1}^{N}$: Predicted obesity types

OUTPUTS: TS: Decision Tree Set

Score: Prediction score

Step-1: $F_0(P) \leftarrow c_1 = mean(T)$

Step-2: Create a subsample $SP = (P_i, t_i)_{i=0}^{\rho 4_i}$ of size $\rho 4_i$ from P

Step-3: Obtain the additive gradient boosted decision trees

For m in $\rho 3_i$

For each sample P_i in SP

$k_i \leftarrow y_i - F_{m-1}(P_i)$ (computed as in Eq.6)

$$r_i \leftarrow \left[\frac{\partial L(y, F(P_i))}{\partial F(P_i)}\right]_{F_{m-1}(P)}$$

Store (P_i, r_i) as instance and create a dataset SP'

$[LS, Tree] \leftarrow DTR(SP', \rho 1_i)$ (Algorithm-3)

$$TS = TS \cup Tree$$

Step-4: Use TS to make final prediction and calculate prediction score

Step-5: Return Score

Algorithm-3: $DTR(SP', \rho 1)$

$SP' = (P_j, r_j)_{j=1}^{\delta}$: Data includes data samples with their associated gradient

$P_j = (p_i)_{i=1}^{d}$: i^{th} input data sample with $d$dimension

Cd: Current depth

LS_i: Leaf space $LS_i = \{LS_{i,j}\}_{j=1}^{L}$, where is the number of samples in j^{th}leaf ofi^{th}tree

Step-1: Cd =1

Step-2: Find minimum MSE

For each P_j in SP'

For each p_i in P_j

Find min(MSE)

Step-3: Continue splitting of data samples until the depth αreached

If $Cd < \rho 1$

Split SP' in to SP_1' and SP_2'

$Cd = Cd + 1$

$DTR(SP_1', \rho 1)$

$DTR(SP_2', \rho 1)$

Else, Make SP'as leaf node

Step-4: Return LS and $Tree$

In Eq. (10.1), m_{θ_i} is the gravitational mass of θ_i. f_{θ_i} is the fitness of θ_i. $f_{\theta_{best}}$ is fitness of the best hyperparameter combination of θ_{best}. $f_{\theta_{worst}}$ is fitness of the worst hyperparameter combination of θ_{worst}.

$$M_{\theta_i} = \frac{m_{\theta_i}}{\sum_{j=1}^{n} m_{\theta_j}} \tag{10.2}$$

In Eq. (10.2), M_{θ_i} is denoted as inertial mass of θ_i.

$$F_{\theta_i,\theta_j} = \begin{Bmatrix} g \times \frac{M_{\theta_i} \times M_{\theta_j}}{\varepsilon d(H_i, H_j)+\delta}\left(\theta_j - \theta_i\right) & i \neq j \\ 0 & i = j \end{Bmatrix} \tag{10.3}$$

In Eq. (10.3), F_{θ_i,θ_j} is force applied on mass of θ_i by the mass of θ_j, g is the gravitational constant, $\varepsilon d(\cdot)$ is the Euclidian distance, and δ is a small value constant.

$$F_{\theta_i} = \sum\nolimits_{j=1, j \neq i}^{n} r \times F_{\theta_i,\theta_j} \tag{10.4}$$

$$a_{\theta_i} = \frac{F_{\theta_i}}{M_{\theta_i}} \tag{10.5}$$

In Eq. (10.4), F_{θ_i} is force applied on mass of θ_i by the all other mass of θ_j, and r is a random number generated in between 0 and 1. In Eq. (10.5), a_{θ_i} is acceleration of θ_i.

The hyperparameter optimization process (Fig. 10.2) begins by generating a random population of hyperparameter sets. These sets represent different configurations of hyperparameters for the GSA. Once the population is obtained, GSA parameters, such as gravitational constant and population size, are defined to guide the optimization process. Subsequently, the fitness of each hyperparameter set is calculated. This involves configuring GBT model based on the values of hyperparameters within the selected set and training the model using the IIoT dataset. The prediction score obtained from the trained model serves as the fitness measure. To identify the best and worst hyperparameter sets in terms of fitness, the algorithm compares the prediction scores across the entire population. The mass of each hyperparameter set is then computed, taking into account its individual fitness, as well as the best and worst fitness values within the population. The gravitational forces between hyperparameter sets are determined by their respective masses. These forces, in turn, influence the acceleration of each hyperparameter set. Acceleration values are employed to calculate velocities, leading to the determination of the next updated hyperparameter set. This iterative process continues until a termination condition is satisfied. If the condition is met, the optimal hyperparameter set is returned. If not, the algorithm recalculates fitness, updates masses, and repeats the entire sequence, dynamically adjusting hyperparameters based on gravitational principles. This comprehensive approach ensures that the algorithm converges towards an optimal configuration, refining hyperparameter sets iteratively until the termination condition is met, and the most effective set is identified.

10.4 Result Analysis

10.4.1 Dataset Information

X-IIoTID [27] was carefully developed by simulating advanced procedures, techniques, authentic behaviors, and recent attacker tactics demonstrated by IIoT systems. This is designed for contemporary IIoT devices operating cohesively with a sophisticated structure. Developed holistically, this dataset encompasses traffic data and

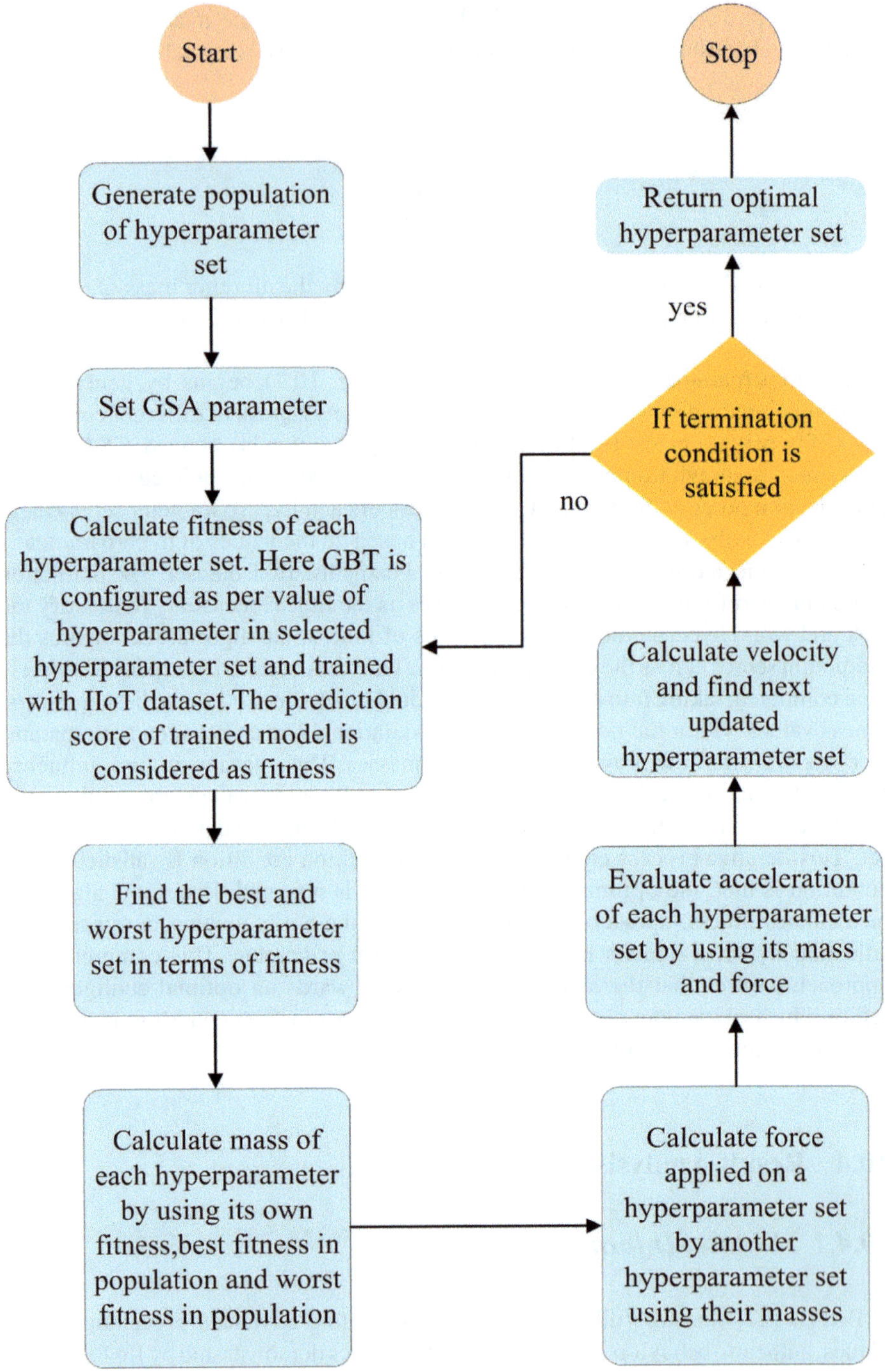

Fig. 10.2 Generalized structure of the proposed model

modifications sourced from protocols and devices within the IIoT network. Additionally, it includes resources, log, and alert features from various connections and devices. The simulation encompassed a diverse array of IoT devices, including controllers, sensors, mobile devices, actuators, edge components, and cloud traffic. Additionally, the dataset encapsulated the intricate dynamics of connectivity protocols such as WebSocket, CoAP, and MQTT. Notably, various communication patterns like Machine-to-Machine (M2M), Human-to-Machine (H2M), and Machine-to-Human (M2H) were integrated, incorporating substantial network traffic and event scenarios. The model under consideration undergoes testing with a simulated X-IIoT ID dataset comprising 820,834 instances, categorized into Normal type (421,417 instances) and Anomalous type (399,417 instances). Each instance in this dataset is characterized by 66 attributes, with the class label represented by the final attribute. To address limitations related to memory and running time, we opted for stratified sampling to resample the dataset, resulting in a total of 82,000 instances, before constructing the model.

10.4.2 Result Analysis

To ensure a comprehensive assessment of overall performance, we have explored a diverse set of machine learning (ML) and ensemble learning (EL) models. Among the ML models considered are Linear Regression (LR), Linear Discriminant Analysis (LDA), Naïve Bayes (NB), Decision Tree (DT), Stochastic Gradient Descent (SGD), Quadratic Discriminant Analysis (QDA), and Multilayer Perceptron (MLP). Additionally, the evaluation encompass EL models such as Bagging, Random Forest (RF), AdaBoost, Gradient Boosting (GBT), and XGBoost for performance comparison. All these models, including the suggested approach, have undergone evaluation and comparison using metrics such as Precision, Recall, F-beta Score, and F1 Score.Table 10.1 reveals that the performance of the proposed GBT + GSA model surpasses that of other models under consideration. It is to be noticed that all the base ML and EL models are implemented and tested with their default parameter setting. The GBT + GSA model demonstrates superior performance across the evaluated metrics. Furthermore, in terms of convergence speed (Fig. 10.3), the proposed approach outperforms both GBT + PSO and GBT + GIO. It is observed that, GSA is found better for hyperparameter optimization of GBT as compared to GIO and PSO. Further it is noticed that GIO is found better on the same context.

Following the experimentation and analysis of simulation results, we have determined the optimal set of hyperparameter values for the considered models from PSO, GIO, and GSA as {0.23120479995090337, 93, 0.8667198035433937, 10}, {0.878742278989701, 24, 0.9286113285064188, 9}, and {0.9138798315367432, 27, 0.9812477028784514, 8}, respectively. Figure 10.4 illustrates the performance comparison of the proposed approach with the considered ML and EL models. It is observed that out of all compared ML model DT is found better than other ML

Table 10.1 Performance evaluation with various metrics

Prediction models	Performance metrics			
	Recall	Precision	F beta score	F1 score
LDA	0.93285306	0.93316747	0.93297800	0.93281265
NB	0.80551957	0.81376202	0.80895771	0.80475843
LR	0.79022623	0.79024652	0.789238	0.79017847
DT	0.99937868	0.99937869	0.99937868	0.99937868
SGD	0.77356843	0.79749694	0.78072535	0.76769301
QDA	0.92455259	0.93278629	0.92813797	0.92403833
MLP	0.94784589	0.95000823	0.94880724	0.94773079
RF	0.999695	0.999695	0.999695	0.999695
AdaBoost	0.99737259	0.99737264	0.99737261	0.99737258
Bagging	0.99946396	0.99946400	0.99946398	0.99946395
GBT	0.99874518	0.99874558	0.99874536	0.99874516
GBT + PSO	0.99979695	0.99979695	0.99979695	0.99979695
GBT + GIO	0.99979695	0.99979695	0.99979695	0.99979695
GBT + GSA	0.99980101	0.99980101	0.99980101	0.99980101

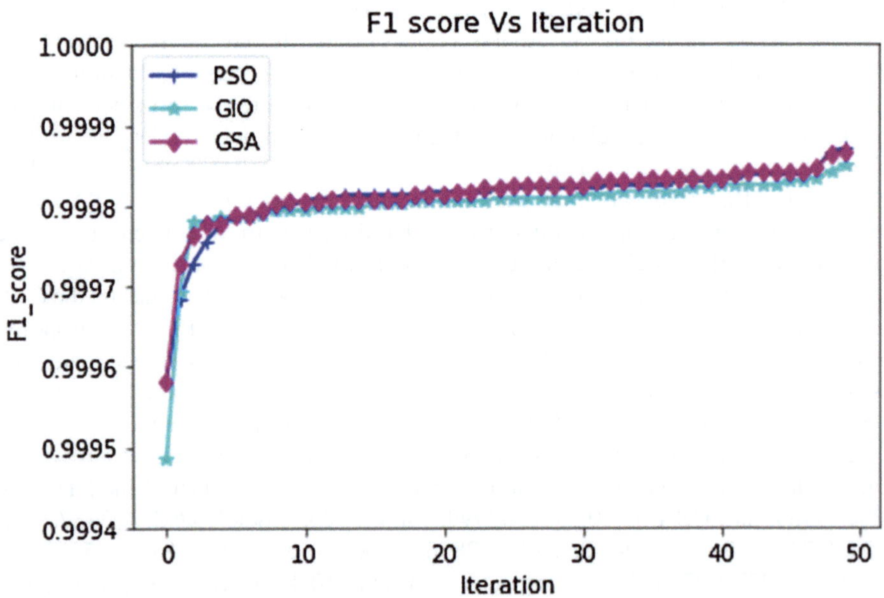

Fig. 10.3 Tracking F1-score of best solution in every generation

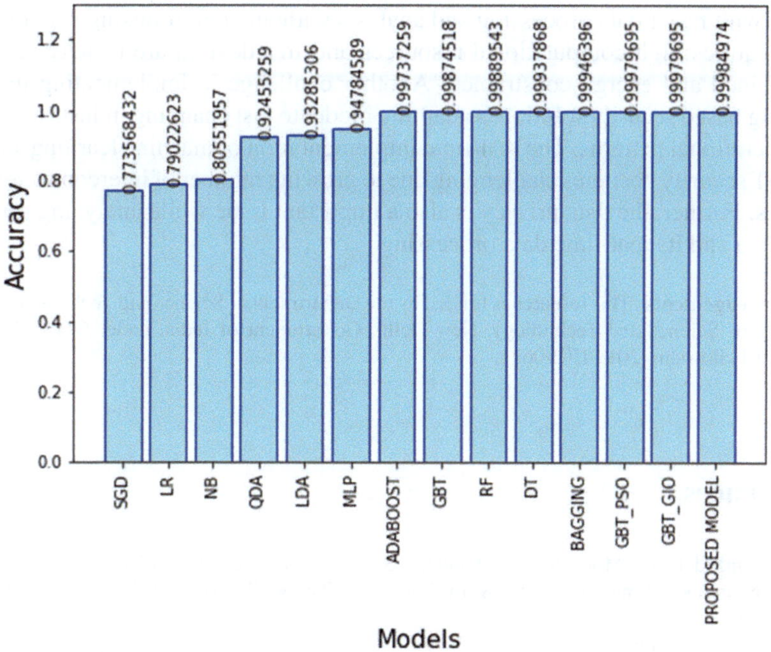

Fig. 10.4 Tracking accuracy of best solution in every generation

models. While comparing all base EL model Bagging approach is found better than all other compared EL models.

10.5 Conclusion

The increasing uses of IoT in industries have witnessed a new efficiency and connectivity in industrial processes. This revolution in IoT technology has enhanced the streamline operations and productivity. At the same time, security challenge becomes a vital issue because of its potential consequences. Security attack to IIoT may interrupt significant industrial processes, leading to financial losses, and operational disruption. IIoT infrastructure requires network of interconnected devices to control and monitor industrial processes. Therefore, ensuring security of IIoT network is essential to avoid unauthorized access. IoT devices interact with each other and they generate and share sensitive data on the network. So, security threats are always a big concern in order to provide safeguard to data confidentiality, and proprietary information related to industrial processes. In this work, an ensemble learning based model is designed to detect anomalies in IIoT network. Here, gradient boosted decision tree is used with its optimized hyperparameters using gravitational search algorithm. The real-time deployment of machine learning security solution demands low-latency model with faster data processing and analysis without compromising performance. This requires high computational resources and IoT devices are usually resource-constrained and energy-constrained. Another challenge in implementing machine learning based security solution is adapting model to fast changing industrial settings and operational patterns. The real-time implementation of machine learning solution for IIoT security becomes challenging due to growing number of interconnected IIoT devices. Further, the data privacy is also a important issue while analyzing sensitive IIoT data and its real-time data processing.

Acknowledgements This research is funded by the Department of Science and Technology (DST), Ministry of Science and Technology, New Delhi, Government of India, under Grant No. DST/INSPIREFellowship/2019/IF190611.

References

1. Hassanzadeh, A., Modi, S., Mulchandani, S.: Towards effective security control assignment in the industrial internet of things. In: Internet of Things (WF-IoT), IEEE 2nd World Forum (2015)
2. Industrial Internet of Things Volume G4: Security Framework, IIC:PUB:G4:V1.0:PB:20160926
3. Muna, A.H., Moustafa, N., Sitnikova, E.: Identification of malicious activities in Industrial Internet of Things based on deep learning models. J. Inf. Secur. Appl. **41**, 1–11 (2018)
4. Defense Use Case. Analysis of the Cyber Attack on the Ukrainian Power Grid. Electricity Information Sharing and Analysis Center (E-ISAC) 388 (2015). https://africautc.org/wp-content/uploads/2018/05/E-ISAC_SANS_Ukraine_DUC_5.pdf. Accessed 7 May 2022
5. Alladi, T., Chamola, V., Zeadally, S.: Industrial control systems: cyberattack trends and countermeasures. Comput. Commun. **155**, 1–8 (2020)
6. Sitnikova, E., Foo, E., Vaughn, R.B.: The power of handson exercises in SCADA cybersecurity education. In: Information Assurance and Security Education and Training. Springer, Berlin/Heidelberg, Germany, pp. 83–94 (2013)
7. Dash, S., Chakraborty, C., Giri, S.K., Pani, S.K., Frnda, J.: BIFM: big-data driven intelligent forecasting model for COVID-19. IEEE Access **9**, 97505–97517 (2021)
8. Koroniotis, N., Moustafa, N., Sitnikova, E.: A new network forensic framework based on deep learning for Internet of Things networks: a particle deep framework. Fut. Gener. Comput. Syst. **110**, 91–106 (2020)
9. Vaiyapuri, T., Binbusayyis, A.: Application of deep autoencoder as an one-class classifier for unsupervised network intrusion detection: a comparative evaluation. PeerJComput. Sci. **6**, e327 (2020)
10. Garcia-Teodoro, P., Diaz-Verdejo, J., Maciá-Fernández, G., Vázquez, E.: Anomaly-based network intrusion detection: techniques, systems and challenges. Comput. Secur. **28**, 18–28 (2009)
11. Gao, X.-C., et al.: Energy-efficient and low-latency massive SIMO using noncoherent ML detection for industrial IoT communications. IEEE IoT J **6**(4), 6247–6261 (2018)
12. Zolanvari, M., Teixeira, M.A., Jain, R.: Effect of imbalanced datasets on security of industrial IoT using machine learning. In: 2018 IEEE International Conference on Intelligence and Security Informatics (ISI). IEEE (2018)
13. Zolanvari, M., et al.: Machine learning-based network vulnerability analysis of industrial Internet of Things. IEEE IoT J **6**(4), 6822–6834 (2019)

14. Latif, S., et al.: A novel attack detection scheme for the industrial internet of things using a lightweight random neural network. IEEE Access **8**, 89337–89350 (2020)
15. Mudassir, M., et al.: Detection of botnet attacks against industrial IoT systems by multilayer deep learning approaches. Wirel. Commun. Mobile Comput. (2022)
16. Qolomany, B., et al.: Particle swarm optimized federated learning for industrial IoT and smart city services. In: GLOBECOM 2020–2020 IEEE Global Communications Conference. IEEE (2020)
17. Ksentini, A., Jebalia, M., Tabbane, S.: Fog-enabled industrial IoT network slicing model based on ML-enabled multi-objective optimization. In: 2020 IEEE 29th International Conference on Enabling Technologies: Infrastructure for Collaborative Enterprises (WETICE). IEEE (2020)
18. Marino, R., et al.: A machine-learning-based distributed system for fault diagnosis with scalable detection quality in industrial IoT. IEEE IoT J **8**(6), 4339–4352 (2020)
19. Taheri, R., et al.: FED-IIoT: A robust federated malware detection architecture in industrial IoT. IEEE Trans. Ind. Informatics **17**(12), 8442–8452 (2020)
20. Yazdinejad, A., et al.: An ensemble deep learning model for cyber threat hunting in industrial internet of things Digital Commun. Netw. **9**(1), 101–110 (2023)
21. Le, T.-T.-H., Oktian, Y.E., Kim, H.: XGBoost for imbalanced multiclass classification-based industrial internet of things intrusion detection systems. Sustainability **14**(14), 8707 (2022)
22. Mohy-Eddine, M., et al.: An ensemble learning based intrusion detection model for industrial IoT security. Big Data Min. Anal. **6**(3), 273–287 (2023)
23. Rashid, Md.M., et al.: A federated learning-based approach for improving intrusion detection in industrial internet of things networks. Network **3**(1), 158–179 (2023)
24. Rafiq, H., Aslam, N., Ahmed, U., Lin, J.C.-W.: Mitigating malicious adversaries evasion attacks in industrial internet of things. IEEE Trans. Industr. Inf. **19**(1), 960–968 (2023). https://doi.org/10.1109/TII.2022.3189046
25. Friedman, J.H.: Greedy function approximation: a gradient boosting machine. Ann. Stat. 1189–1232 (2001)
26. Rashedi, E., Nezamabadi-Pour, H., Saryazdi, S.: GSA: a gravitational search algorithm. Inf. Sci. **179**(13), 2232–2248 (2009)
27. Al-Hawawreh, M., Sitnikova, E., Aboutorab, N.: X-IIoTID: a connectivity-agnostic and device-agnostic intrusion data set for industrial internet of things. IEEE Internet Things J. **9**, 3962–3977 (2022)

Chapter 11
Machine Learning Based Intelligent Diagnosis of Brain Tumor: Advances and Challenges

Surendra Kumar Panda, Ram Chandra Barik, Danilo Pelusi, and Ganapati Panda

Abstract One of the fatal diseases that kills a large number of humanity across the globe is brain tumor. If the brain tumor detection is delayed, then the patient has to spend a large amount of money as well as to face severe suffering. Therefore, there is an essential need to detect the brain tumor so that money and life can be saved. The conventional examination of brain images by doctors does not reveal the presence of a tumor in a reliable and accurate manner. To overcome these issues, early and accurate brain tumor identification is of prime importance. A short while ago, methods employing machine learning (ML) and artificial intelligence (AI) were utilized to properly diagnose other diseases using test attributes, electrocardiogram (ECG), electromyography (EMG), Heart Sounds, and other types of signals obtained from the human body. This chapter presents a complete overview of the detection of patient-provided brain MR pictures and classifying patients' brain tumor using AI and ML approaches. For this pose, brain images obtained from kaggle.com website have been employed for developing various AI and ML classifiers. Through simulation-based experiments conducted on the AI and ML classifiers, performance matrices have been obtained and compared. From the analysis of results reported in the different articles, it is observed that Random Forest exhibit superior detection of brain tumor. There is still further scope for improving the performances as well as developing affordable, reliable, and robust AI-based brain tumor classifiers.

S. K. Panda (✉) · R. Chandra Barik
Department of Computer Science and Engineering, C. V. Raman Global University, Odisha, India
e-mail: surendrapanda000@gmail.com

R. Chandra Barik
e-mail: ram.chandra@cgu-odisha.ac.in

D. Pelusi
Faculty of Communication Sciences, Teramo, Italy
e-mail: dpelusi@unite.it

G. Panda
Department of Electronics and Communication Engineering, C. V. Raman Global University, Odisha, India

J. Nayak et al. (eds.), *Machine Learning for Cyber Physical System: Advances and Challenges*, Intelligent Systems Reference Library 60,
https://doi.org/10.1007/978-3-031-54038-7_11

Keywords Brain tumor classification · Magnetic Resonance Imaging (MRI) · Machine learning · Feature extraction · Performance analysis

11.1 Introduction

With thousands of instances discovered each year, brain tumors are a major health problem on a global scale. Conventional methods to identify brain tumors involve the use of Computed Tomography(CT) scans, MR images, etc.. Medical professionals use MR images to diagnose patients and analyze brain tumor. Thus, the MR images continues to play an important role in brain tumor detection [1]. The accuracy and time requirements of conventional diagnostic techniques are constrained. For successful treatment and better patient outcomes, brain tumor must be identified as early as possible. The use of computer-aided techniques(CAT) might lead to more accurate brain tumor detection [2]. The internet of medical things (IoMT), ML, and DL among other technological breakthroughs, provide potential options to improve brain tumor detection and diagnosis. As a result of their impressive performance in image analysis and pattern recognition tasks, ML and DL algorithms are excellent options for detecting brain tumor from MR images and CT scans. Various ML based techniques are constantly advancing to enhance the precision of detection. Identifying the features is very important in the process of brain tumor detection. For instance, Ghassemi et al. [3] have proposed a DL technique in which a generative adversarial network (GAN) is trained on multiple datasets so to make it capable of extracting strong and required features from the MR images to make brain tumor detection easier. Along with features extraction, segmentation also plays a very essential role in brain tumor detection. Many methods have been introduced to improve the segmentation process. In [4], a grab cut method is introduced which helps in accurate segmentation. It also uses VGG-19 for tuning to extract features. Duan et al. [5] have discussed the importance of deformable registration. They proposed a tuning-free 3D image registration model which provided very accurate registration. In [6], the automated segmentation of MR images has been proposed. The approach has also taken the noisy and inconsistent data into consideration. Such data might lead to unexpected and inaccurate results. Hence, they must be handled to achieve accurate results. Brain tumors are categorized as benign and malignant. Malignant tumors are more harmful as compared to benign tumors. Brain tissue is differentiated using a hybrid technique [7], according to whether it is normal, has a benign tumor, or has a malignant tumor. Further brain tumor has two variant one is glioma and another is meningioma, pituitary tumors, etc based on their position. Anaraki et al. [14] performed multiple case studies to identify the type of brain tumour present. In order to identify the kind of brain tumour that is present and to do it extremely early on, they have used a hybrid genetic algorithm (GA) combined with convolutional neural networks (CNNs). A brain tumor can further have different stages or grades such as grade I,II,III and IV etc. The rate of recovery greatly depends on the grade of the tumor. Sultan et al. [16] processed two distinct datasets using DL methods. Their

objectives were to identify the type of brain tumor in one dataset and the grade of glioma in another dataset. They achieved very promising results for the two datasets. Consequently, the employment of DL and ML technologies has a notable positive effect on the identification and diagnosis of brain tumor. Brain tumor detection and treatment might be revolutionized by their capacity to accurately and quickly analyze complicated medical imaging data, which would ultimately improve patient care and results.
Contribution of the chapter:

- This chapter provides ML based methods for early and accurate detection of brain tumor using standard brain images obtained from kaggle.com website.
- The chapter presents a generalized ML based method for brain tumor detection.
- Seven different ML methods have been proposed and performance matrices have been obtained and compared.
- It is demonstrated that in ML categories, standard ML based classifiers shows improved performance compared to other methods.

The organization of this chapter is represented as below:

The Sect. 11.2 consists of system under study, while the Sect. 11.3 explains the material and methodology used in the chapter. The Sect. 11.4 provides a detailed analysis of the results obtained, and the Sects. 11.5 and 11.6 consist of the discussion and conclusion, respectively.

11.2 System Under Study

Numerous methodologies have been put forth to increase the effectiveness and precision of brain tumor detection. El-Melegy et al. [6] have formulated a new fuzzy method based on the traditional fuzzy algorithm. It helps in the automatic segmentation of MR images, by taking into consideration the noisy data as well. As a result, the performance of the Fuzzy C-means (FCM) method is notably improved. An amalgam of GA and the support vector machine (SVM) has been introduced by Kharrat et al. [7], which is used to classify tumor in brain MR images. The GA model is used to classify the wavelet's texture feature, which is provided as an input to the SVM. In this instance, the accuracy percentage ranges from 94.44 to 98.14. It analyzed the ML-based back propagation neural networks (MLBPNN) by using an infrared sensor imaging technique. They used the fractal dimension algorithm (FDA) to extract features and the multi-fractal detection (MFD) to select the most essential features. The data is then transferred to a clinician via a wireless infrared imaging sensor, which is reported in [8]. The average specificity was 99.8%, while the average sensitivity was 95.103%. Kanmani et al. [9] proposed an approach for classifying brain tumor using threshold-based region optimization (TBRO). It helps to overcome the limitations of traditional CAT. It had 96.57% percent accuracy. An attentive residual U-Net (AResU-Net) is being used to segment the ROI area from two well-known 2-D image datasets, according to a novel segmentation approach proposed by Zhang et al. [10]. Their second experiment with BRATS 2018 had them obtain the biggest Dice score in the case of enhancing tumor and place second in the case of core tumor, whereas their first experiment with BRATS 2017 saw them reach the largest dice score in the case of entire along with enhancing tumor. Kumar et al. [11] presented the Dolphin-SCA-based Deep CNN DL method, using this strategy, the MR images were first pre-processed before being segmented, feature-extracted, and then classified. In this case, the segmentation was accomplished by hybridizing the fuzzy deformable fusion model with the Sine Cosine Algorithm in conjunction with the Dolphin equalization classification of informative features, which is derived from Deep CNN. The experimentation was performed on two datasets namely BRATS and SimBRATS. It achieved an accuracy of 95.3% in the case of the former, and 96.3% in the case of the latter. Alam et al. [12] introduced an improved FCM and template-based K means (TK-means) model. At first, the segmentation is initialized by the use of the TK-means algorithm and by using FCM algorithm the distance between centroid to data points is calculated.Lastly, the position of the tumor is detected using the improved FCM algorithm. The proposed model reached an accuracy of 97.5%. Islam et al. [13] exploited the TK-means algorithm along with the superpixels and used principle component analysis (PCA) to quickly and accurately identify brain tumor. In the method, the important features are first extracted by the use of superpixels and PCA. Then the image's quality is improved and finally segmented with the TK-means algorithm. It achieved an accuracy of 95% in 35-60 seconds, which is much lower compared to other methods. Anaraki et al. [14] outlined a method based on CNNs and GA that is used to classify the various stages of glioma from MR images. In this technique, the structure of CNN is developed by the use of GA. In the case of classification of the stage of Glioma, it achieved an accuracy of 90.9%, and in the case of classification of the type of tumor, it achieved an accuracy of 94.2%. The Berkeley wavelet transformation (BWT) along with SVM is intended to improve the efficiency of the segmentation process as well as reduce its complexity. It was found to have a 96.51% accuracy, 94% specificity, and 97.72% sensitivity as reported in [15]. Sultan et al. [16] for the categorization of various forms of brain tumor, a CNN-based DL model has been presented. Two datasets were utilized in the experiment; one was used to categorize the different types of brain tumor into meningioma, glioma, and pituitary tumor, while the other dataset was used to identify the glioma's grade. Nanda et al. [17] have introduced a K-means cluster-based segmentation technique combined with a hybrid saliency map to divide the tumor area. On three distinct datasets, it obtained accuracy rates of 96%, 92%, and 94%. The extension of 2D-CNNs into multimodal 3D-CNNs, which may create brain lesions in three-dimensional space under different modal characteristics, is a technique used by Li et al. [18]. Tumor lesions may be effectively identified using the recommended method, which also produced higher correlation coefficient, sensitivity, and specificity values. In [19], ensemble classifier was used for the detection of brain tumor. The ensemble classifier consisted of six different ML algorithms which were compared to find the best among them. In order to diagnose brain tumors, Ahmad et al. [20] looked at a range of DL methods based on transfer learning and several conventional classifiers. The

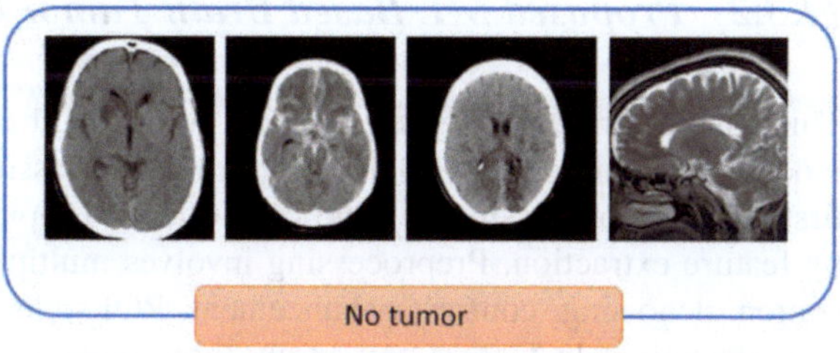

Fig. 11.1 MR images having no tumor

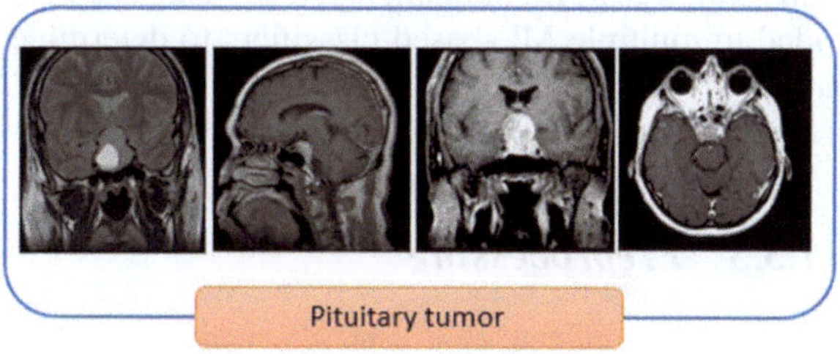

Fig. 11.2 MR images having pituitary tumor

investigation's conclusions are based on a labeled dataset including images of brain tissue that is both normal and abnormal.

11.3 Materials and Methodology

11.3.1 Dataset

A total of 3757 MR images are utilized in this chapter. Eighty percent of the dataset is utilised for training, while twenty percent is used for testing. Figure 11.1. shows brain images with no tumor and Fig. 11.2. shows brain images having a pituitary tumor. A total of 3004 MR images are included in the training set, while the testing set holds 751 MR images. Table 11.1 describes the dataset. The link of dataset is given below:
https://www.kaggle.com/datasets/masoudnickparvar/brain-tumor-mri-dataset.

Table 11.1 Dataset description

Total MR images: 3755			
Classes	Total MR images	Having tumor	No tumor
Training set	3004	1450	1554
Testing set	751	350	401

11.3.2 Proposed ML Based Brain Tumor Classifier

This chapter presents the development of an ML-based brain tumor classification model, as shown in Fig. 11.3, for the purpose of classifying brain tumors. In the very first step, pre-processing is done on the loaded input brain MR images, followed by feature extraction. Preprocessing involves multiple techniques such as normalization, denoising, contrast enhancement, ROI segmentation, and derivative-based edge detection. In feature extraction, three types of features are obtained namely statistical, transform domain, and technical. The feature vector obtained is then provided to multiple ML-based classifiers to determine if a brain tumor is present or not.

11.3.3 Preprocessing

Preprocessing is a crucial step in identifying brain tumor from MR images. It uses a number of approaches to improve picture quality, lower noise, and get the data ready for proper analysis and interpretation. Preprocessing aims at improving the efficiency of the classification algorithms and increasing the suitability of the images for further analysis. Image resizing and standardization, intensity normalization, noise reduction, contrast enhancement, and derivative-based edge detection are all included in preprocessing.

11.3.4 Features Extraction

Feature extraction entails the conversion of unprocessed picture data into a set of representative features that effectively record essential details about the underlying structures. To distinguish between areas of tumor and normal brain tissue, several characteristics are retrieved from MR images. These features are designed to identify the distinctive traits of tumor and facilitate precise identification. Instead of directly providing the MR image data to the model, the required extracted features act as feature vectors that are provided to the model.Transform domain, statistical, and technical characteristics are retrieved as three different categories of features.

The average intensity of the image pixels is known as the mean intensity. It is mathematically represented as given in Eq. (11.1):

$$\mu = \frac{\sum I}{np} \tag{11.1}$$

where μ represents the mean intensity, I denotes the intensity of each pixel's value, and the sum of the pixels indicated by np.

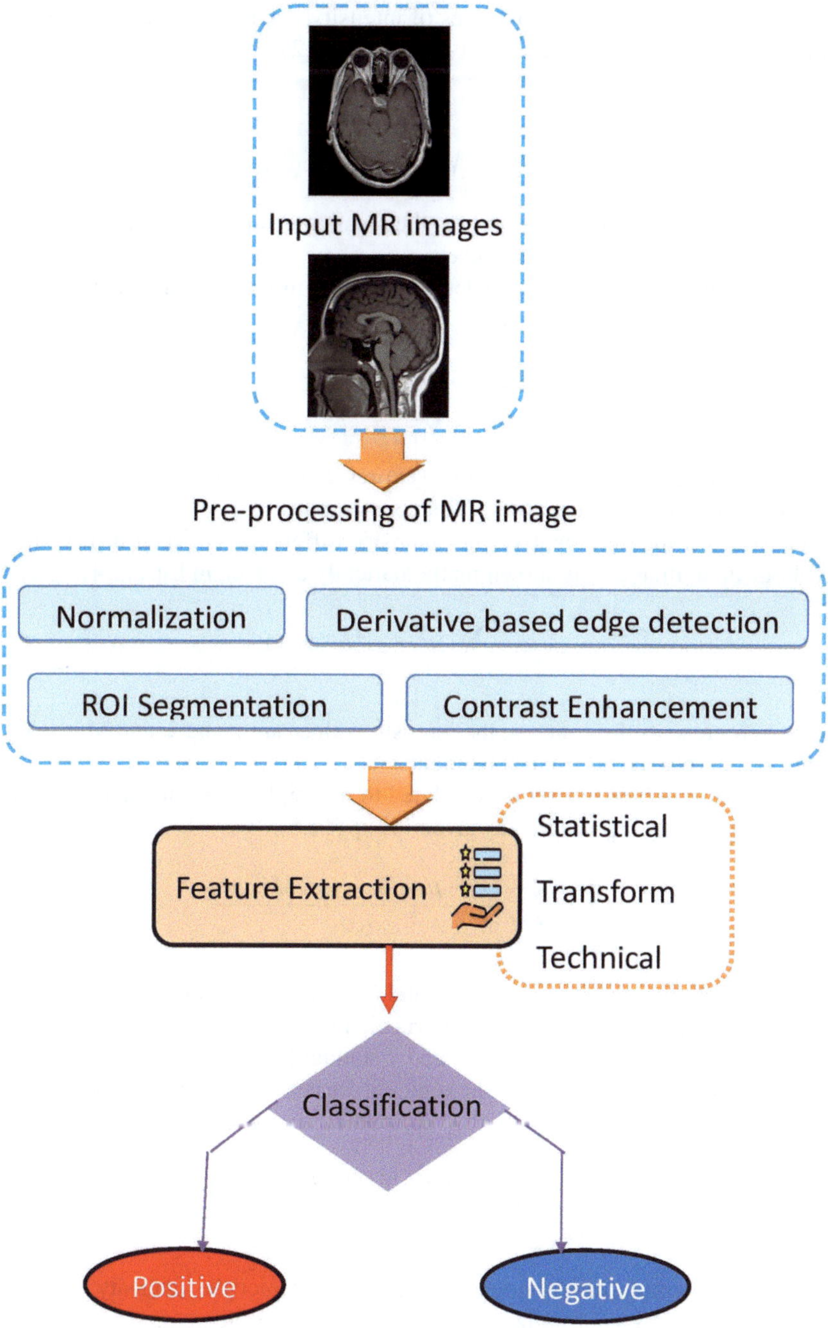

Fig. 11.3 Proposed workflow of ML-based brain tumor classification

The measure of the spread of values of intensity is known as standard deviation. It is mathematically represented as given in Eq. (11.2):

$$\sigma = \sqrt{\frac{\sum (I - \mu)^2}{np}} \tag{11.2}$$

where σ depicts standard deviation, μ is the mean intensity, I indicate pixel intensity value, and *np* represents all of the pixels.

The energy of an image is used to determine its homogeneity. It is mathematically represented as given in Eq. (11.3):

$$En = \sum_{x,y} M(x, y)^2 \tag{11.3}$$

where En represents the energy, both x, y expresses the intensity values, and M(x, y) shows the normalized co-occurrence matrix element.

Contrast is a measurement of the intensity difference between a pixel and its neighbour in an image. It is shown mathematically as given in Eq. (11.4):

$$Cn = \sum_{x,y} |x - y|^2 M(x, y) \tag{11.4}$$

where Cn represents the contrast, the intensity values are x and y, and M(x, y) is the normalized co-occurrence matrix element.

The measure of how a pixel is correlated to its neighbor is known as correlation. It is mathematically represented as given in Eq. (11.5):

$$Cr = \sum_{x,y} \frac{(x - \mu x)(y - \mu y) M(x, y)}{\sigma_x \sigma_y} \tag{11.5}$$

where Cr represents the correlation, x, and y exhibit the intensity values, M(x, y) denote the normalized co-occurrence matrix element, μ represents the mean intensity, and σ_x and σ_y indicates the relevant x and y standard deviations.

Homogeneity offers details on the regional variation or coarseness of the texture of an image area. It is mathematically represented as given in Eq. (11.6):

$$Hm = \sum_{x,y} \frac{M(x, y)}{1 + |x - y|} \tag{11.6}$$

where Hm represents the homogeneity, x, and y depict the intensity values, and M(x, y) represents the normalized co-occurrence matrix element.

The measure of complexity of an image texture is known as entropy. It is mathematically represented as given in Eq. (11.7):

$$Et = -\sum_{x,y} M(x, y)log(M(x, y)) \tag{11.7}$$

where Et represents the entropy, x, and y act as the intensity values, and M(x, y) imitate the normalized co-occurrence matrix element.

The measure of the asymmetry of intensity distribution is known as skewness. It is mathematically represented as given in Eq. (11.8):

$$\gamma = \frac{\frac{\sum(I-\mu)^3}{np}}{\sigma^3} \tag{11.8}$$

where γ represents the skewness, μ represents the mean intensity, I indicate the value of the intensity of each pixel, *np* illustrates all pixels, and the standard deviation is denoted by σ

The measure of the peak point of the intensity distribution is known as kurtosis. It is mathematically represented as given in Eq. (11.9):

$$kt = \frac{\frac{\sum(I-\mu)^4}{np}}{\sigma^4} \tag{11.9}$$

where kt represents the kurtosis, μ represents the mean intensity, I indicate the value of the intensity of each pixel, *np* represents the full count of pixels, and σ indicates the standard deviation.

11.3.5 ML Based Classifiers

A total of seven algorithms are used for the simulations study. They are SVM, RF, AdaBoost, Decision Tree (DT), LDA, ANN and RBF. A detailed description of the algorithms along with their limitations are described in Table 11.1.

11.3.5.1 Support Vector Machine

SVM is a unique ML model that is focused on classification and regression approach based on the labelled data. Mathematically it establishes an optimal decision function as a hyperplane whose objective is to maximize the margin of the training set, which inherently minimizes the generalization error. The architecture of SVM is shown in Fig. 11.4. The equation is given in (11.10). The architecture of SVM given in Fig. 11.4 was made by referring to the architecture given in [19].

$$c_0 + (c_1 * d_1) + (c_2 * d_2) = 0 \tag{11.10}$$

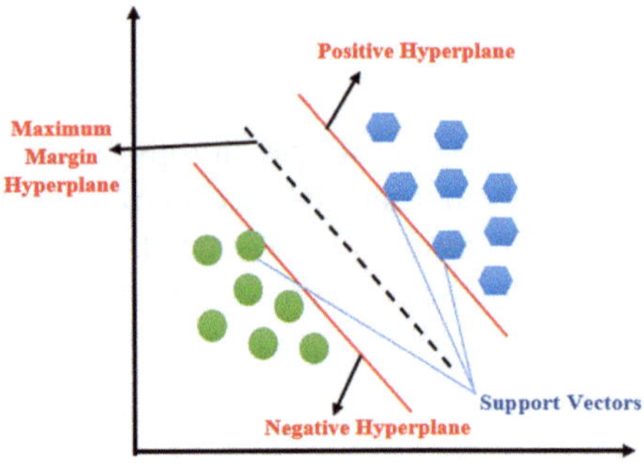

Fig. 11.4 Architecture of support vector machine

where c_1 and c_2 determines the slope of the line, c_0 represents the intercept and d_1 and d_2 represents the input variables.

11.3.5.2 Random Forest

Random Forest is the finest variant of a combination of decision trees which is also focused towards classification and regression approach based on the labeled data. Rather than concentrating on a unified decision tree prediction, it predicts the result from a series of decision tree for forecasting as an ensemble learning approach. The architecture of RF is shown in Fig. 11.5. The equation is given in (11.11). The architecture of RF given in Fig. 11.5 was made by referring to the architecture given in [19].

$$GI = 1 - [(pr_+)^2 + (pr_-)^2] \quad (11.11)$$

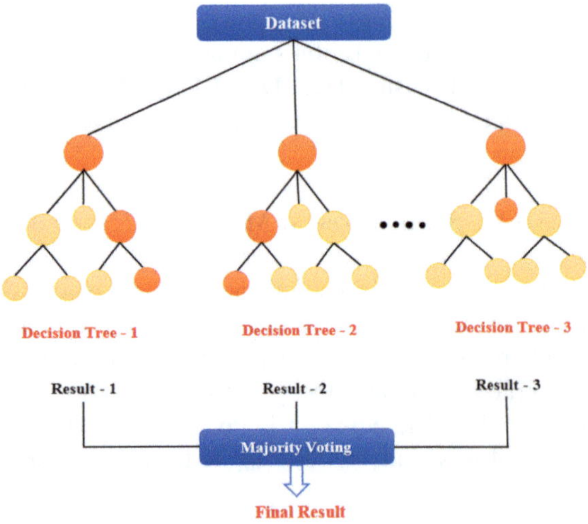

Fig. 11.5 Architecture of random forest

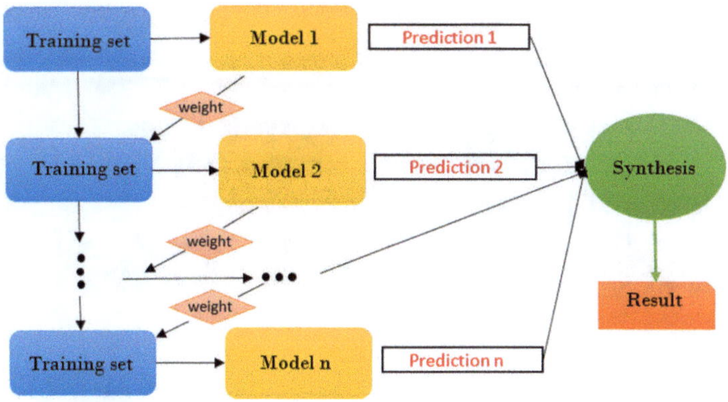

Fig. 11.6 Architecture of AdaBoost

where GI represents Gini Index, pr_+ and pr_- represents the probability of positive and negative class respectively.

11.3.5.3 AdaBoost

Boosting a series of weak learner to generate a strong learner step by step is the policy of AdaBoost algorithm. The value of the alpha parameter's is indirectly proportional to the mistakes of the weak learners. The architecture of AdaBoost is shown in Fig. 11.6. The equation is given in (11.12).

$$h(x) = sg(\sum_{y=1}^{Y} \alpha_y o_y(i)) \tag{11.12}$$

where h(x) represents the hypothesis function of a value x, sg represents the sign, α_y represents the weight given to the classifier, and $o_y(i)$ exhibit the output of weak classifier y for input i.

11.3.5.4 Decision Tree

A decision tree is a conventional technique used for regression and classification that focuses on supervised, labeled data. Every leaf or terminal node preserves the labeled class whereas every branch defines the results of the test. Internal nodes signify attribute-based tests. The architecture of DT is shown in Fig. 11.7. The equation is given in (11.13).

$$E_v = (F_{po}L_o) + (S_{po}L_o) - Cost \tag{11.13}$$

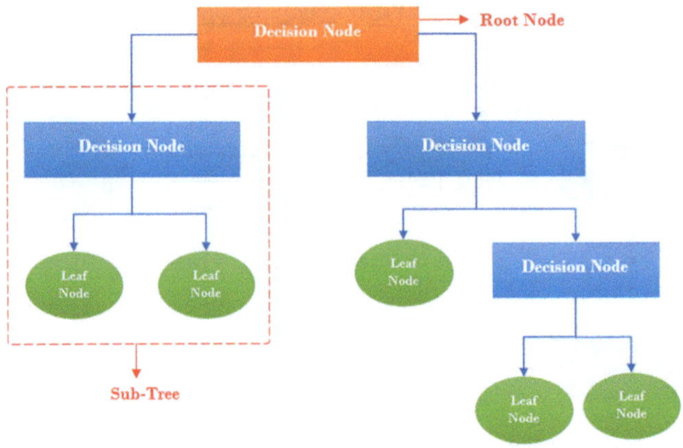

Fig. 11.7 Architecture of decision tree

where E_v represents the expected value, F_{po} and S_{po} represents the first and second possible outcome respectively, and L_o represents the likelihood of outcome.

11.3.5.5 Linear Discriminant Analysis

One supervised learning approach for classification tasks in machine learning is linear discriminant analysis (LDA). A linear feature combination that best distinguishes the classes in a dataset is found using this method. LDA works by projecting the data into a smaller-dimensional space where the distance between the classes is maximized. The architecture of LDA is shown in Fig. 11.8. The equation is given in (11.14).

$$\beta^T(m - (\frac{\mu_1 + \mu_2}{2})) > -log\frac{pr(cl_1)}{pr(cl_2)} \tag{11.14}$$

where β^T represents the coefficient vectors, m represents the data vector, μ_1 and μ_2 represent the mean vector, and $pr(cl_1)$ and $pr(cl_2)$ represent the class probability.

11.3.5.6 Artificial Neural Networks

ANN are created using a model of a human brain's neuronal network. Units, sometimes known as artificial neurons, are found in artificial neural networks. These components are stacked in a number of layers. The input, hidden, and output layers comprise the three tiers of this layout. The input layer is where the neural network gets data from the outside world that it needs to evaluate or learn about. After thereafter, the inputs are processed by one or more hidden layers into information that

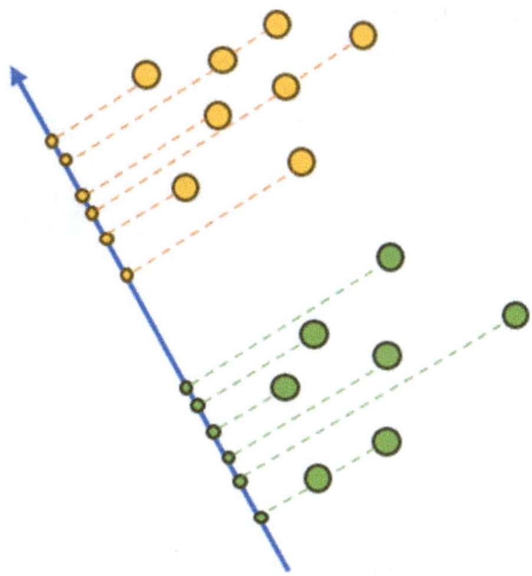

Fig. 11.8 Architecture of linear discriminant analysis

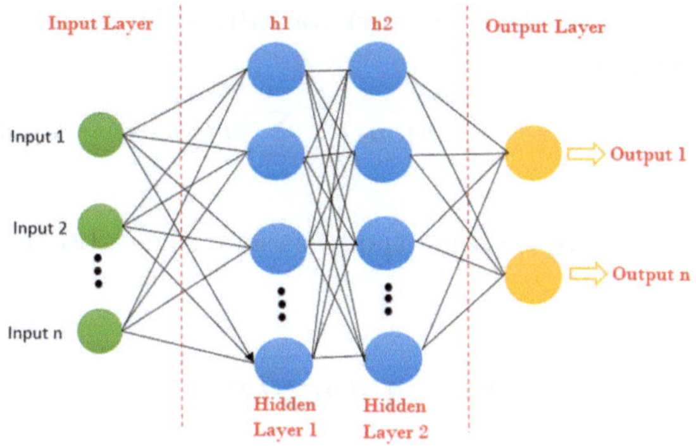

Fig. 11.9 Architecture of artificial neural networks

may be utilised by the output layer. The architecture of ANN is shown in Fig. 11.9. The equation is given in (11.15).

$$Z = Bias + y_1 i_1 + y_2 i_2 + \ldots + y_n i_n \tag{11.15}$$

where Z represents the sum of bias along with the product of input node and its weight, y represents the weights of the beta coefficients, The intercept is represented by Bias, while the independent variables are marked by i.

11.3.6 Radial Basis Functions

It is composed of three distinct layers, an input layer, a hidden layer, and an output layer. Radial basis functions are a unique class of feed-forward neural networks. This

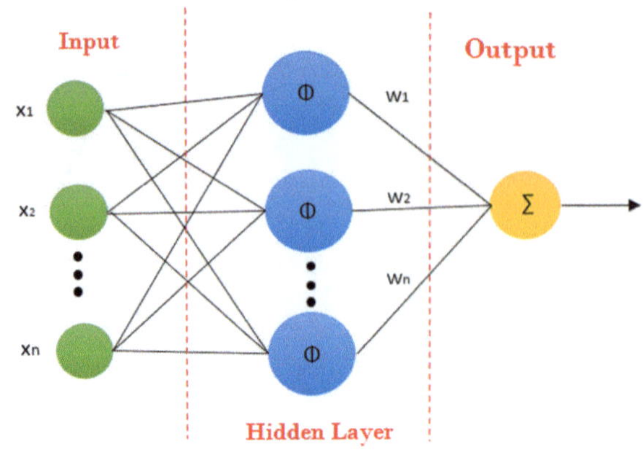

Fig. 11.10 Architecture of Radial basis functions

is fundamentally distinct from the majority of neural network topologies, which have several layers and produce non-linearity by repeatedly using nonlinear activation functions. The architecture of RBF is shown in Fig. 11.10. The equation is given in (11.16).

$$h(obs) = \sum_{n=1}^{N} wt_n * exp(-\gamma ||obs - obs_n||^2) \tag{11.16}$$

where h(obs) is the hypothesis set for new observation obs.

11.3.7 Activation Function

A neural network or other computational model with artificial neurons is incomplete without an activation function. As a result, the network gains non-linearity, which enables it to learn and approximatively represent complicated connections in data. A neuron's activation functions decide what it will produce based on its input or combination of inputs. The different types of activation functions are Linear Activation Function, Unit Step Heaviside Term Activation Function, Sign(Signum) Activation Function, Piecewise Linear Activation Function, Logistic(Sigmoid) Activation Function, Hyperbolic Tangent (tanh) Activation Function, and Rectified Linear Unit (ReLU) Activation Function. A detailed description of the activation functions along with their limitations are described in Table 11.3.

11.3.7.1 Linear

Often referred to as the linear function, the linear activation function is one of the most basic activation functions utilised in neural networks and other computational models. The function's result is the same as its input. The linear activation function

Table 11.2 ML-based classifiers

Techniques	Advantages	Limitations
SVM	It is appropriate for a variety of machine learning problems due to their stability in high-dimensional spaces, efficiency in addressing non-linear interactions through adaptable kernel functions, and global optimality.	It is susceptible to noise and outliers
RF	It is a flexible and effective ensemble learning approach that excels in high accuracy, can handle big datasets with high dimensionality, and provides robustness against overfitting	It may be prone to overfitting noisy data, especially if the forest has an excessively high tree population
AdaBoost	This effective ensemble learning technique uses the strengths of weak learners to increase the accuracy of the model. It is also less prone to overfitting and adapts well to different types of data	Although it is less likely to overfit than a single weak learner, it can still overfit in cases when the noise or outliers in the data are present or the base learners are somewhat complicated
DT	It is a versatile tool for machine learning problems related to regression and classification because of their interpretability, simplicity, and capacity to handle both numerical and categorical input	It is prone to overfitting, particularly in cases when the training data contains noise and the models are deep
LDA	It is a useful tool for feature extraction, classification, and enhancing model performance in supervised learning tasks since it reduces dimensionality while maintaining class separability	Assuming that the data has a normal distribution, it is susceptible to outliers
ANN	It is incredibly effective in machine learning applications because of its capacity to recognize intricate links in data, learn from a wide range of patterns, and adjust to a variety of tasks	For efficient training, ANNs need a lot of data. Insufficient data might result in overfitting or poor generalization to new, unobserved samples
RBF	It is useful for jobs involving complicated connections in data because they provide flexibility in capturing non-linear and complex patterns	Interpretability of the model may be compromised by the RBF kernel's transformation

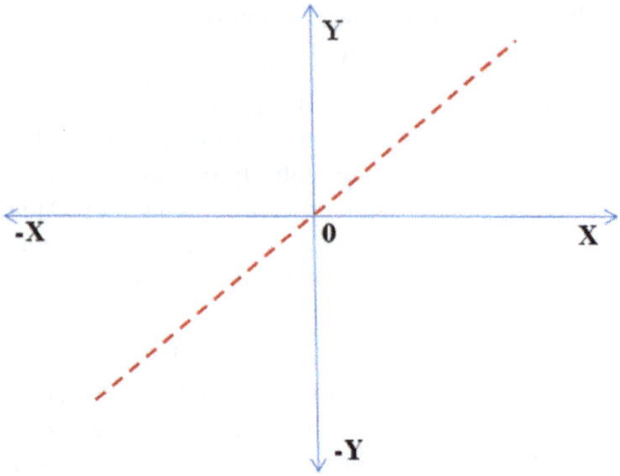

Fig. 11.11 Linear activation function

does not cause the network to become non-linear. Figure 11.11. shows the graphical representation of the linear activation function. It is mathematically represented as given in Eq. (11.17):

$$\phi(z) = z \tag{11.17}$$

where $\phi(z)$ denotes the output of the activation function for an input z.

Example: Adaline Linear Regression.

11.3.7.2 Unit Step Heaviside Term

The Heaviside step function is a binary activation function that is used in various applications. It can incorporate a threshold behavior into a neural network when applied to the output of a neuron or layer, where outputs below a given value are set to one value and outputs beyond that value are assigned to another. The graphical depiction of the unit step heaviside term activation function is displayed in Fig. 11.12. It is mathematically represented as given in Eq. (11.18):

$$\phi(z) = \begin{cases} 0 & \text{if } z < 0 \\ 0.5 & \text{if } z = 0 \\ 1 & \text{if } z > 0 \end{cases} \tag{11.18}$$

where $\phi(z)$ is equal to 0 if input z is less than z0, 0.5 if z is equal to 0, and 1 if z is greater than 0.

Example: Perceptron Variant.

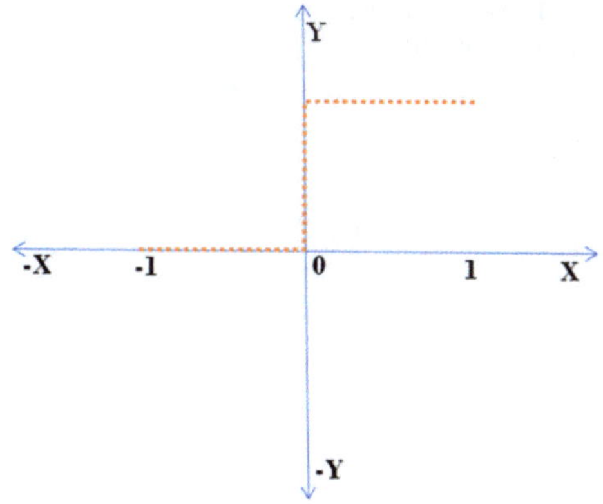

Fig. 11.12 Unit step heaviside term activation function

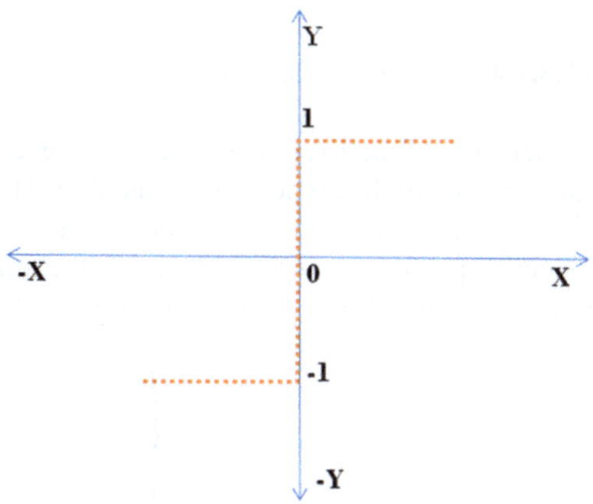

Fig. 11.13 Sign(Signum) activation function

11.3.7.3 Sign(Signum)

The sign activation function generates a binary output that encodes the polarity of the input (positive or negative). In situations where the magnitude of the input doesn't matter and the value's direction is more concerned than its magnitude, this function may be helpful. Figure 11.13 shows the graphical representation of the sign activation function. It is mathematically represented as given in Eq. (11.19):

$$\phi(z) = \begin{cases} -1 & \text{if } z < 0 \\ 0 & \text{if } z = 0 \\ 1 & \text{if } z > 0 \end{cases} \tag{11.19}$$

where $\phi(z)$ is equal to -1 if input z is less than 0, 0 if z is equal to 0, and 1 if z is greater than 0.
Example: Perceptron Variant.

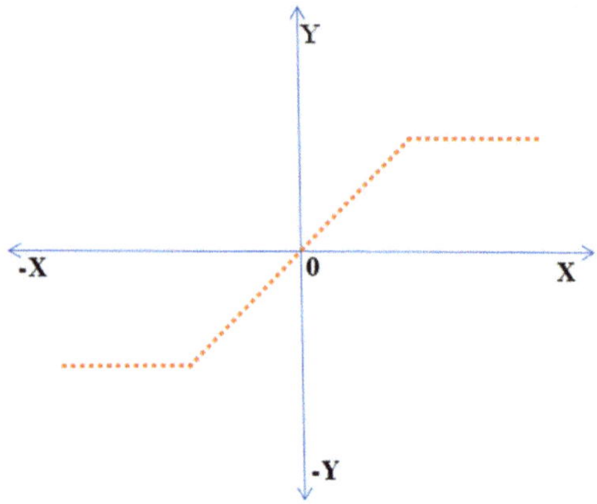

Fig. 11.14 Piecewise linear activation function

11.3.7.4 Piecewise Linear

An activation function that is made up of several linear sections or segments is known as the piecewise linear activation function. The function is generated by joining these linear segments together at particular places, with each segment being described by a linear equation. Figure 11.14. shows the graphical representation of the piecewise linear activation function. It is mathematically represented as given in Eq. (11.20):

$$\phi(z) = \begin{cases} 0 & \text{if } z < -\frac{1}{2} \\ z + \frac{1}{2} & \text{if } -\frac{1}{2} < z < \frac{1}{2} \\ 1 & \text{if } z >= \frac{1}{2} \end{cases} \tag{11.20}$$

where $\phi(z)$ is equal to 0 if input z is less than $-\frac{1}{2}$, $z + \frac{1}{2}$ if $-\frac{1}{2}$ is less than z and z is less than$\frac{1}{2}$, and 1 if z is greater than or equal to $\frac{1}{2}$.
Example: SVM.

11.3.7.5 Logistic(Sigmoid)

An activation function class referred to as the logistic or sigmoid activation function is defined by the logistic or sigmoid curve. In neural networks and other machine learning models, especially for binary classification problems, it is a frequently employed activation function. A graphical illustration of the logistic activation function is presented in Fig. 11.15. It is mathematically represented as given in Eq. (11.21):

$$\phi(z) = \frac{1}{1 + e^{-z}} \tag{11.21}$$

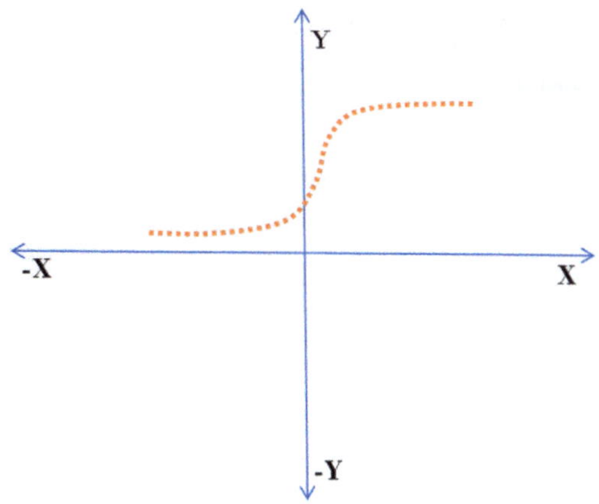

Fig. 11.15 Logistic(Sigmoid) activation function

where $\phi(z)$ is equal to the inverse of 1 + exponent of -z.
Example: Logistic Regression, Multilayer Neural Network.

11.3.7.6 Hyperbolic Tangent (tanh)

The hyperbolic tangent activation, often known as the tanh activation function, is a mathematical operation that is frequently employed in neural networks and a variety of machine learning methods. Figure 11.16. shows the graphical representation of the tanh activation function. It is mathematically represented as given in Eq. (11.22):

$$\phi(z) = \frac{e^z - e^{-z}}{e^z + e^{-z}} \tag{11.22}$$

where e^z and e^{-z} represents the exponent of z and -z respectively.
Example: Multilayer Neural Network, Radial Neural Network.

11.3.7.7 Rectified Linear Unit (ReLU)

Among all, most often utilized activation functions in contemporary neural networks is the ReLU activation function. DL models have been successful in large part because of ReLU. An illustration of the ReLU activation function is presented in Fig. 11.17. It is mathematically represented as given in Eq. (11.23):

$$\phi(z) = \begin{cases} 0 & \text{if } z < 0 \\ z & \text{if } z > u \end{cases} \tag{11.23}$$

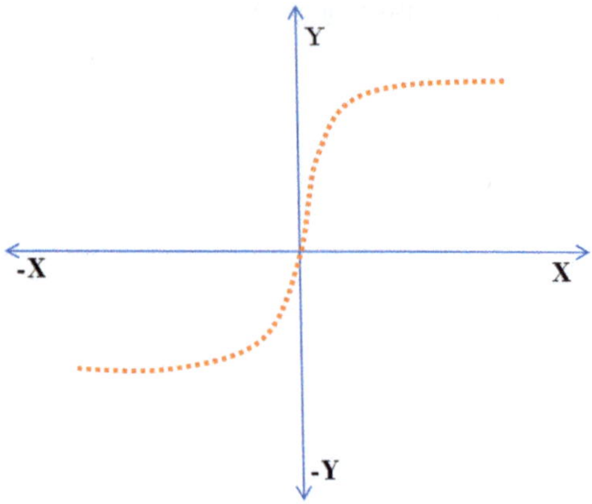

Fig. 11.16 Hyperbolic Tangent (tanh) activation function

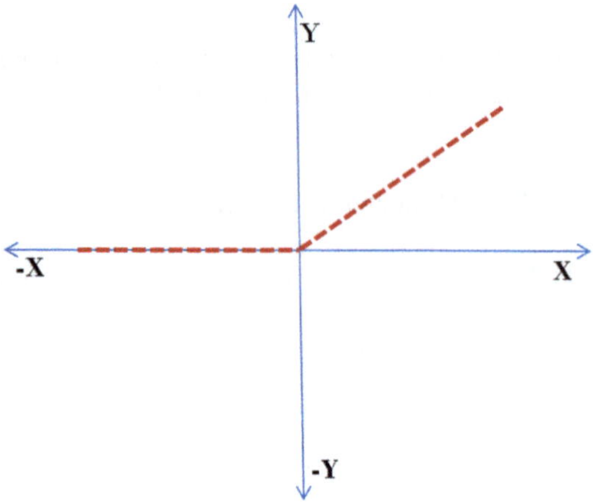

Fig. 11.17 Rectified Linear Unit (ReLU) activation function

where $\phi(z)$ denotes the output of the activation function for an input z.
Example: Multilayer Neural Network, Convolutional Neural Network.

11.3.8 K-Fold Cross-Validation

In order to assess a prediction model's performance and solve overfitting and bias issues, k-fold cross-validation is a technique that is widely used in machine learning and statistics. A dataset is divided into K folds, each of a comparable size. Following that, K iterations of the training and evaluation procedure are carried out, each time utilizing a new fold as the validation set and the remaining folds as the training set. Table 11.4 lists the benefits and drawbacks of each kind.

Table 11.3 Activation functions

Techniques	Advantages	Limitations
Linear	It is basic and comprehensible	For projects requiring the capture of intricate, non-linear patterns in the data, it might not be appropriate
Unit Step Heaviside Term	It is simple and understandable	Its restricted expressiveness, smoothness, and lack of differentiability make it unsuitable for neural network training using contemporary optimization approaches
Sign	It is easy to compute and generates binary output	It is less suited for training contemporary neural networks with gradient-based optimization techniques due to its non-differentiability and restricted expressiveness
Piecewise Linear	It is adaptable for a range of activities because it provides a balance between linearity and non-linearity	To fully utilize it's benefits, breakpoint selection, and careful design considerations are essential
Logistic	It is extensively utilised, particularly in applications involving binary classification at the output layer. Its appeal is attributed to its smoothness, interpretability, and avoidance of the vanishing gradient issue	For some applications, its ability to handle saturation, zero-centering, and output range may be limited
tanh	It is more successful in some situations because of its zero-centered output and larger output range when compared to the logistic function	The range of output is limited and is not used for regression
ReLU	The simplicity and effectiveness of this technique make it popular in practise, particularly for deep neural networks' hidden layers	ReLU-activated neurons may become dead during training, and it is sensitivity to outliers

11.3.8.1 Leave One Out Cross-Validation (LOOCV)

A rigorous approach used in model evaluation and parameter adjustment is known as Leave-One-Out Cross-Validation (LOOCV). In LOOCV, the model is trained using a subset of the data while each data point is alternately held out as a validation set.

Table 11.4 K-fold cross validation

Techniques	Advantages	Limitations
LOOCV	It ensures optimum data utilisation by excluding one data point for validation, providing an unbiased and low-variance assessment of model performance	For big datasets, LOOCV can be computationally costly, and its efficacy could be reduced if the dataset contains outliers or is prone to high fluctuation
5-fold validation	It provides a solid assessment of performance with less computation than LOOCV by striking a compromise between computational economy and trustworthy model evaluation	The dataset attributes, the number of folds, and the specific data split can all have an impact on the method's performance
10-fold validation	It divides the data into 10 subgroups to provide a dependable assessment of performance while striking a balance between computational economy and robust model evaluation	Large datasets may need a lot of processing power, and the features of the dataset may affect how effective it is

11.3.8.2 5-Fold Validation

A popular method for assessing model performance is 5-fold cross-validation. This method splits the dataset into five subsets, or folds, and trains the model on four of them while holding off on the fifth for validation.

11.3.8.3 10-Fold Validation

A reliable and popular method for evaluating the effectiveness of models is 10-fold cross-validation. The dataset is splitted into ten subsets or folds, and the model goes through ten iterations of training and testing.

11.4 Analysis of Result

Table 11.5 presents the classifiers' accuracy, sensitivity, specificity, f1 score, and precision utilized in the article. Table 11.6 compares the suggested work with other methods already in use.

The disparity of the f1 score between the categorization models is shown graphically in Fig. 11.18, and the comparison with other methods already in use is shown graphically in Fig. 11.19.

Table 11.5 Evaluation metrics

Model	Precision (%)	Sensitivity (%)	Specificity (%)	F1 Score (%)	Accuracy (%)
SVM	98.86	98.86	99	98.86	98.93
RF	99.14	99.14	99.25	99.14	99.20
AdaBoost	98.86	98.30	99	98.58	98.67
DT	98.29	98.29	98.50	98.29	98.40
LDA	98	98	98.25	98	98.14
RBF	99.43	98.31	99.50	98.86	98.93
ANN	98.29	98.85	98.29	98.57	98.67

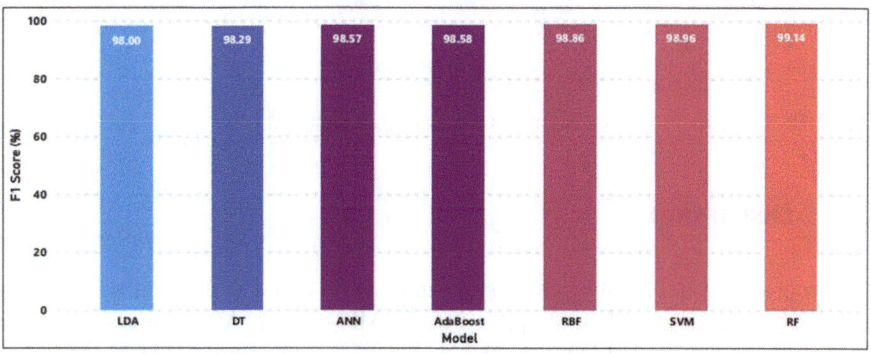

Fig. 11.18 Comparison of F1 score between the different proposed models

Table 11.6 Comparison of proposed model with existing techniques

References	Model used	Accuracy (%)
Kanmani and Marikkannu [9]	TBRO-segmentation	96.57
Kumar and Mankame [11]	Dolphin-SCA	95.3 & 96.3
Alam et al. [12]	TK-FCM	97.5
Islam et al. [13]	TK-means	95
Anaraki et al. [14]	CNN & GA based	90.9 & 94.2
Bahadure et al. [15]	BWT	96.51
Sultan et al. [16]	CNN based	96.13 & 98.7
proposed ML based classifiers	LDA	98.14
	DT	98.40
	AdaBoost	98.67
	SVM	98.93
	RF	99.20
	ANN	98.67
	RBF	98.93

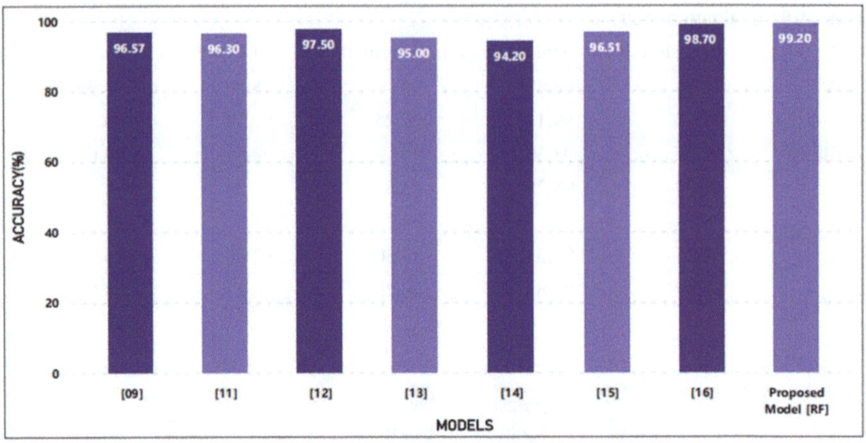

Fig. 11.19 Comparison of accuracy with other existing techniques

11.5 Discussion

The proposed model involves three steps in total. To enhance the quality of the MR pictures, pre-processing is first applied to the MR images. After the completion of pre-processing, important features such as statistical, transform domain, and technical features were extracted to obtain the desired feature vector. Subsequently, each feature vector is fed to each of the proposed models, for training and validation purposes.

Each model is trained to a reasonable degree, and then its performance is assessed and contrasted. Subsequently, the best two models have been identified as RBF, and RF respectively. To demonstrate the robustness of the developed model, performance needs to be evaluated using other standard datasets. Further, the potentiality of each of the proposed models needs to be assessed using imbalanced feature data. The suggested approach and models may be applied to different types of disease recognition and classification which requires an image dataset as input.

11.6 Conclusion

This chapter presents a set of ML-based classifiers that may be used to identify brain tumor using standard MR-based image input. Subsequently, requisite features have been extracted from these raw images and fed to each of these models for achieving satisfactory training.

In the second stage, each of the developed models has undergone different validation schemes. The performance of every model that was generated has been evaluated and compared in the third stage. It is demonstrated that, among the seven study models, the suggested RF model has the greatest accuracy (99.20%) and F1 score (99.14%).

References

1. Li, H., Li, A., Wang, M.: A novel end-to-end brain tumor segmentation method using improved fully convolutional networks. Comput. Biol. Med. **108**, 150–160 (2019)
2. Zacharaki, E.I., Wang, S., Chawla, S., Soo Yoo, D., Wolf, R., Melhem, E.R., Davatzikos, C.: Classification of brain tumor type and grade using MRI texture and shape in a machine learning scheme. Magn. Reson. Med. Off. J. Int. Soc. Magn. Reson. Med. **62**(6), 1609–1618 (2009)
3. Ghassemi, N., Shoeibi, A., Rouhani, M.: Deep neural network with generative adversarial networks pre-training for brain tumor classification based on MR images. Biomed. Signal Process. Control **57**, 101678 (2020)
4. Saba, T., Mohamed, A.S., El-Affendi, M., Amin, J., Sharif, M.: Brain tumor detection using fusion of hand crafted and deep learning features. Cogn. Syst. Res. **59**, 221–230 (2020)
5. Duan, L., Yuan, G., Gong, L., Fu, T., Yang, X., Chen, X., Zheng, J.: Adversarial learning for deformable registration of brain MR image using a multi-scale fully convolutional network. Biomed. Signal Process. Control **53**, 101562 (2019)
6. El-Melegy, M.T., Mokhtar, H.M.: Tumor segmentation in brain MRI using a fuzzy approach with class center priors. EURASIP J. Image and Video Process. **2014**(1), 1–14 (2014)
7. Kharrat, A., Gasmi, K., Messaoud, M.B., Benamrane, N., Abid, M.: A hybrid approach for automatic classification of brain MRI using genetic algorithm and support vector machine. Leonardo J. Sci. **17**(1), 71–82 (2010)
8. Shakeel, P.M., Tobely, T.E.E., Al-Feel, H., Manogaran, G., Baskar, S.: Neural network based brain tumor detection using wireless infrared imaging sensor. IEEE Access **7**, 5577–5588 (2019)
9. Kanmani, P., Marikkannu, P.: MRI brain images classification: a multi-level threshold based region optimization technique. J. Med. Syst. **42**, 1–12 (2018)
10. Zhang, J., Lv, X., Zhang, H., Liu, B.: AResU-Net: attention residual U-Net for brain tumor segmentation. Symmetry **12**(5), 721 (2020)
11. Kumar, S., Mankame, D.P.: Optimization driven deep convolution neural network for brain tumor classification. Biocybern. Biomed. Eng. **40**(3), 1190–1204 (2020)
12. Alam, M.S., Rahman, M.M., Hossain, M.A., Islam, M.K., Ahmed, K.M., Ahmed, K.T., Singh, B.K., Miah, M.S.: Automatic human brain tumor detection in MRI image using template-based K means and improved fuzzy C means clustering algorithm. Big Data Cogn. Comput. **3**(2), 27 (2019)
13. Islam, M.K., Ali, M.S., Miah, M.S., Rahman, M.M., Alam, M.S., Hossain, M.A.: Brain tumor detection in MR image using superpixels, principal component analysis and template based K-means clustering algorithm. Mach. Learn. Appl. **5**, 100044 (2021)
14. Anaraki, A.K., Ayati, M., Kazemi, F.: Magnetic resonance imaging-based brain tumor grades classification and grading via convolutional neural networks and genetic algorithms. Biocybern. Biomed. Eng. **39**(1), 63–74 (2019)
15. Bahadure, N.B., Ray, A.K., Thethi, H.P.: Image analysis for MRI based brain tumor detection and features extraction using biologically inspired BWT and SVM. Int. J. Biomed. Imaging (2017)
16. Sultan, H.H., Salem, N.M., Al-Atabany, W.: Multi-classification of brain tumor images using deep neural network. IEEE Access **7**, 69215–69225 (2019)
17. Nanda, A., Barik, R.C., Bakshi, S.: SSO-RBNN driven brain tumor classification with Saliency-K-means segmentation technique. Biomed. Signal Process. Control **81**, 104356 (2023)
18. Li, M., Kuang, L., Xu, S., Sha, Z.: Brain tumor detection based on multimodal information fusion and convolutional neural network. IEEE Access **7**, 180134–180146 (2019)

19. Panda, S.K., Barik, R.C.: MR Brain 2D image tumor and cyst classification approach: an empirical analogy. In 2023 IEEE International Students' Conference on Electrical, Electronics and Computer Science (SCEECS), pp. 1–6. IEEE (2023)
20. Ahmad, S., Choudhury, P.K.: On the performance of deep transfer learning networks for brain tumor detection using MR images. IEEE Access **10**, 59099–59114 (2022)

Chapter 12
Cyber-Physical Security in Smart Grids: A Holistic View with Machine Learning Integration

Bhaskar Patnaik, Manohar Mishra, and Shazia Hasan

Abstract Cyber-physical attacks are become more challenging in each passing days owing to the continuous advancement of smart-grid systems. In the present industrial revolution, the smart grid is integrated with a wide-range of technologies, equipment/ devices and tools/software to make the system more trustworthy, reliable, efficient, and cost-effective. Regardless of achieving these objectives, the peril area for the critical attacks has also been stretched owing to the add-on cyber-layers. In order to detect and mitigate these attacks, the machine learning (ML) tools are being reliably and massively used. In this chapter, the authors have reviewed several state-of-the-art related researches comprehensively. The advantages and disadvantages of each ML based schemes are identified and reported in this chapter. Finally, the authors have presented the shortcomings of the existing researches and possible future research direction based on their investigation.

Keywords Cyber-physical · Attacks · Machine learning · Industrial revolution · Smart grid

B. Patnaik
Nalla Malla Reddy Engineering College, Hyderabad, Telangana, India

M. Mishra (✉)
Department of Electrical and Electronics Engineering, Siksha O Anusandhan University, Bhubaneswar, India
e-mail: manohar2006mishra@gmail.com

S. Hasan
Department of Electrical and Electronics Engineering, Birla Institute of Technology & Science, Dubai Campus, Dubai, United Arab Emirates
e-mail: shazia.hasan@dubai.bits-pilani.ac.in

J. Nayak et al. (eds.), *Machine Learning for Cyber Physical System: Advances and Challenges*, Intelligent Systems Reference Library 60,
https://doi.org/10.1007/978-3-031-54038-7_12

12.1 Introduction

The International Energy Agency (IEA), established in 1974 and collaborating with governments and industry to forge a secure and sustainable energy future for all, characterizes a smart grid as:

> Smart grids are electrical networks incorporating digital technologies, sensors, and software to efficiently synchronize electricity supply and demand in real-time. This is achieved by minimizing costs and ensuring the stability and reliability of the grid [1].

A fairly more explicit definition of smart grid can be found as furnished by the "National Smart Grid Mission, Ministry of Power, Government of India" [2], which is:

> A Smart Grid refers to an electrical grid equipped with automation, communication, and IT systems that oversee power distribution from generation points to consumption points, including individual appliances. It can regulate power flow and adjust loads in real-time or near-real-time to align with current generation levels. Realizing Smart Grids involves implementing efficient transmission and distribution systems, improving system operations, integrating consumers effectively, and seamlessly incorporating renewable energy sources.

Smart grid solutions play a crucial role in monitoring, measuring, and controlling power flows in real-time, enabling the identification of losses. This functionality allows for the implementation of suitable technical and managerial measures to mitigate these losses. The deployment of smart grid solutions can significantly contribute to reducing transmission and distribution (T&D) losses, managing peak loads, enhancing service quality, improving reliability, optimizing asset management, integrating renewable energy sources, and increasing electricity accessibility. Furthermore, smart grids have the potential to create self-healing grids. In essence, smart grid solutions provide a comprehensive approach to addressing various challenges within the electrical grid, fostering more efficient and sustainable energy management. A smart grid is a futuristic electrical power grid which is aimed to evolve in order to be able to address the varying needs of global consumer and global concerns. A general architecture of smart grid in block diagram form is represented in Fig. 12.1. With increasing population there has been tremendous increase in the demand of power. With changing lifestyle needs and awareness, consumers have become more discerning in context of quality of power. While fast depleting natural source of energy is a growing concern, arresting the environmental degradation by limiting the carbon emission has also become a global concern. All these factors have hastened up the search for solutions that should help generate more electrical power by sustainable means, should be environmentally friendly, should facilitate cost effective quality power with highly reliable, stable, resilient servicing, and last but not the least, should ensure data privacy and security in the event of increasing consumer participation in the process. Although the above-mentioned advantages of smart grids are substantial, it is crucial to acknowledge the existence of significant challenges, including advanced system complexities, monitoring and control intricacies, and the paramount issue of cybersecurity.

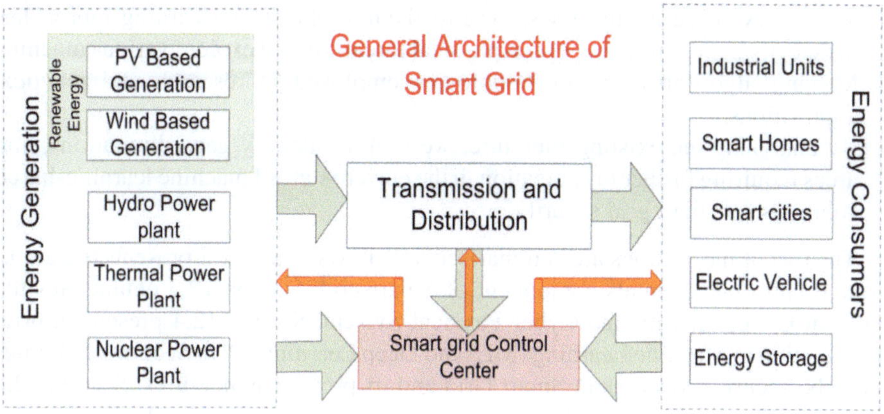

Fig. 12.1 A conceptual architecture of smart grid

The increasing importance of cybersecurity in smart grids highlights a pressing necessity for robust measures to safeguard these sophisticated electrical systems. Smart grids heavily rely on automation, communication, and information technology, rendering them more susceptible to cyber threats. It is crucial to implement stringent measures to protect against malicious activities in order to ensure the integrity, reliability, and security of smart grid operations. Given the central role of interconnected devices and communication networks in smart grids, the potential attack vectors for cyber threats are on the rise. A breach in cybersecurity could compromise data integrity, disrupt power flow controls, and result in unauthorized access, posing serious risks to the overall functionality of the grid. As smart grids continue to advance with the incorporation of more sophisticated technologies, the need for proactive cybersecurity measures becomes paramount. Investments in cutting-edge security protocols, continuous monitoring, and the development of resilient frameworks are essential to counteract cyber threats and uphold the trust and dependability of smart grid systems. Recognizing and addressing cybersecurity concerns is fundamental, especially in light of the growing interconnectivity of critical infrastructure, to ensure the long-term viability and success of smart grids.

The primary objective of this chapter is to offer a comprehensive examination of the role of machine learning in the context of cyber-physical attack detection and mitigation within smart grid systems. As the smart grid evolves into a critical component of modern energy infrastructure, the increasing integration of digital technologies exposes it to various cybersecurity threats. Recognizing the significance of these threats, the purpose of this review is to synthesize existing knowledge and advancements in leveraging machine learning techniques to safeguard smart grids from cyber-physical attacks. The major objectives are highlighted as follows:

- Here, the authors aim to provide an in-depth understanding of the cybersecurity challenges faced by smart grids, emphasizing the unique nature of cyber-physical attacks that exploit the interconnectedness of digital and physical components.

- This review will critically assess the effectiveness of machine learning approaches in detecting and mitigating cyber-physical threats. It will explore various machine learning algorithms and methodologies employed in research and practical applications.
- By analyzing the existing literature, we seek to identify gaps, limitations, and areas requiring further investigation in the current state of machine learning-based solutions for smart grid security.

The rest of the sections are summarised as follows: Sect. 12.2 presents the Background and Fundamentals component of Smart-grid. Section 12.3 enumerates the basic of Cyber security and Cyber Physical System. Section 12.4 presents a brief introduction to Machine Learning (ML) and Deep Learning (DL). Section 12.5 states the cybersecurity concerns in Smart Grid and its protective measures. Section 12.6 deals with the associated challenges and future directions. Section 12.7 concludes with overall concluding remarks.

12.2 Background and Fundamentals Component of Smart-Grid

While technologies and infrastructures like, microgrid, smart metering, advanced communication systems, distributed renewable/non-renewable energy sources, electric vehicles etc. are considered as smart grid enabler, the compositional architecture of a smart grid can be divided into three major subcomponents; namely Operational Technology (OT), Information Technology (IT), and Advanced Metering Infrastructure [3].

12.2.1 Advanced Metering Infrastructure

Figure 12.2 illustrates the structure of the Advanced Metering Infrastructure (AMI). At the core of the AMI are smart meters installed at both small- and large-scale consumer locations. These smart meters, distinguished from traditional energy meters, are fully digital devices equipped with a range of additional features and functionalities, as detailed in Table 12.1.

The AMI functions as a wireless network comprising smart meters, enabling various smart services such as remote billing, monitoring of supply–demand management, integration and oversight of distributed energy sources, consumer engagement, and energy conservation, among others. Essentially, the AMI structure forms a communication network that facilitates interaction among the smart grid central control server, aggregators, and power consumers. In a smart grid environment, a smart home connects all its appliances through a Home Area Network (HAN), transmitting data to the smart meter via Wi-Fi, ZigBee, or Wide Area Network

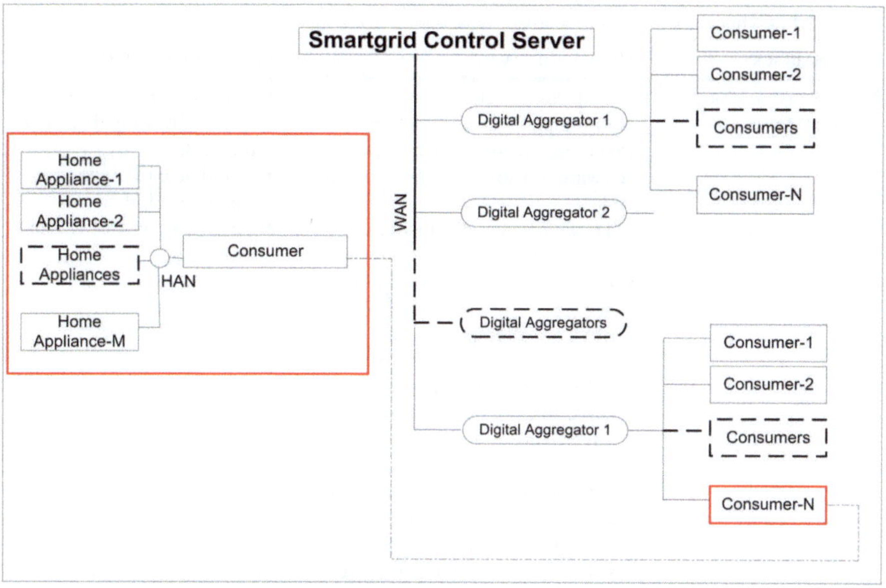

Fig. 12.2 Layout of advanced metering infrastructure [3]

(WAN). Smart meters, strategically placed in homes and diverse consumer locations (e.g., factories, offices, social infrastructures), convey crucial information, including power consumption and related data, to the aggregators through a Neighborhood Area Network (NAN). The collected data is then forwarded to the central control server.

The smart grid leverages this data to make informed decisions and implement necessary measures to ensure a stable power supply, considering the fluctuating power demand from consumers. Specific Smart Grid (SG) enabling devices, such as Electric Vehicles (EVs) in a Vehicle-to-Grid (V2G) network, utilize the AMI network based on technologies such as WiMAX, LTE, Wi-Fi, or WAN. Furthermore, power plants and generators communicate their status data to the Smart Grid through Power Line Communication (PLC).

12.2.2 Operational Technology Component

The operational technology (OT) in a smart grid structure can be visualized as a multi layered structure consisting of Industrial Control System (ICS), Power Line Communication (PLC), Distributed Control System (DCS), and Supervisory Control and Data Acquisition (SCADA) system as illustrated in Fig. 12.3 [3–5]. ICS is the controlling network that operate and automate industrial processes. SCADA is involved in gathering data through Remote Terminal Units (RTUs) from PLCs. It

Table 12.1 Cyber threats and preventive measures

Type of attack	How do they work?	Prevention methods
Malware Attack (also, Fileless Malware)	A wide spread and most commonplace virus which includes malicious software such as worms, spyware, ransomware, adware, and trojans • Trojan appears disguises as a legitimate software • Ransomware attempts to block access to a network's key features • Spyware attempts to steal confidential data stealthily • Adware is a nuisance (at sometimes) that pops up advertising contents on a user's display • Malware exploits the vulnerability of a network, such as when a naïve user clicks a dangerous link solicitated to be a useful one or when an infected pen drive is used	• Use antivirus software • Use of firewalls that filters the traffic • Avoidance of clicking on suspicious links • Regular update of OS and browsers
Phishing Attack [Can be an Identity based attack] (Whale-Phishing Attacks, Spear-Phishing Attacks, Angler Phishing Attacks, Spamming) and Spoofing	It is often manifest in the form of widespread social engineering attacks, where the attacker assumes the identity of a trusted contact and sends deceptive emails to the victim. In this scenario, the unsuspecting recipient may disclose personal information or perform actions as directed by the hacker, granting them access to confidential data and account credentials. Phishing is a common method employed in such attacks, wherein malware can be surreptitiously installed Various types of phishing attacks exist; for instance, Whale-Phishing targets high-profile individuals, while Spear-Phishing is directed at specific individuals or groups within an organization. These attacks leverage social engineering techniques to extract sensitive information	• Thorough scrutiny of emails to look mails which may have significant errors or uncharacteristic format indicating to a possible phishing email • Use of an anti-phishing toolbar • Regular password update
Password Attack (Corporate Account Takeover (CATO), Dictionary Attacks)	Hacker attempts to crack the password using various programs and password cracking tools There exist several password attacks types, such as brute force, dictionary, and keylogger attacks	• Use of strong passwords • Avoiding repeated use of same password for multiple accounts • Regular update of passwords

(continued)

Table 12.1 (continued)

Type of attack	How do they work?	Prevention methods
Man-in-the-Middle Attack (MITM) (Eavesdropping) [Can be an Identity based attack]	A Man-in-the-Middle Attack (MITM) is the type of, wherein the attacker comes in between a two-party communication, i.e., instead of a direct client server communication, the communication line gets routed through the hacker	• Remain careful of the security of the website being used. Use of encrypted devices • Abstain from use of public Wi-Fi networks
SQL Injection Attack (Code Injection Attacks)	An attack that is done on a database-driven website wherein the attacker manipulates a standard SQL query The hacker becomes capable of viewing, editing, and deleting tables in the databases. Attackers can also get administrative rights through this	• Use of intrusion detection system, that are designed to detect unauthorized access to a network • Carry out a validation of the user-supplied data in order to keep the user input in check
Denial-of-Service Attack	DoS pose a significant threat to organizations and businesses, particularly those dealing with extensive data and offering critical, time-sensitive services. In a DoS attack, the target systems, servers, or networks are flooded with traffic to the point of exhausting their resources and bandwidth. This results in the inability of servers to respond to incoming requests, leading to the potential shutdown or slowdown of the host server. In many cases, legitimate service requests go unaddressed When attackers employ multiple compromised systems to execute a DoS attack, it is termed a Distributed Denial-of-Service (DDoS) attack. Bots and botnets, which are software programs, can facilitate the execution of DDoS attacks	• Running of traffic analysis to identify malicious traffic • Understand the warning signs like network slowdown, intermittent website shutdowns, etc. and actions must be taken immediately • Outsourcing of DDoS prevention to cloud-based service providers
Insider Threat	The attacker could be an individual from within the organization privy of significant amount of information, potential enough to cause tremendous damages	• Organizations should have a good culture of security awareness and must have limited staff having access to the IT resources

(continued)

Table 12.1 (continued)

Type of attack	How do they work?	Prevention methods
Cryptojacking	The hacker's objective is to illicitly access an individual's computer for the purpose of cryptocurrency mining This unauthorized access is achieved by tricking the victim into using an infected website, clicking on a malicious link, or interacting with JavaScript-encoded online advertisements. While the victim waits for the execution of any of the above events (which unusually takes longer time to get executed), the crypto mining code keeps working in the background	• Update software and all the security apps regularly • Use of ad blocker as ads are a primary source of cryptojacking scripts • Use of extensions like Miner Block, to identify and block crypto mining scripts
. Zero-Day Exploit	When certain vulnerability is made known by a user, which evidently comes to the knowledge of not only all users but also the hackers, the vulnerability period (i.e., the time in between happening and fixing up the loophole) provides the hacker to exploit the situation	• Well organized patch management processes, automated management solutions, and incident response plan must be in force by the organization in order deal with such type of attacks
Watering Hole Attack	The target of the hacker in this case happens to be a particular group of an organization, region, etc. who frequent specific websites. The hackers infect these websites with malware, which in turn infects the victims' systems. The hacker not only gains access to the victim's personal information but also has remote access to the infected computer	• Regular software update • Use of network security tools • Use of intrusion prevention systems (IPS) • To go for concealed online activities
DNS Tunnelling & DNS Spoofing [can be Backdoors type]	The assailant leverages the Domain Name System (DNS) to circumvent security measures and establish communication with a remote server This involves a cyberattack where the attacker manipulates the DNS records of a website, gaining control over its traffic and potentially redirecting it for malicious purposes	Regular monitoring of DNS traffic for: • Anomaly detection • Payload analysis • Rate limiting (limiting of DNS queries) • Intrusion Detection System
IoT-Based Attacks	Leveraging weaknesses in Internet of Things (IoT) devices, such as smart thermostats and security cameras, is employed to unlawfully pilfer data	Maintaining separate networks for IoT devices Using security tools to ensure IoT devices spoofing proof

(continued)

Table 12.1 (continued)

Type of attack	How do they work?	Prevention methods
URL Interpretation	Competent attackers use URL (web address), rewriting in several programming languages in order to achieve malicious objectives	Web developers need to enforce security measures, such as input validation and proper data sanitization
Birthday Attack	Is sort of a cryptographic attack, based on the mathematics behind the birthday problem in probability theory	• Use of secure hash functions with large hash code length • Implement slated hashing • Regular update of hash algorithms
Protocol Attacks:	Capitalizing on vulnerabilities within network protocols, an attacker seeks unauthorized entry into a system or disrupts its normal operation. Illustrative instances encompass the Transmission Control Protocol (TCP) SYN Flood attack and the Internet Control Message Protocol (ICMP) Flood attack	To thwart protocol attacks, organizations should deploy firewalls, intrusion prevention systems, and encryption. Network segmentation limits the impact, while regular updates and patching address vulnerabilities. Anomaly detection, access controls, and user education enhance security. Deep packet inspection and monitoring aid in detecting irregularities, and a well-defined incident response plan ensures a prompt and coordinated reaction
Application Layer Attacks	This focuses on the application layer of a system, with the objective of capitalizing on vulnerabilities within applications or web servers	Prevent application layer attacks by employing Web Application Firewalls, secure coding practices, regular audits, input validation, session management, Content Security Policies, Multi-Factor Authentication, rate limiting, and timely software updates
AI-Powered Attacks	Employing artificial intelligence and machine learning techniques to circumvent conventional security measures	To defend against AI-powered attacks, organizations should implement robust cybersecurity strategies, incorporating advanced threat detection systems, regular updates, anomaly detection, user awareness training, and adaptive security measures

(continued)

Table 12.1 (continued)

Type of attack	How do they work?	Prevention methods
Rootkits	Granting attackers privileged access to a victim's computer system, rootkits serve as tools to conceal various types of malware, including spyware or keyloggers. Their ability to evade detection and removal poses a significant challenge	Guard against rootkits by utilizing advanced anti-malware tools, conducting regular system scans, implementing secure boot processes, practicing principle of least privilege, and maintaining up-to-date software and firmware to address vulnerabilities
Advanced Persistent Threat (APT)	Is a cyberattack characterized by long-term, persistent access to a victim's computer system. APT attacks are highly sophisticated and difficult to detect and remove	To thwart Advanced Persistent Threats (APTs), organizations should deploy sophisticated threat detection systems, conduct regular security audits, enforce robust access controls, employ encryption, educate users on phishing risks, and establish an incident response plan for swift and effective mitigation

processes the gathered data and sends action messages back to the PLCs, which in turn makes the devices connected to it to conduct the pre-defined procedures. DCS is the controlling mechanism that operates machines under the ambit of SCADA infrastructure.

It may be inferred from above descriptions that operation technology (OT) is the terminology refers to a large system that encompasses the monitoring and control of the whole gamut of activities that is executed by its sub-component ICS.

In order to understand the operational methodologies of ICS, it may be visualized as composition of several layer. Systems responsible for infrastructure operations make for the supervisory layer. The physical components in a given facility make for the physical layer [4]. As each of these systems or physical components are designed and manufactured by different companies and they invariably use different protocols of their choice, it is imperative that each layer will have reliance on different network types resulting in data and signal incompatibility amongst them. In this situation an Open Platform Communication (OPC) server id engaged to provision for a common platform of interface of these layers with the management server of the enterprise management layer. Servers engaged in the field layer record the historical and real time data pertaining to its connected devices, which are used to enable the system to bounce back from an abnormal state to the normal one. The field layer also uses the services of data acquisition layer, which manages multiple RTUs, PLCs, MTUs and IEDs, and ensures synchronization of communications amongst them. IED is another important device which helps protect the smart grid through

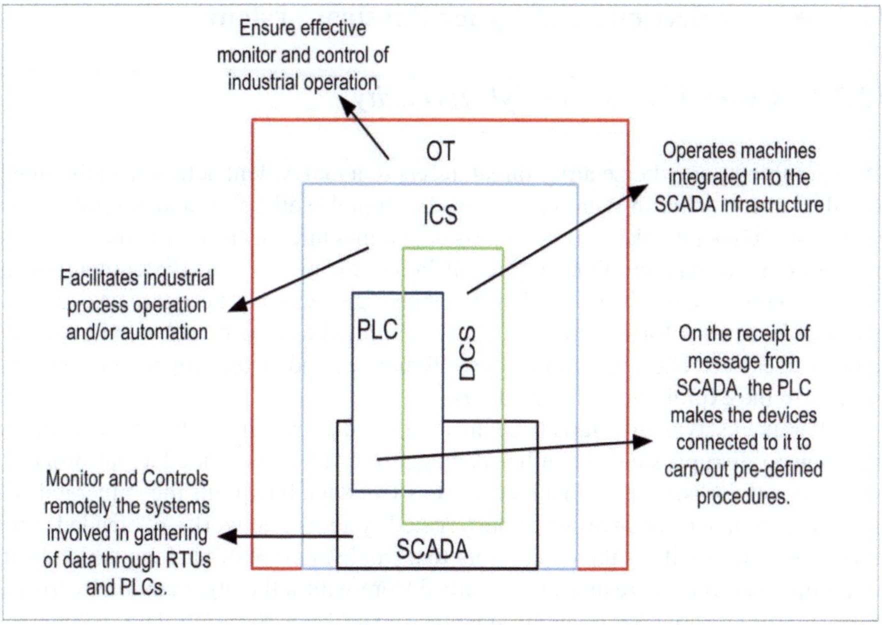

Fig. 12.3 Components of operational technology [3]

blocking procedures before the incident of any critical system failure. The OT is also equipped with authentication servers and application servers, which are part of SCADA infrastructure and facilitate authenticated user access and enable system systems-devices compatibility. A Human Machine Interface (HMU) also makes a part of OT to facilitate operator's interface with the linked apparatuses.

12.2.3 Information Technology Component

Information technology is pivotal to the management of Smart Grid business enterprise. It helps the managers in decision making, resource management, monitoring of production, logistics, stocks, purchases, accounting, and sales etc. It supports monitoring of work process for efficient manufacturing. Multiple IT based servers and software systems, such as ERP. MES. MIS etc. help serve to this end [6].

12.3 Cyber Security and Cyber Physical System

12.3.1 Cyber Threat and Cybersecurity

"A cyber threat, or cybersecurity threat, refers to a malevolent action with the intent to either pilfer or harm data, or disrupt the digital well-being and stability of an enterprise." Cyberattacks can encompass both unintentional and deliberate activities, posing potential dangers that may result in significant harm to the computational systems, networks, or other digital assets of an organization. These threats or attacks manifest in various forms, including but not limited to data breaches, viruses, and denial of service. The spectrum of cyber threats extends from trojans, viruses, and hackers to the exploitation of back doors.

Cyberattacks typically target the unauthorized acquisition of access, with the intention to disrupt, steal, or inflict damage upon IT assets, intellectual property, computer networks, or any other form of sensitive data. It exploits the vulnerabilities in a system to launch an invasion of the targeted system or network. A "blended cyber threat", which usually is the case, refers to a single attempt of hacking which leads to multiple exploits. Threats can be sourced from within the organization by trusted users or from remote locations by unknown external parties. While a cyberattack of the type adware may have inconsequential effect, an attack of type denial-of-service can have catastrophic effect on an organization. The impact of cyberattacks can be as severe as electrical blackouts, malfunctions in military equipment, or the compromise of national security secrets. In short, it affects each aspect of our life. The Table 12.1 provides an exhaustive list of cyber threats, how they act and plausible counter measures in some of these cases.

The significance of cybersecurity in this context can be succinctly outlined as follows [7]:

- Guards sensitive information against unauthorized access or theft.
- Provides protection from cyber threats like malware, viruses, and ransomware.
- Ensures the integrity and confidentiality of digital systems and networks.
- Averts disruptions to critical services and operations.
- Mitigates financial losses and preserves the reputation of businesses.
- Assists in compliance with legal and regulatory standards.
- Builds trust and confidence among customers and users.
- Enhances secure communication and collaboration within organizations.
- Facilitates the safe integration of emerging technologies such as cloud computing and the Internet of Things (IoT).

The goals of cybersecurity can be succinctly outlined as follows [7]:

- **Confidentiality of Data**: Ensuring protection against unauthorized access.
- **Integrity of Data and Information**: Safeguarding against unauthorized alterations.

- **Data Availability**: Ensuring data is accessible when needed.
- **Authentication**: Verifying the identity of users or systems.
- **Authorization**: Granting appropriate access permissions.
- **Auditing and Monitoring**: Continuous surveillance for suspicious activities.
- **Incident Response**: Swift and effective actions in response to security incidents.
- **Non-repudiation**: Preventing denial of actions by parties involved.
- **Security Awareness and Training**: Educating users to recognize and address security threats.
- **Compliance**: Adhering to legal and regulatory requirements.
- **Continuous Improvement**: Ongoing enhancements to adapt to evolving cyber threats.

However, the three major objectives of cyber security, commonly referred as CIA, stands on the three pillars—Confidentiality, Integrity, and Availability. People, Processes, and Technology come together to attain these objectives of cyber security and ensure effective security system [8].

12.3.2 Cyber Physical System: Smart Grid

Overall, a Smart Grid, as smart as it gets, heavily relies upon a behemoth of digital network comprising of fast communication channels dealing with humongous flow of data, which are processed, filtered, and applied with intelligent computational methods in order to come out with instantaneous solutions that help not only to operate, maintain and protect the physical devices in the smart grid but also help run the smart grid enterprise and increased consumer participation. In fact, this digital layer over the physical entities of the grid system, intricately connected exchanging data and information with each other, makes for what is so called as a Cyber Physical System (CPS).

> The smart grid is recognized as a quintessential cyber-physical system (CPS), embodying an integration of physical power systems with cyber components. This fusion encompasses elements such as sensing, monitoring, communication, computation, and control within the smart grid framework. [9, 10]

Needless to say that such a vast network of networks that the cyber physical system a smart grid is definitely is vulnerable to threats that a cyber system is susceptible to notwithstanding the usual share of protection issues that is generally accrued to the physical components of a smart grid [11, 12]. Any kind of infringement in the cyber layer of the smart grid can have colossal damaging effect and a smart grid needs to be smart enough to shield itself from such cyber infringements or cyber threats; an additional and most important technical challenge that can be ascribed to the evolution of smart grid.

12.4 A Brief Introduction to Machine Learning (ML) and Deep Learning (DL)

Before delving into the deliberation on the significance of machine learning (ML) in enhancing cybersecurity, it is important to understand what machine learning is all about and its context in view of Artificial Intelligence (AI), as too often it is observed that both AI and ML are used to refer to the same activity. While ML is considered a subset of AI, the subtle difference between them in terms of their deployment is generally misunderstood.

ML can be viewed as a class of statistical tools which identifies the relationships and patterns in a given set of data and the process build up to a ML model which represents the event or phenomena that the data pertains to. In the same coin the AI can be viewed as a software that aligns the tool (i.e., ML in this case) with a controller that takes action based on the tool's output. The tools can be any other suitable algorithm, such as a logic or am expert system to implement the AI [13]. To put it in a simpler way, the ML tool initiates a training phase wherein the ML model learns automatically analyzing the available data set (training data set). Such a model developed through training on existing data implements a function to make decisions on future data. The performance of the ML model is assessed before deploying it into the intended operational environment, an exercise known as validation. In pursuit of this objective, the machine learning (ML) model processes designated "validation" data, and the resulting predictions undergo analysis by humans or are compared against established ground truth. Consequently, a machine learning method is delineated as "the process of constructing a machine learning model through the application of ML algorithms on specific training data" [14].

Based on the data type, labelled or non-labelled, training of ML methods can be either supervised or unsupervised respectively. Labelled training data usually available naturally. If not, labels can be attributed to the training data through manual verification. In contrast unsupervised training do not require labelled data and may involve a feedback process, acquiring the labels automatically as the ML model develops. The ML model based on reinforcement learning is such an instance of unsupervised learning.

The ML methods, on the other hand, can be classified as shallow and deep learning type. Deep learning methods bank upon neural networks and require greater computational power with larger training dataset in comparison to shallow ML methods (based on structures/algorithms/logics other than neural networks). It is important to note that deep learning performs much better than shallow methods when it is needed to handle large datasets with high complexities, whereas shallow methods perform equally well while data available has small number of features. Nevertheless, deep learning ML methods stand out while dealing with large dataset with varied complexities involving images, unstructured text, temporal dependencies etc. besides being trained both supervised or unsupervised manner [15–17]. The Fig. 12.4 enumerates some of the popular ML algorithms under the categories as discussed above. A brief description of the ML algorithms as depicted in the Fig. 12.4 is as follows:

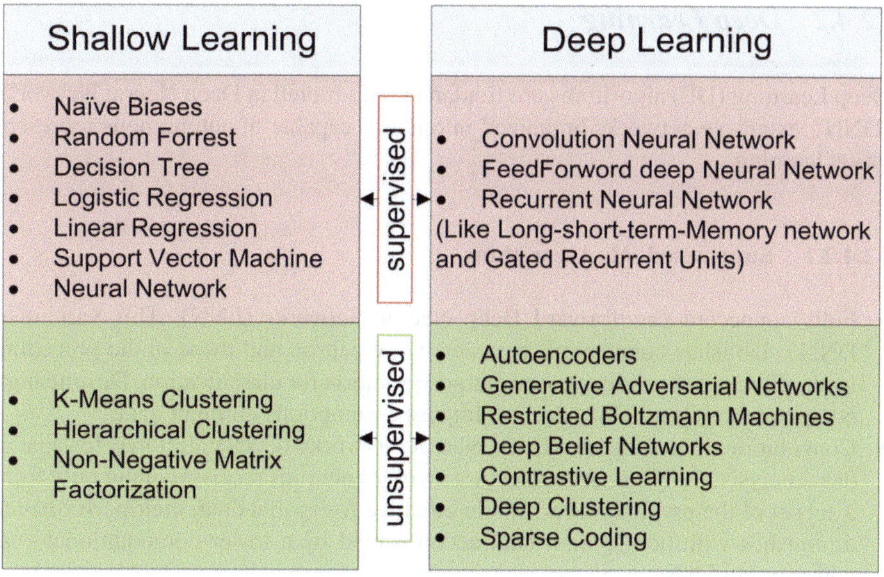

Fig. 12.4 Taxonomy of machine learning techniques

12.4.1 Shallow Learning

Several popular shallow machine learning algorithms that rely on supervised learning are described below. Naïve Bayes (NB) is a probabilistic classifier that assumes a priori independence among input features, making it efficient for small datasets. Logistic Regression (LR), a categorical classifier, shares a similar a priori assumption as NB but is increasingly reliant on larger datasets for effective training. Support Vector Machines (SVM) are highly effective binary classifiers but face challenges with scalability and extended processing times. Random Forest (RF) comprises a collection of decision trees, each acting as a conditional classifier. The final RF output integrates the results of individual trees, making it beneficial for large datasets and multiclass problems but susceptible to overfitting.

Hidden Markov Models (HMM) represent a set of states producing outputs with distinct probabilities, aiming to determine the sequence of states that can produce observed outputs. HMMs can be trained on both labeled and unlabeled datasets. K-Nearest Neighbor (KNN), like RF, is useful for solving multiclass problems, but the computational intensity of training and testing poses challenges. Shallow Neural Network (SNN) belongs to a class of algorithms based on neural networks.

Moving to unsupervised learning, some popular shallow machine learning algorithms are highlighted below. Clustering involves grouping data with similar characteristics, with k-means and hierarchical clustering being prominent examples. Association, another unsupervised learning method, aims to identify patterns between data, making it particularly suitable for predictive purposes.

12.4.2 Deep Learning

Deep Learning (DL) algorithms are fundamentally rooted in Deep Neural Networks (DNN), extensive networks organized into layers capable of autonomous representation learning.

12.4.2.1 Supervised DL Algorithms

- Fully-connected Feedforward Deep Neural Networks (FNN): This variant of DNN establishes connections between every neuron and those in the preceding layer, offering a flexible, general-purpose solution for classification. Despite high computational costs, FNN doesn't impose assumptions on input data.
- Convolutional Feedforward Deep Neural Networks (CNN): Tailored for spatial data analysis, CNN's unique structure involves neurons receiving input only from a subset of the previous layer. While effective for spatial data, their performance diminishes with non-spatial data, accompanied by a lower computational cost compared to FNN.
- Recurrent Deep Neural Networks (RNN): Differing from FNN, RNN allows neurons to send output to previous layers. Though more challenging to train, they shine as sequence generators, especially the long short-term memory variant.

12.4.2.2 Unsupervised DL Algorithms

- Deep Belief Networks (DBN): Comprising Restricted Boltzmann Machines (RBM), these networks lack an output layer. Ideal for pre-training due to superior feature extraction, DBN excels with unlabeled datasets, requiring a dedicated training phase.
- Stacked AutoEncoders (SAE): Comprising multiple Autoencoders, where input and output neuron numbers match, SAE excels in pre-training tasks akin to DBN. They demonstrate superior results on smaller datasets, highlighting their efficacy.

12.5 Cybersecurity in Smart Grid

Considering the smart grid as a cyber-physical system, where the interconnection between physical and cyber components is intricate, a comprehensive study of cyber threats and their implications becomes essential from both the cyber network and physical infrastructure perspectives. Initially, the vulnerability of devices to specific cyber threats within various components of the smart grid architecture is emphasized. Subsequently, potential countermeasures to prevent these threats are explored, aiming to enhance the overall cybersecurity posture of the smart grid.

12.5.1 Cyber Threats in Smart Grid: Smart Grid Devices Vulnerable to Cyber Attacks

The vulnerability of the devices in the AMI component of the smart grid infrastructure to different types of cyberthreats and the security objectives compromised thereof is enlisted in the Table 12.2.

Table 12.3 similarly enumerates the devices belonging to OT component of the smart grid infrastructure that are susceptible to cyberattacks and the impacted cyber security objectives.

Table 12.4 in the similar line shows devices in the IT component of the smart grid architecture that are prone to cyberattacks and the related cyber security objectives compromised.

Table 12.2 AMI Devices vulnerable to cyberattack [3, 4, 18–29]

Device name	Device description	Vulnerable to attacks	Security goal compromised
Smart Meter	Measures and records electricity consumption	Yes (Data manipulation, DoS, Firmware vulnerabilities)	A, NR, I, C
Phasor Measurement Units (PMUs)	Measures and records voltage and current data by specific time synchronised operation	Yes (False data injection, Time synchronization attacks, Spoofing attacks)	I
Meter Data Management System (MDMS)	Manages data from smart meters	Yes (Data breaches, Man-in-the-middle attacks, Ransomware attacks)	A, I, C
AMI Head-End System (AMI-HE)	Centralizes management of the AMI network	Yes (Zero-day attacks, Distributed denial-of-service (DDoS), Unauthorized access)	I, A, C
Communication Network	Enables data exchange between devices	Yes (Eavesdropping, Jamming attacks, Man-in-the-middle attacks)	I, A, C
In-Home Displays (IHDs)	Provides energy consumption information to consumers	Yes (Phishing attacks, Malware injection, Physical attacks)	I, A, C
Vehicle-to-Grid (V2G) Devices	Enables communication between electric vehicles and the grid	Yes (Malicious charging, Rogue charging, Man-in-the-middle attacks)	I, A, C

I: Integrity A: Availability C: Confidentiality NR: Non-Repudiation

Table 12.3 OT Devices vulnerable to cyberattack [3, 4, 18–29]

Device Name	Device Description	Vulnerable to attacks	Security goal compromised
Generator	A device designed to produce electrical power	Yes, That could manipulate power generation output, leading to blackouts or grid instability	A
Transmission line	A physical medium for data communication	Yes, Vulnerable to physical attacks that could damage the line and disrupt power flow	A
Transformer	An apparatus responsible for altering voltage levels	Yes, Vulnerable to attacks that could overload the transformer or cause it to malfunction	A
Load	A device that regulates impedance within an electrical circuit	Yes, Vulnerable to attacks that could manipulate the amount of electricity used by consumers	A
State estimator	Monitors devices by assessing their feedback and status	Yes, Vulnerable to attacks that could manipulate the state data, leading to incorrect grid operations and potential blackouts	I, C
WAPMC	Equipment that furnishes precise phasor and frequency data to Phasor Measurement Units (PMUs)	Yes, Vulnerable to attacks that could manipulate data collected from across the grid, leading to incorrect situational awareness and grid management decisions	I, C
Physical system component	A general device falling under Operational Technology (OT) components	Yes, Vulnerable to physical attacks that could damage or destroy critical infrastructure	I, C
Local sensor	A small gadget designed to measure specific attributes like light, sound, and pressure	Yes, Vulnerable to attacks that could manipulate sensor data, leading to incorrect grid operations and potential blackouts	A, C

(continued)

Table 12.3 (continued)

Device Name	Device Description	Vulnerable to attacks	Security goal compromised
Synchronous generator	Similar to a conventional generator, it produces electricity at a constant rate	Yes, Vulnerable to attacks that could manipulate the generator's operation, leading to grid instability or blackouts	C
Controller	A device responsible for operating other sensors and actuators based on control messages	Yes, Vulnerable to attacks that could manipulate control signals, leading to incorrect grid operations and potential blackouts	I, C
HVAC	A system managing environmental conditions	Yes, Vulnerable to attacks that could manipulate the voltage or frequency of the power transmission system, leading to equipment damage and blackouts	A, I
Generator Controls	Devices that regulate and control the operation of generators	Yes	I, A, C, NR
Transmission Line Monitoring Systems	Systems designed to monitor the condition of transmission lines	Yes	Transmission Line Monitoring Systems

I: Integrity A: Availability C: Confidentiality NR: Non-Repudiation

12.5.2 Cyber Threats in Smart Grid: Proactive Measures

The various cyberattacks that affects the devices in the AMI, OT, and IT Components of a Smart Grid and the cyber security objects that remain unattained or compromised have been enumerated in a table M, N, and O and this section dwells upon the methods, process, tools, or practices that can adopted to detect, prevent and throttle the impending cyberattacks. While the tabular enumeration of cyber threats has specific countermeasures, there are certain techniques or approaches of cyber security which are generic and are applicable to devices across the smart grid depending upon the threat type threat perception. As such smart grid provides a very large attack surface for attackers to make an entry, and it is not feasible to deploy equal level of security measures throughout the infrastructure. A minor loophole in the security setup could jeopardize the entire power grid infrastructure and unfortunately information related

Table 12.4 IT Devices vulnerable to cyberattack [3, 4, 18, 19, 21–29]

Device Name	Device Description	Vulnerable to attacks	Security goal compromised	Specific Threat Examples
Server system	Stores and processes data, runs applications	Yes, Malware, SQL injection	A	Ransomware, Cryptojacking, Zero-day attacks
Router	Directs network traffic	Yes, Malware, DDoS	A	DNS hijacking, Port scanning, Packet sniffing
Network node	Connects devices on a network such as client, router, switch, or hub	Yes, Malware, DDoS	A, I	Botnets, ARP spoofing, Denial-of-service attacks
Storage system	Stores data	Yes, Malware, Phishing	C	Data breaches, Data manipulation, Ransomware attacks
System memory	Physical RAM (Stores temporary data and programs)	Yes, Malware, Phishing	A	Buffer overflows, Memory scraping, Data breaches
CPU	A core unit for a computer that processes instructions	Yes, Malware, DDoS	A	Resource exhaustion attacks, CPU hijacking, Code injection
Network hardware resources	Includes switches, firewalls, and load balancers	Yes, Malware, DDoS	A	Configuration errors, Firmware vulnerabilities, Denial-of-service attacks
Wireless signal	A radio frequency that is used to send and receive data that enables wireless communication between devices	Yes, Phishing, Man-in-the-middle	A, I, C	Evil twin attacks, WEP cracking, WPA vulnerabilities
Authentication server	Manages user access and authentication	Yes, Phishing, Man-in-the-middle	C, I	Credential stuffing, Brute force attacks, Password spraying

I: Integrity A: Availability C: Confidentiality NR: Non-Repudiation to grid devices and systems are already commonplace in online search engines like Shodan [30]. These engines possess the capability to gather data from Internet of Things (IoT) devices, including Industrial Control Systems (ICSs) associated with electrical grids. For example, as of April 2019, they have indexed over 1,200 devices supporting the IEC 60,870–5-104 protocol and nearly 500 devices supporting the DNP3 protocol. Both of these protocols are widely used for the remote control and monitoring of modernized power grid systems. Moreover, considering additional protocols utilized in broader Industrial Control Systems, such as Modbus, the total number of indexed devices is even more extensive [31]. In such an environment it is very much essential that the security protocol of organizations of infrastructure of the proportion of a smart grid must have the first line of defence to learn as to who is accessing and scanning which exposed devices rather than waiting for development of a threat.

Honeypot is such a generic cybersecurity approach which is also deployed for cybersecurity in smart grid. Honeypots appear as the targets likely to be attacked by the hackers, such as a vulnerable device and its associated networks in a smart grid. Hackers get lured to these honeypots assuming them to be legitimate targets, and in the process the security analysts get the chance to detect and deflect cyber criminals from hacking the intended targets.

A smart grid, given its nature of complexity and cost involved in its implementation and operation, effective or optimal utilization of its resources, including the defence mechanism against cyber threats, is of paramount importance. In this context Game Theories could be highly helpful as they are widely used to predict an attack method that is most likely to take place. Game theories are deployed to make out the process of a specific attack scheme and a tree-structure-based game theory defence mechanism is best suited to a smart grid scenario as it analyses the path generated from the tree-structure model to predict the line of attack or its procedures [25].

12.5.2.1 Proactive Measures Against Cyber Threats: AMI Infrastructure Component

AMI essentially involves communication protocols for sharing of data between smart meters and SG control centre (V2G), between EV to EV and EV to EV charging stations, and so forth. All these communication channels can be target of the hackers and securing them against cyber attacks is the primary aspect of cyber security in the AMI component of a smart grid infrastructure.

One of the distinct features of smart meters is the embedded encryption algorithms or encryption key which is very vital for secure smart meter communication and for proper coordination amongst them encryption key management system is also very much essential [32]. Efficient key management and frequent auto updating of these keys can be taken up as an intrusion detection measure against data injection attack [33]. Similarly, authors in [34] have suggested an hash-based encryption with bidirectional authentication for secure V2G communication. In a smart grid scenario, it is very essential to have a robust key agreement and subscriber authentication for protected communication and the same has been pointed out by authors in [35, 36] highlighting the consequences of a feeble authentication and key agreement algorithm which provide the leeway to tampering of smart meters by the adversaries. Authors in [36–38] have proposed countermeasures against may include location and data stamp information. The above has been reiterated for EV-to-EV communication in [39]. The Table 12.5 enumerates some of the counter measures against cyber threats to the AMI component of a smart grid infostructure.

12.5.2.2 Proactive Measures Against for Cyber Threats: IT Component

The certificate of authenticity is a very crucial instrument for secured communication in an IT infrastructure. Each and every smart grid CPCs, including devices, users, keys, servers, and clients need to be authenticated. Generally, the authentication certificates are stored in something called "Certificate Authority" (CA), and if the CA itself is compromised in the event of a cyberattack, then it may lead to devastating consequences in an infrastructure of the size smart grid. Essentially the whole Public Key Infrastructure (PKI) that relies upon certificate authentication scheme will be jeopardised. Countermeasures to address this vital security concern needs to put in force in the IT component of a smart grid. In this aspect, authors in [42] have proposed to decentralise the CA applying a certificate-based authentication scheme, called Meta-PKI, in order to achieve effective monitoring of the authentication processes. Use of auditable logs for trusted certificate authentication between a server and a client can be a very effective countermeasure against Homograph attacks. Use of latest firmware version can also help prevent malware contamination [26]. Implementation of processes such as cross-domain authentication, risk assessment model etc. also help ensure smart grid IT security [43, 44].

Another important aspect of IT security is the security of device-to-device (D2D) communication, wherein the wireless signals exchanged between the devices carry both data and electricity alternately through a base station while getting themselves charged. These wireless networks are inherently susceptible to breaches, necessitating protection against eavesdroppers. To address this, a game theory approach incorporating cooperative jamming techniques has been developed for secure wireless charging communication, as detailed in reference [45]. There are several other countermeasures against varied types of attacks and listed in the Table 12.6 below.

12.5.2.3 Proactive Measures Against for Cyber Threats: OT Component

Countermeasures concerning DoS, FDI, message replay, and TSA attacks are similar to as used in AMI and IT component of smart grid. Quantification of the intensity of cyberattack through its impact on the smart grid is a necessary countermeasure to discern a traffic flooded by an attacker. As suggested in [47] the effect of a cyberattack can be measured by using channel bandwidth. A DoS attack on the DNP3 (Distributed

Table 12.5 Proactive Measures: AMI [3]

Sub-Component/ Devices of AMI under threat	Type and nature of attack	Proposed Counter/Preventive Measures	Strength (S) and Weakness (W) of the Counter/ Preventive Measure
AMI	DDoS attacks [21]	Bayesian honeypot game model	S: It has the potential to improve energy consumption for defense and increase the accuracy of attack detection W: Attackers can bypass the honeypot by employing anti-honeypot techniques
AMI		The dynamic honeypot defence mechanism	S: It tackles the bypass issue by scrutinizing interactions between attackers and defenders S: This approach aids in forecasting optimal strategies within the Advanced Metering Infrastructure (AMI) network
AMI	Abnormal activity detection	It relies on the Kullback–Leibler divergence (KL Distance), utilizing this metric to detect a compromised smart meter by assessing the relative entropy between the historical distribution model and the current model. [28]	
AMI	Message replay attacks	The results from state estimators, including the Kalman filter, Minimum Mean Square Error (MMSE), and Generalized Likelihood Ratio Test (GLRT), are juxtaposed with real-time measured data to verify the integrity of messages [18]	Additionally help prevent metre manipulation and theft attacks from compromised smart metres

(continued)

Table 12.5 (continued)

Sub-Component/ Devices of AMI under threat	Type and nature of attack	Proposed Counter/Preventive Measures	Strength (S) and Weakness (W) of the Counter/ Preventive Measure
AMI	Session key exposure attacks	In the context of session key exposure attacks, the recommended preventive measures involve regularly changing the random number utilized in generating a session key or securely sharing this number, as suggested by reference [35]	–
AMI	substitution attacks	it is advisable to utilize a cryptographic algorithm that produces randomized encryption patterns. This measure enhances security by making it challenging for attackers to deduce the plaintext easily, as outlined in reference [35]	–
AMI	TSA attacks	To address TSA attacks, a recommended countermeasure involves employing an algorithm capable of computing average estimation with error covariance for attack detection. Additionally, implementing a shift-invariant transmission policy can help minimize the impact of TSA attacks. [40]	–
EV Charging Infrastructure		Adherence to NISTIR 7628 framework that defines security objectives and requirements for a EV charging system [41]	–

(continued)

Table 12.5 (continued)

Sub-Component/ Devices of AMI under threat	Type and nature of attack	Proposed Counter/Preventive Measures	Strength (S) and Weakness (W) of the Counter/ Preventive Measure
Smart Meter	Tampering and physical manipulation	AES-CBC and SHA1-HMAC	The security measures in place should be sufficiently robust to prevent tampering with a smart meter and physical manipulation
Smart Meter	Data falsification attacks	Detection schemes, classified as additive, deductive, and camouflage methods use a statistical classifiers (such as Arithmetic Mean (AM), Geometric Mean (GM), and Harmonic Mean (HM)) to infer whether the falsification attacks have occurred or not [28]	–

Table 12.6 Proactive Measures Against: IT

Type and nature of attack	Proposed Counter/Preventive Measures
DoS and DDoS attacks	A fusion based defence mechanism based on analysis of feedback data of the smart grid network nodes.
Desynchronisation attacks	Deployment of suitable fault diagnosis scheme in the smart grid infrastructure
FDI attacks,	Dynamic measurement of a sample of the data flow in the smart grid network in order to make out abnormal data packets.
Eavesdropping attacks:	Countermeasures: message encryption, access control, anti-virus programs, firewall, VPN, and IDS [39]
Brute force attacks	Adoption of encryption mechanism with a large key size and robust authentication process [26, 28]
PDoS & botnets	Countermeasures as suggested bin the [46] enumerates detection model based on Poisson signalling game. The same can also detect botnets

Network Protocol 3) can be ascertained by analysing the attack intensity. The high intensity indicates network flooding and the attacker stands chance to get exposed easily.

It is observed that an attacker usually adopts an attack method involving less cost and the cost involved in unceasing a cyberattack depend on the strength of the defence mechanism in force at the targeted SG. For such reason it is good to calculate the cost involved in setting up an attack or defence mechanism. Such a scheme based on game-tree based Markov Model is proposed in [48] which calculates the cost involved in launching an attack or deploying a defence mechanism for SCADA used in a smart grid.

The cybersecurity of smart grids involving nuclear power stations is another major area of concern as a compromised network can have colossal damaging effects. Countermeasures against cyberattacks in such cases must be robust and full proof. The Feedback Linearisation Control as a part of system synchronism could be considered as an effective solution to ensure grid resiliency against cyberattacks with severe negative impact [22].

There are many cyber threats which are capable of compromising multiple security objectives and are termed as hybrid attacks. An IDS meant to uphold integrity objective may not be up to the task against these hybrid attack types, necessitating quantification of the impact of attacks as a part of preventive measure. In this aspect, a Mixed Integer Linear Programing (MILP) based security metrics to assess a smart grids vulnerability against hybrid attacks has been proposed in [49]. As IoT happens to be an integral part of any smart grid scheme, safeguard of IoT devices by security model designed on blockchain technology is suggested in [50].

12.6 Application of ML and DL Algorithms in Smart Grid Cybersecurity

An IDS model based on ML model with random forest (RF) as the classifier is proposed in [51] where the data collected by PMUs across the smart grid is used to detect **data injection threats** with very high accuracy and detection rate. Another such intelligent IDS modelled on multi-layer deep algorithm for detection of cyber threats in smart meter communication network is proposed in [52]. The accuracy and speed of detecting several cyber attacks of type and nature, such as **benign, DoS, PortScan, Web Attack, Bot, FTP Parator, and SSH Parator**, in a cyber physical system like smart grid claimed to be very high. A deep neural network (DNN) model has been proposed in [53] which proves to be highly accurate in classifying smart grid cyber attacks into types, namely Probe, DoS, U2R, R2L. The cyberattack type, False Data Injection (FDI) in a smart grid can be mitigated by a ML model which is designed based on Convolutional Neural Network (CNN) and a Long Short Term Memory (LSTM) network [54]. The model makes a time series anomaly search encompassing all the evolved features of a FDI attack [55]. A two-staged DL based threat detection and localization is proposed in [56] wherein 2D images of the encoded correlations of the measured variables in the smart grid is used to develop a deep Convolutional Neural Network (CNN)-based classifier to detect FDI attacks with high accuracy. Also, the authors in [56] have proposed a two-layered sequential auto detector of cyber threats in a smart grid. The first layer indicates the presence of cyberattack followed by the second layer which classifies the cyberattacks. For both the layers the ML algorithm RF is chosen for the intended purposes. A pattern recognition approach for detection of cyber threats in the physical layer of smart grid is proposed in [57] which relies on an ensemble of ML classifiers and neural networks for enhanced pattern recognition.

The study [58] proposed to view the task of anomaly detection in the data traffic of a digital network as a partially observable Markov decision process (POMDP) problem. To handle such problem, it suggests a universal robust online detection algorithm based on the framework of model-free reinforcement learning (RL) for POMDPs. The anomaly detection scheme aims to identify attacks of types, jamming, FDI, and DoS. The article [59] proposes an anomaly detection and classification model for smart grid architectures using Modbus/Transmission Control Protocol (TCP) and Distributed Network Protocol 3 (DNP3) protocols. The model adopts an Autoencoder-Generative Adversarial Network (GAN) architecture for (a) detecting operational anomalies and (b) classifying Modbus/TCP and DNP3 cyberattacks.

The increased adoption of Internet of Things (IoT) and digital communication networks for Monitoring and controlling of industrial control systems (ICS) in a smart grid, exposes the CPS to many cyber threats with devastating consequences. While traditional IDSs are found inadequate, the intelligent IDSs proposed in various literatures do not take into account the imbalanced observed in the ICS datasets. The model proposed in [60] is based on Deep Neural Network (DNN) and Decision Tree (DT) classifiers, taking advantage of the inherent capabilities of the intelligent algorithms in handling imbalanced datasets and providing high accuracy of classification.

The authors in [61] have proposed a DL based IDS specifically to address the FDI attack on Supervisory Control and Data Acquisition (SCADA) system of a smart grid in order to ensure data integrity of the data collected by the system. Stacked autoencoder (SAE) based deep learning framework for mitigation of threats against transmission SCADA is proposed in [62] which also counts on the inherent capacity of DL in unsupervised feature learning in complex security scenario. A two stage IDS model with each stage deployed with an agent-based model is proposed [63] for preserving data integrity in the physical layers of a smart grid. The first stage comes out with attack exposure matric while the second stage explores decentralization of security in the system. The study [64] takes in to account the varied attack strategies likely to be adopted by the hackers based on factors such as cost, time, availability of information, and level of vulnerability of the system chosen to be attacked. In this context a scenario based two-stage sparse cyber-attack models for smart grid with complete and incomplete network information are proposed, which works on DL based interval state estimation (ISE).

With means of advanced technology and vulnerabilities of smart grids due to heavy reliance on IT, a rudimentary act of electricity theft has also gone digital and is very much a cyber threat concern. The authors in [65] have highlighted this aspect in context of electricity theft in a distributed generation (DG) scenario, wherein the consumers with malicious intent hack into the smart meters deployed with their own grid-tied-DG units to claim higher than the actual supply of energy. The authors in their study have proposed a deep convolutional-recurrent neural network-based model to address such an issue.

12.7 Challenges and Future Directions

AI has the potential to revolutionize cybersecurity in smartgrids by automating threat detection, response, and incident analysis. However, several challenges need to be addressed for AI to reach its full potential:

Challenges

- Data Availability and Quality: Training and evaluating AI models require large amounts of labeled data, which can be scarce and challenging to collect in the cybersecurity domain.
- Explainability and Transparency: AI models are often opaque, making it difficult to understand their decision-making process and ensure they are unbiased and fair.
- Adaptability and Generalizability: Cyberattacks are constantly evolving, and AI models need to be adaptable to detect and respond to new threats.
- Integration with Existing Systems: Integrating AI solutions with existing cybersecurity infrastructure and workflows can be complex and require significant resources.
- Privacy and Security Concerns: AI applications themselves can be vulnerable to attacks, and it's crucial to ensure the privacy and security of collected data.

Future Directions

- Federated Learning: This approach allows training AI models on decentralized datasets, addressing data privacy concerns and improving data availability.
- Explainable AI (XAI): Techniques like LIME and SHAP can help explain AI model decisions, making them more transparent and trustworthy.
- Generative Adversarial Networks (GANs): These neural networks can be used to generate synthetic data for training AI models, addressing data scarcity challenges.
- Multi-agent Systems: Collaborative AI agents can work together to detect and respond to cyberattacks more effectively.
- Homomorphic Encryption: This encryption technique allows computations to be performed on encrypted data, preserving privacy while enabling AI analysis.

12.8 Conclusion

In conclusion, this comprehensive review has delved into various facets of smart grid cybersecurity, providing a nuanced understanding of the challenges and potential solutions in this critical domain. Here, the authors scrutinized the intricacies of smart grid infrastructure, highlighting challenges that range from data availability and quality to the integration complexities of associated devices such as AMI, IT, and OT. Our exploration extended to the diverse landscape of cyber threats, encompassing the types of attacks and the specific devices susceptible to these threats within a smart grid framework. By elucidating effective countermeasures, we underscored the importance of securing smart grid components against potential vulnerabilities. Moreover, the study explored the integration of artificial intelligence, encompassing both machine learning (ML) and deep learning (DL), as a transformative approach to fortify smart grid cybersecurity. This work discussed the application of ML and DL techniques, recognizing their potential to automate threat detection and response. In acknowledging the evolving nature of cyber threats, the work outlined challenges associated with AI adoption in this context. Looking ahead, we proposed future directions, including federated learning, explainable AI, generative adversarial networks, multi-agent systems, and homomorphic encryption, as promising avenues to enhance the resilience of smart grids against cyber threats. This holistic examination contributes to the collective knowledge base, offering insights that can inform future research, policy development, and practical implementations in the ever-evolving landscape of smart grid cybersecurity.

References

1. https://www.iea.org/energy-system/electricity/smart-grids [Accessed on June 28 2023]
2. https://www.nsgm.gov.in/en/smart-grid [Accessed on July 12, 2023]
3. Kim, Y., Hakak, S., Ghorbani, A.: Smart grid security: Attacks and defence techniques. IET Smart Grid **6**(2), 103–123 (2023)
4. Canadian Institute for Cybersecurity (CIC): Operational Technology (OT) Forensics, pp. 1–141. University of New Brunswick (2019)
5. Stouffer, K., Falco, J., Scarfone, K.: Guide to Industrial Control Systems (ICS) Security (No. NIST Special Publication (SP) 800-82 (Retired Draft)). National Institute of Standards and Technology
6. Wang, Y., et al.: Analysis of smart grid security standards. In: Proc. Int. Conf. Computer Science and Automation Engineering, Shanghai, China, June 2011, pp. 697–701
7. https://www.nwkings.com/objectives-of-cyber-security [Accessed on July 15, 2023]
8. https://sprinto.com/blog/cyber-security-goals/# What_are_Cyber_Security _Goals_or_Objectives [Accessed on July 18, 2023]
9. Chen, B., Wang, J., Shahidehpour, M.: Cyber–physical perspective on smart grid design and operation. IET Cyber-Physical Systems: Theory & Applications **3**(3), 129–141 (2018)
10. Guo, Q., Hiskens, I., Jin, D., Su, W., Zhang, L.: Editorial: cyberphysical Systems in Smart Grids: security and operation. IET Cyber-Physical Systems: Theory & Applications **2**(4), 153–154 (2017)
11. Patnaik, B., Mishra, M., Bansal, R.C., Jena, R.K.: AC microgrid protection–A review: Current and future prospective. Appl. Energy **271**, 115210 (2020)
12. Mishra, M., Patnaik, B., Biswal, M., Hasan, S., Bansal, R.C.: A systematic review on DC-microgrid protection and grounding techniques: Issues, challenges and future perspective. Appl. Energy **313**, 118810 (2022)
13. Spring, J. M., Fallon, J., Galyardt, A., Horneman, A., Metcalf, L., & Stoner, E. (2019). Machine Learning in Cybersecurity: A Guide. SEI Carnegie Mellon Technical Report CMU/SEI-2019-TR-005.

14. Apruzzese, G., Laskov, P., Montes de Oca, E., Mallouli, W., Brdalo Rapa, L., Grammatopoulos, A.V., Di Franco, F.: The role of machine learning in cybersecurity. Digital Threats: Research and Practice **4**(1), 1–38 (2023)
15. Giovanni Apruzzese, Michele Colajanni, Luca Ferretti, Alessandro Guido, and Mirco Marchetti. 2018. On the effectiveness of machine and deep learning for cybersecurity. In Proceedings of the IEEE International Conference on Cyber Conflicts. 371–390.
16. Giovanni Apruzzese, Michele Colajanni, Luca Ferretti, and Mirco Marchetti. 2019. Addressing adversarial attacks against security systems based on machine learning. In Proceedings of the IEEE International Conference on Cyber Conflicts. 1–18.
17. Kasun Amarasinghe, Kevin Kenney, and Milos Manic. 2018. Toward explainable deep neural network based anomaly detection. In Proceedings of the IEEE International Conference Human System Interaction. 311–317.
18. Baig, Z.A., Amoudi, A.R.: An analysis of smart grid attacks and countermeasures. J. Commun. 8(8), 473–479 (2013). https://doi.org/10. 12720/jcm.8.8.473-479
19. Bou-Harb, E., et al.: Communication security for smart grid distribution networks. IEEE Commun. Mag. 51(1), 42–49 (2013). https://doi.org/10. 1109/mcom.2013.6400437
20. Hansen, A., Staggs, J., Shenoi, S.: Security analysis of an advanced metering infrastructure. Int. J. Crit. Infrastruct. Protect. **18**, 3–19 (2017). https://doi.org/10.1016/j.ijcip.2017.03.004
21. Wang, K., et al.: Strategic honeypot game model for distributed denial of service attacks in smart grid. IEEE Trans. Smart Grid. **8**(5), 2474–2482 (2017). https://doi.org/10.1109/tsg.2017.2670144
22. Farraj, A., Hammad, E., Kundur, D.: A distributed control paradigm for smart grid to address attacks on data integrity and availability. IEEE Trans. Signal Inf. Process. Netw. 4(1), 70–81 (2017). https://doi.org/10. 1109/tsipn.2017.2723762
23. Chen, P.Y., Cheng, S.M., Chen, K.C.: Smart attacks in smart grid communication networks. IEEE Commun. Mag.Commun. Mag. **50**(8), 24–29 (2012). https://doi.org/10.1109/mcom.2012.6257523
24. Sanjab, A., et al.: Smart grid security: threats, challenges, and solutions. arXiv preprint arXiv: 1606.06992
25. Liu, S.Z., Li, Y.F., Yang, Z.: Modeling of cyber-attacks and defenses in local metering system. Energy Proc. 145, 421–426 (2018). https://doi.org/10.1016/j.egypro.2018.04.069
26. Sun, C.C., et al.: Intrusion detection for cybersecurity of smart meters. IEEE Trans. Smart Grid. 12(1), 612–622 (2020). https://doi.org/10. 1109/tsg.2020.3010230
27. Bansal, G., Naren, N., Chamola, V.: RAMA: real-time automobile mutual authentication protocol using PUF. In: Proc. Int. Conf. Cloud Computing Environment Based on Game Theory, Barcelona, Spain, January 2020, pp. 265–270
28. Bhattacharjee, S., et al.: Statistical security incident forensics against data falsification in smart grid advanced metering infrastructure. In: Proc. Int. Conf. Data and Application Security and Privacy, Scottsdale, USA, March 2017, pp. 35–45
29. Wei, L., et al.: Stochastic games for power grid protection against co-ordinated cyber-physical attacks. IEEE Trans. Smart Grid. **9**(2), 684–694 (2018). https://doi.org/10.1109/tsg.2016.2561266
30. "Shodan," https://www.shodan.io/. [Accessed on August 8, 2023]
31. Mashima, D., Li, Y., & Chen, B. (2019, December). Who's scanning our smart grid? empirical study on honeypot data. In 2019 IEEE Global Communications Conference (GLOBECOM) (pp. 1–6). IEEE.
32. Liu, N., et al.: A key management scheme for secure communications of advanced metering infrastructure in smart grid. IEEE Trans. Ind. Electron. **60**(10), 4746–4756 (2012). https://doi.org/10.1109/tie.2012.2216237
33. Liu, X., et al.: A collaborative intrusion detection mechanism against false data injection attack in advanced metering infrastructure. IEEE Trans. Smart Grid. **6**(5), 2435–2443 (2015). https://doi.org/10.1109/tsg.2015.2418280
34. Lee, S.: Security and privacy protection of vehicle-to-grid technology for electric vehicle in smart grid environment. J. Convergence Culture Technol. **6**(1), 441–448 (2020)

35. Park, K.S., Yoon, D.G., Noh, S.: A secure authentication and key agreement scheme for smart grid environments without tamper-resistant devices. J. Korea Inst. Inf. Secur. Cryptol. **30**(3), 313–323 (2020)
36. Kaveh, M., Martín, D., Mosavi, M.R.: A lightweight authentication scheme for V2G communications: a PUF-based approach ensuring cyber/physical security and identity/location privacy. Electronics **9**(9), 1479 (2020). https://doi.org/10.3390/electronics9091479
37. Zhang, L., et al.: A lightweight authentication scheme with privacy protection for Smart Grid communications. Future Generat. Comput. Syst. **100**, 770–778 (2019). https://doi.org/10.1016/j.future.2019.05.069
38. Go, Y.M., Kwon, K.H.: Countermeasure of SIP impersonation attack using a location server. J. Korea Contents Assoc. **13**(4), 17–22 (2013). https://doi.org/10.5392/jkca.2013.13.04.017
39. Roberts, B., et al.: An authentication framework for electric vehicle-to- electric vehicle charging applications. In: Proc. Int. Conf. Mobile Ad Hoc and Sensor Systems, Orlando, USA, November 2017, pp. 565–569
40. Guo, Z., et al.: Time synchronization attack and countermeasure for multisystem scheduling in remote estimation. IEEE Trans. Automat. Control. **66**(2), 916–923 (2020). https://doi.org/10.1109/tac.2020.2997318
41. Chan, A.C.F., Zhou, J.: A secure, intelligent electric vehicle ecosystem for safe integration with smart grid. IEEE Trans. Intell. Transport. Syst. **16**(6), 3367–3376 (2015). https://doi.org/10.1109/tits.2015.2449307
42. Kakei, S., et al.: Cross-certification towards distributed authentication infrastructure: a case of hyperledger fabric. IEEE Access. **8**, 135742–135757 (2020). https://doi.org/10.1109/access.2020.3011137
43. Li, Q., et al.: A risk assessment method of smart grid in cloud computing environment based on game theory. In: Proc. Int. Conf. Cloud Computing and Big Data Analytics, Chengdu, China, April 2020, pp. 67–72
44. Shen, S., Tang, S.: Cross-domain grid authentication and authorization scheme based on trust management and delegation. In: Proc. Int. Conf. Computational Intelligence and Security, Suzhou, China, December 2008, pp. 399–404
45. Chu, Z., et al.: Game theory based secure wireless powered D2D communications with cooperative jamming. In: Proc. Int. Conf. Wireless Days, Porto, Portugal, March 2017, pp. 95–98
46. Pawlick, J., Zhu, Q.: Proactive defense against physical denial of service attacks using Poisson signaling games. In: International Conference on Decision and Game Theory for Security, October 2017, pp. 336–356. Springer, Cham
47. Lu, Z., et al.: Review and evaluation of security threats on the communication networks in smart grid. In: Proc. Int. Conf. Military Communications, San Jose, USA
48. Hewett, R., Rudrapattana, S., Kijsanayothin, P.: Cyber-security analysis of smart grid SCADA systems with game models. In: Proc. Int. Conf. Cyber and Information Security Research, New York, USA, April 2014, pp. 109–112
49. Pan, K., et al.: Combined data integrity and availability attacks on state estimation in cyber-physical power grids. In: Proc. Int. Conf. Smart Grid Communications, Sydney, Australia, November 2016, pp. 271–277
50. Jeong, Y.S.: Probability-based IoT management model using blockchain to expand multilayered networks. J. Korea Convergence Soc. **11**(4), 33–39 (2020)
51. Wang, D., Wang, X., Zhang, Y., Jin, L.: Detection of power grid disturbances and cyber-attacks based on machine learning. Journal of information security and applications **46**, 42–52 (2019)
52. Vijayanand, R., Devaraj, D., & Kannapiran, B. (2019, April). A novel deep learning based intrusion detection system for smart meter communication network. In 2019 IEEE International Conference on Intelligent Techniques in Control, Optimization and Signal Processing (INCOS) (pp. 1–3). IEEE.
53. Zhou, L., Ouyang, X., Ying, H., Han, L., Cheng, Y., & Zhang, T. (2018, October). Cyber-attack classification in smart grid via deep neural network. In Proceedings of the 2nd international conference on computer science and application engineering (pp. 1–5).

54. Niu, X., Li, J., Sun, J., & Tomsovic, K. (2019, February). Dynamic detection of false data injection attack in smart grid using deep learning. In 2019 IEEE Power & Energy Society Innovative Smart Grid Technologies Conference (ISGT) (pp. 1–6). IEEE.
55. Mohammadpourfard, M., Genc, I., Lakshminarayana, S., & Konstantinou, C. (2021, October). Attack detection and localization in smart grid with image-based deep learning. In 2021 IEEE international conference on communications, control, and computing technologies for smart grids (SmartGridComm) (pp. 121–126). IEEE.
56. Farrukh, Y. A., Ahmad, Z., Khan, I., & Elavarasan, R. M. (2021, November). A sequential supervised machine learning approach for cyber attack detection in a smart grid system. In 2021 North American Power Symposium (NAPS) (pp. 1–6). IEEE.
57. Sakhnini, J., Karimipour, H., Dehghantanha, A., Parizi, R.M.: Physical layer attack identification and localization in cyber–physical grid: An ensemble deep learning based approach. Physical Communication **47**, 101394 (2021)
58. Kurt, M.N., Ogundijo, O., Li, C., Wang, X.: Online cyber-attack detection in smart grid: A reinforcement learning approach. IEEE Transactions on Smart Grid **10**(5), 5174–5185 (2018)
59. Siniosoglou, I., Radoglou-Grammatikis, P., Efstathopoulos, G., Fouliras, P., Sarigiannidis, P.: A unified deep learning anomaly detection and classification approach for smart grid environments. IEEE Trans. Netw. Serv. Manage.Netw. Serv. Manage. **18**(2), 1137–1151 (2021)
60. Al-Abassi, A., Karimipour, H., Dehghantanha, A., Parizi, R.M.: An ensemble deep learning-based cyber-attack detection in industrial control system. IEEE Access **8**, 83965–83973 (2020)
61. He, Y., Mendis, G.J., Wei, J.: Real-time detection of false data injection attacks in smart grid: A deep learning-based intelligent mechanism. IEEE Transactions on Smart Grid **8**(5), 2505–2516 (2017)
62. Wilson, D., Tang, Y., Yan, J., & Lu, Z. (2018, August). Deep learning-aided cyber-attack detection in power transmission systems. In 2018 IEEE Power & Energy Society General Meeting (PESGM) (pp. 1–5). IEEE.
63. Sengan, S., Subramaniyaswamy, V., Indragandhi, V., Velayutham, P., Ravi, L.: Detection of false data cyber-attacks for the assessment of security in smart grid using deep learning. Comput. Electr. Eng.. Electr. Eng. **93**, 107211 (2021)
64. Wang, H., Ruan, J., Wang, G., Zhou, B., Liu, Y., Fu, X., Peng, J.: Deep learning-based interval state estimation of AC smart grids against sparse cyber attacks. IEEE Trans. Industr. Inf.Industr. Inf. **14**(11), 4766–4778 (2018)
65. Ismail, M., Shaaban, M.F., Naidu, M., Serpedin, E.: Deep learning detection of electricity theft cyber-attacks in renewable distributed generation. IEEE Transactions on Smart Grid **11**(4), 3428–3437 (2020)

Chapter 13
Intelligent Biometric Authentication-Based Intrusion Detection in Medical Cyber Physical System Using Deep Learning

Pandit Byomakesha Dash, Pooja Puspita Priyadarshani, and Meltem Kurt Pehlivanoğlu

Abstract The current generation of technology is evolving at a rapid speed and gaining a prominent position in the hearts of individuals. For instance, when internet-connected gadgets link to other devices, they create a large system that solves complicated issues and makes people's lives simpler and longer. The Cyber Physical System (CPS) is a very essential advancement in technology. The rapid and important growth of CPS influences several facts of people's lifestyles and allows a more comprehensive selection of services and applications, including smart homes, e-Health, smart transport, and e-Commerce, etc. In the advanced medical field, a medical cyber-physical system (MCPS) is a one-of-a-kind cyber-physical system that integrates networking capability, embedded software control devices and the complicated health records of patients. Medical cyber-physical data are digitally produced, electronically saved, and remotely accessible by medical personnel or patients through the process of MCPS's interaction between communication, devices, and information systems. MCPS is based on the concept that biometric readings can be used as a way to verify a user's identity in order to protect their security and privacy. Several studies have revealed that Machine Learning (ML) algorithms for CPS technology have achieved significant advancements. Interactions between real-time physical systems and dynamic surroundings have been significantly simplified by the use of more effective ML techniques in CPS. In this study, we have suggested a convolutional

P. B. Dash (✉)
Department of Information Technology, Aditya Institute of Technology and Management (AITAM), Tekkali, Andhra Pradesh 532201, India
e-mail: byomakeshdash2000@gmail.com

P. P. Priyadarshani
Department of Computer Science, Maharaja Sriram Chandra Bhanja Deo University (MSCBU), Baripada, Odisha 757003, India

M. K. Pehlivanoğlu
Department of Computer Engineering, Kocaeli University, Kocaeli, Türkiye
e-mail: meltem.kurt@kocaeli.edu.tr

J. Nayak et al. (eds.), *Machine Learning for Cyber Physical System: Advances and Challenges*, Intelligent Systems Reference Library 60,
https://doi.org/10.1007/978-3-031-54038-7_13

neural network (CNN)-based intrusion detection system for identifying anomalies in MCPS. The ECU-IoHT dataset has been used for our research. The experimental findings outperform the conventional ML baseline models, demonstrating the efficacy of our proposed approach.

Keywords CPS · MCPS · Deep learning · IDS · CNN · ML · Health Care

13.1 Introduction

The phrase “cyber-physical system” was first used by Helen Gill at the National Science Foundation (NSF) in the United States in 2006. Cyber-Physical System (CPS) serves as platforms that facilitate the coordination of computing device, internet connectivity, and physical activities. This integration enables smooth interaction between web activities and real-world components [1]. A CPS refers to a device that employs computer-based algorithms to manage and supervise a particular process. The interconnection between physical and software components is a prominent feature of cyber-physical systems, which operate across multiple spatial and temporal dimensions. These systems exhibit diverse and unique modes of behavior and engage in communication within varied environments. The increasing popularity of the CPS is being attributed to the fast growing nature of the internet. The CPS paradigm has been increasingly used in the development of intelligent applications including smart robots, smart transportation, smart healthcare, smart agriculture, smart manufacturing, smart distribution of water, and smart homes, including several technical domains and associated services. Figure 13.1 represents some applications of CPS. Water supply networks are dynamic due to climate change and customer demand uncertainty. The rapid advancement of technology makes water supply system improvements possible. Thus, communication and networking, sensing and instrumentation, computation, and control technology is linked with water delivery system infrastructures to improve operations [2]. Cyber physical manufacturing systems provide distinct advantages over conventional production methods. Five methods, smart supply chains, including production line monitoring, predictive analysis, asset monitoring, predictive analysis, and personalized goods, demonstrate the superiority of cyber manufacturing systems over conventional approaches [3]. Transport networks affect national productivity, environment, and energy consumption. Developing innovative, efficient transport systems requires overcoming technological hurdles related to the cyber-physical characteristics of current systems [4]. A significant proportion of the global population is relocating to metropolitan areas. Countries are actively pursuing smart city initiatives to enhance the overall welfare of their residents. CPS is the fundamental basis of smart city infrastructures. Practically every part of a smart city’s infrastructure makes use of CPS [5].

MCPS is designed to combine a variety of intelligent health care sensor gadgets and effectively gather signal information. The data is securely stored inside cloud storage architecture. MCPS conducts monitoring and surveillance activities that

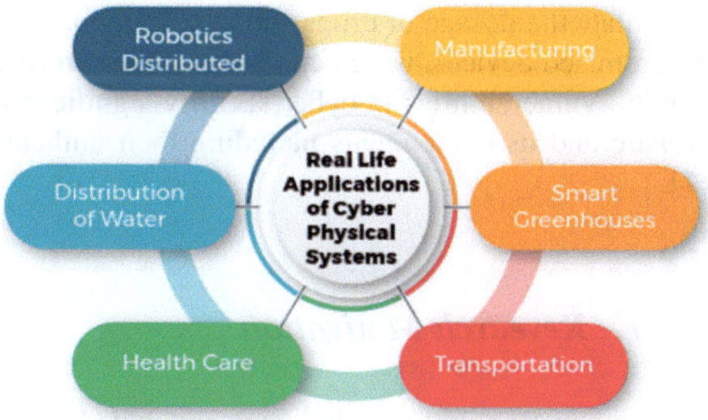

Fig. 13.1 Application of CPS

monitor the functioning of several integrated smart sensor devices throughout different time frames. Subsequently, the collected data is sent to the medical professional. The use of the MCPS environment is widely implemented in several healthcare facilities to present an accurate and streamlined examination of the patient's overall health condition. Ensuring the secure preservation of a patient's health information is of the highest priority. The potential consequences of an attack on patient information include the theft or alteration of recorded data, which might result in the misdiagnosis of a medical condition [6]. The Internet of Things (IoT) is the backbone of a complex healthcare system's cloud storage, health care sensor gadgets, and wireless connectivity to transmit patient's medical data through mobile application [7].

A good example is the biosensor which may be utilized with or without human interaction to link individuals to the healthcare system [8]. A wide range of biosensors may be employed for monitoring a patient's vital signs, including their movement, breathing, temperature, eyesight, heart rate, and more health information. This kind of sensor can be implanted within a human being and produces massive amounts of data in real time [9]. A forecast by International Data Corporation (IDC) estimates that there will be 41.6 billion IoT devices in 2025. Therefore, there is a need for advancements in data preservation methodologies, data analysis approaches, and the solution of security-related challenges to keep up with the rising needs and growth of MCPS systems. The accelerating growth of MCPS devices and infrastructure has been accompanied by a variety of cyber-attacks, which have identified the vulnerable aspects within the MCPS ecosystem. According to skilled professionals, a significant number of MCPS devices have been identified as susceptible to cyber-attacks, which might possibly compromise the health and security of patients. MCPS uses open wireless connection methods for its equipment. In addition, the healthcare industry has established a digitalized and interconnected network of clinical devices, which constantly transfer unstructured and possibly insecure data therefore exposing them susceptible to cyber-attacks [10]. It is possible for an intruder to gain access to MCPS network as a result of insecure architecture and insufficient authentication protocols. Another security risk that arises is the potential for unauthorized access to occur without being identified, which may be attributed to the lack of capacity to identify and prevent such assaults. Consequently, an attacker has the capability to remotely manipulate the dosage of drugs and may transform MCPS sensors into networks of compromised devices, which can be used for carrying out Denial-of-Service (DoS) assaults. Vulnerabilities in cyber security significantly compromise the security of software and its components including their authenticity, privacy, and availability [11].

13.1.1 Research Motivation

Based on existing studies, it has been concluded that there are many security-related difficulties and challenges present in applications of the MCPS. The increasing frequency of cyber-attacks in the MCPS environment poses a significant threat to the whole healthcare ecosystem, exposing it susceptible to hackers. The following are the primary difficulties that need to be addressed:

1. The fluctuating nature of MCPS networks (IoT devices, fog, and the cloud) makes it difficult to design distributed security architecture for distributed MCPS applications. Furthermore, with MCPS, the transmission network may be interrupted by the change of attacker behavior.
2. The decentralized framework for analyzing the huge amount of data generated by MCPS devices presents a significant challenge to the security mechanisms designed to protect devices.
3. It is difficult to build an intrusion detection system (IDS) that can discriminate between an attack and ordinary observations in an MCPS environment. Thousands of devices and sensors are linked together in such a network environment, which suffers from poor architecture and inadequate authentication procedures.

We have presented a DL based IDS for addressing these issues. The detection system employs complex CNN architecture to minimize the impact of cyber-attacks in an MCPS environment.

13.1.2 Research Contribution

Following are the key contributions of this study:

1. The MCPS ecosystem includes a wide variety of sensors and devices, which provide huge quantities of information. To manage massive amounts of data in real-time and enable quick and efficient decision-making, the DL technique has been employed as a framework. It is essential to use robust data processing techniques.
2. The suggested CNN approach trains and evaluates the model using a novel dataset called ECU-IoHT [12] from the healthcare sector.

3. State-of-the-art methodologies including Random Forest (RF), Decision Tree (DT), Adaptive Boosting (AdaBoost), Gradient Boosting (GBoost), Extreme Gradient Boosting (XGBoost), and CatBoost are compared with the suggested CNN model.
4. The CNN technique has shown superior performance compared to conventional approaches achieving an accuracy rate of 99.74% with the comprehensive analysis of large datasets.

13.1.3 Organization of Paper

The following is the outline for the rest of the paper. Section 13.2 provides an overview of the vulnerabilities present in health care security, as well as the current solutions available to address cyber-attacks. Additionally, it examines a specific attack scenario using medical sensors inside the IoT health care ecosystem. Section 13.3 of the article extends into a study of several ML methodologies, with a specific focus on CNN which are implemented in this research. The design and architecture of the suggested model has been described in Section 13.4. Section 13.5 of the document explores into a detailed analysis of the dataset and the setup of the experimental environment. The evaluation of the proposed model's experimental findings and performance is presented in Section 13.6; In conclusion, Section 13.7 serves as the last section of this article.

13.2 Related Works

The increasing popularity of MCPS devices has created new security concerns, such as increased network traffic across the MCPS environment. Attacks have been employed against the MCPS environment including network spoofing, DoS and DDoS attacks. Several studies have shown the enhancement of security and protection in MCPS by using ML and DL approaches. These approaches have been effective in improving the accuracy and efficiency of security threat detection in MCPS, enabling early prevention measures to be implemented before any potential damage occurs. This section provides an overview of some researches that have used IDS based on ML and DL approaches in MCPS.

Begli et al. [13] have been suggested a secure support vector machine (SVM) based IDS specifically for remote healthcare environments. This approach employs to address and prevent Denial of Service (DoS) assaults and User to Root (U2R) assaults. The anomaly identification system has been evaluated by implementing the NSL-KDD data samples, which achieved an accuracy detection rate of 95.01% for identifying abnormalities. Newaz et al. [14] introduced Health Guard, a safety framework designed specifically for smart healthcare systems (SHS). The present approach examines the vital signals of diverse sensors inside a SHS and establishes correlations between these signals to determine alterations in the patient's activities in the body. The objective is to differentiate between regular and pathological activities. The suggested framework employs DT, RF, Artificial Neural Network (ANN) and the k-Nearest Neighbor (k-NN) methodologies for the purpose of detecting harmful events. The performance of the DT model is superior by achieving an accuracy rate of 93%.

He et al. [15] introduced a novel stacked auto encoder based intrusion detection system (SAE-IDS) specifically for healthcare environments. The model employs a stacked auto encoder for the purpose of feature selection. They have implemented several ML algorithms such as Naive Bayes (NB), SVM, k-NN, and XGBoost for the purpose of identifying harmful activity. The implementation of the XGBoost algorithm exhibits superior performance by achieving an accuracy rate of 97.83%. The primary emphasis of this approach is on optimizing parameters for performance rather than giving importance to security considerations. Alrashdi et al. [16] have been proposed the introduction of fog-based ensemble detection system (FBAD) for effectively recognize both attack and normal events. This study employs the online sequential extreme learning machine (OS-ELM) model to identify attacks in the healthcare domain. The study used the NSLKDD dataset and achieved a classification accuracy of 97.09%. Hady et al. [17] have implemented an Enhanced Healthcare Monitoring System (EHMS) capable of real-time monitoring of patients' biometrics and collection of network traffic measurements. They have compared several ML techniques to train and evaluate the dataset to identify and mitigate several types of attacks. Thus, an efficient IDS has been developed on dynamic collection of data. Through the process of analyzing the 10-fold accuracy score comparison, the SVM algorithm shown superior performance compared to other algorithms by achieving an accuracy rate of 92.44%.

Susilo et al. [18] introduced DL based IDS for the IoT environment. Their findings indicate that as the number of IoT devices increases, consequently increases the associated security risk and vulnerability. They have conducted a comparison study of the proposed CNN and other ML algorithms including MLP, and RF utilizing the Bot-IoT data samples. CNN demonstrated the best level of accuracy of 91.27%. This result was obtained by having 128 batch sizes with 50 iterations. The total time taken to complete the experiment is about 3267 seconds. The minimum accuracy is being achieved 88.30% while using 32 batch sizes with 50 iterations. The total time taken for this training was 227 minutes and 21 seconds. The augmentation of the batch size resulted in a corresponding improvement in accuracy. The accuracy of the suggested model was found to be inferior compared to RF model that achieved 100% accuracy rate in detecting DOS and DDoS attacks. Ibitoye et al. [19] performed an analysis on application of DL approaches for security detection in the presence of adversarial attacks. Specifically, they employed a feed forward neural network (FNN) and self-normalizing neural network (SNN) as alternatives to the standard methods. Those methodologies were inadequate and inefficient in countering dynamic assaults. They have employed Bot-IoT data samples in their study. The experimental outcomes indicated that FNN approach attained a maximum accuracy of 95.1%. Additionally, the average recall, F1- score, and precision were observed to be 0.95%. However, it was shown that SNN with a 9% higher sensitivity exhibited more robustness compared to FNN when it came to feature normalization in the context of adversarial attacks.

Hizal et al. [20] have implemented CNN model-based IDS that were executed on a Graphics Processing Unit (GPU) runtime environment. They have achieved a classification accuracy of 99.86% for a 5-class classification task using the NSL-KDD data samples. Gopalakrishnan et al. [21] introduced a system called DLTPDO-CD, which integrates DL for traffic prediction, data offloading, and cyber-attack detection. The model incorporates three primary operations, namely traffic prediction, data unloading, and attack detection. The detection of attacks in mobile edge computing involves the use of a deep belief network (DBN) that has been optimized using the barnacles mating optimizer (BMO) method, referred as BMO-DBN. They have achieved an accuracy of 97.65% using BMO-DBN. In contrast, it was found that the use of DBN resulted in a slightly decreased accuracy rate of 96.17%. Xun et al. [22] performed tests using CNN and LSTM networks to build several models for evaluating driving behavior in the context of edge network aided vehicle driving. The training data in CNN exhibits an accuracy rate of 96.7% and a loss value of 0.189. Conversely, the LSTM model achieves an accuracy rate of 98.5% and a loss value of 0.029. The accuracy rates for the CNN and LSTM models on the test dataset are 90.2% and 95.1% respectively. Table 13.1 presents more studies that specifically address the identification of attacks within the context of the IoT in healthcare environments.

Based on the above-mentioned studies, several issues have been observed. Firstly, it has been shown that anomaly detection by statistical approaches requires an adequate amount of repetitions to effectively train the model. Additionally, the threshold used for detecting complex attacks may not be appropriate for real-world scenarios. Another limitation of different proposed approaches pertains to the decline in performance observed in IDS when the network experiences high levels of traffic congestion. Several frameworks demonstrate suboptimal performance when it comes to identifying complex attack. One of the primary limitations in the field of the MCPS is the scarcity of publically available data that can accurately represent cyber-attacks targeting this domain. To address the existing research gap, this study introduces a cyber-attack detection system that utilizes a DL approach. The system is designed to identify a variety of cyber-attacks, including DoS attacks, ARP Spoofing, Smurf attacks and, Nmap PortScan within the context of the MCPS. Additionally, the system incorporates the capability to perform multi-class classification, enabling it to determine the specific type of attack associated with a given malicious event.

13.3 Basic Preliminaries

This section provides a comprehensive analysis of the operational framework of ML methodologies used in the development of robust IDS for MCPS environments.

Table 13.1 Summary of related works

Model used	Dataset used	Performance of model	Limitations of research	Year	Refs.
CNN	Bot-IoT	Accuracy: 91.27% (with batch size 128)	Accuracy of the model falls down when trained with less batch size like 32 and 64	2020	[18]
FNN	Bot-IoT	Accuracy: 95.10% F1-score: 95%	The feature normalization of the Bot-IoT dataset indicates that the accuracy would decrease to less than 50%	2019	[19]
Bidirectional-LSTM	UNSW-NB15 and Bot-IoT	Accuracy: 99.41% (UNSW-NB15) Accuracy: 98.91% (Bot-IoT)	The suggested approach has the drawback of IDS' effectiveness failing under excessive network traffic and failing to properly alert against and detect a complicated attack	2020	[23]
FNN	Bot-IoT	Accuracy: 99.41%	The suggested approach has been found to be inappropriate for providing security to data theft and key logging attacks for binary classification, and it also achieved a low accuracy of 88.9% in multi-class classification	2019	[24]

(continued)

Table 13.1 (continued)

Model used	Dataset used	Performance of model	Limitations of research	Year	Refs.
LSTM	N_BaIoT-2018	Accuracy: 99.85% F1-score: 99.12%	The suggested approach requires extensive training time and large data sets	2020	[25]
LSTM	N_BaIoT	Accuracy: 97.84%	New attack detection is not possible with the recommended approach	2020	[26]
RF	SmartFall dataset	Accuracy: 99.9%	LSTM's accuracy is poor when compared to that of other approaches. The effectiveness of LSTMs may be improved	2021	[27]
RNN	IOTPOT	Accuracy: 98.71%	Multiclass categorization and category malware on system calls might be used to enhance it LSTM and other DL techniques may potentially help it improve	2020	[28]

13.3.1 Decision Tree

The decision tree (DT) approach is a frequently used technique in the field of data mining, utilized for the creation of classification systems that rely on multiple variables. Furthermore, it is used for the development of prediction algorithms that seek to anticipate results for a certain target variable. The suggested approach entails categorizing a given population into segments that resemble branches, leading to the creation of an inverted tree structure. This structure has a primary node, intermediary nodes, and terminal nodes. The method used in this study is non-parametric in nature, allowing it to effectively handle extensive and complex datasets without requiring for a complex parametric framework. When the sample size reaches a certain size, it becomes possible to partition research data into distinct training and validation datasets. The training dataset is used for constructing a DT model, whereas the validation dataset is used to find the right tree size for the best possible model.

The first step in the DT classifier involves the calculation of the entropy of the given database. The metric provides an indication of the level of uncertainty present in the database. A decrease in the magnitude of the uncertainty value corresponds to an improvement in the quality of the categorization outcomes. The information gain of each feature is computed. This subsequently provides the uncertainty that diminishes after the partitioning of the database. At last, the calculation of information gain is performed for each feature, and thereafter, the database is partitioned based on the features that exhibit significant information gain. The procedure mentioned above is iteratively executed until all nodes have been successfully organized. Equation (13.1) provides a mathematical representation of the DT.

$$f(X) = \sum_{k=1}^{N_{leaf}} Y_k * I_{leaf}(X, k) \tag{13.1}$$

where, N_{leaf} = number of leaf nodes in the DT, Y_k = outcome associated with the kth leaf, $I_{leaf}(X, k)$ = indicator function.

13.3.2 Random Forest

Random forests (RF) are an ensemble learning approach that combines several tree predictors. Every tree inside the forest is created by using random vectors which is independently generated from a uniform population for all trees. The convergence of the generalization error of random forests gets more clear and precise as the total number of trees in the forest approaches infinity. The generalization error of a forest consisting of tree classifiers is influenced by the performance of each individual tree within the forest and the degree of connection among them. The use of a stochastic selection of characteristics for partitioning each node results in the occurrence of error rates. The RF model is represented in Equation (13.2).

$$\hat{Z} = mode(f_1(x), f_2(x), \ldots, f_n(x)) \tag{13.2}$$

where, $\hat{Z}$ = Final prediction of RF, $f_n(x)$ = Prediction of nth decision trees.

13.3.3 Adaptive Boosting

AdaBoost is referred to as Adaptive Boosting. It is an ensemble-based ML method that demonstrates adaptability in its application to various classification and regression applications. This supervised learning employs the combination of numerous weak or base learners, such as decision trees to form a robust learner capable of classifying data properly. The AdaBoost algorithm operates by assigning weights to instances in the training dataset, which are determined by the accuracy of prior classification iterations. Equation (13.3) shows the working principle of AdaBoost.

$$\hat{Z}= sign\left(\sum_{k=1}^{K} \alpha_k \cdot h_k(x)\right) \tag{13.3}$$

where, $\alpha_k = \frac{1}{2} ln\left(\frac{1-\varepsilon_k}{\varepsilon_k}\right)$ weight importance of kth weak learner, $h_k(x) =$ prediction of kth weak learner with input x, $\varepsilon_k =$ weight error of weak learner.

13.3.4 GBoost

Gradient boosting is referred to as GBoost employ a learning approach that iteratively fits additional models in order to enhance the precision of the estimated output value. The foundational idea of this methodology is to generate newer base-learners that exhibit the strongest correlation with the negative gradient of the loss function, which has connection to the whole group performance. The choice of loss functions may be arbitrary. However, it provides a clearer understanding if the error function is the traditional squared-error loss which includes iteratively minimizing the errors. Equation (13.4) shows the working principle of GBoost.

$$\hat{Z}= \sum_{k=1}^{K} \eta . h_k(x) \tag{13.4}$$

where, $h_k(x) =$ prediction of kth weak learner with input x, $\eta =$ learning rate hyper parameter controlling the step size during each update.

13.3.5 XGBoost

It has been proven that XGBoost is a superior ML method due to its effective implementation of gradient-boosted decision trees. It has been implemented to make the most efficient use of memory and the available processing power. XGBoost reduces execution time while improving performance when compared to other ML approaches and even DL approaches. The primary goal of boosting is to construct sub-trees from a parent tree in such a way that error rates of each successive tree are less than those of the parent tree. In this method, the new sub trees will revise the old residuals to lower the cost function's inaccuracy. Equation (13.5) shows the working principle of XGBoost.

$$Xgb(\theta) = \sum_{i=1}^{N} L(y_i, p_i) + \sum_{k=1}^{T} \Omega(f_k) \tag{13.5}$$

where, $L(y_i, p_i)$= Loss function with y_i, p_i denoting actual target value and predicted value from weak learner respectively, $\Omega(f_k)$ = regularization term for kth trees.

13.3.6 CatBoost

The Catboost method is a notable example of a gradient boosting technique that has gained popularity in recent years. Catboost is a versatile ML algorithm that addresses both regression and classification tasks. It has gained attention due to its inclusion in a recently developed open-source gradient boosting library, which is freely available and compatible with several platforms. The Catboost algorithm and gradient boosting use DTs as a primary weak learner, using a sequential fitting approach. The use of inconsistent permutation of gradient learning information has been proposed as a means to improve the performance of the Catboost model and mitigate the issue of over fitting. Equation (13.6) shows the working principle of XGBoost.

$$F_k(x) = F_{k-1}(x) + \gamma_k h_k(x) \tag{13.6}$$

where, $F_k(x)$= Ensemble Prediction at iteration K, $F_{k-1}(x)$= prediction from previous iteration, γ_k= learning rate at iteration k, $h_k(x)$= weak learner added in iteration k.

13.4 Proposed Method

In recent years, CNN models have been widely used in the field of computer vision, image classification, segmentation, detection, and natural language processing. This is primarily due to the higher performance shown by these models, which may be attributed to the effective utilization of multiple DL methodologies. CNN has been extensively used in the healthcare domain. CNN often known as covnet is a specific type of neural network that may possess shared parameters. CNN is composed of several layers, through which each layer has the ability to convert one volume into another volume using differentiable functions. The CNN design consists of consecutive layers of convolution and pooling, with the inclusion of at least one fully linked layer at the final stage.

The input layer stores the original image data. To compute the output value, the convolution layer performs a dot product calculation between each filter and each patch of the image. The activation function can be implemented in convolution layer as well as in other layers of CNN. Various activation functions may be used, such as Rectified Linear Unit (ReLU), Sigmoid, Softmax, Leaky ReLU, and Hyperbolic Tangent (Tanh) among others. The pool layer is responsible for decreasing the volume

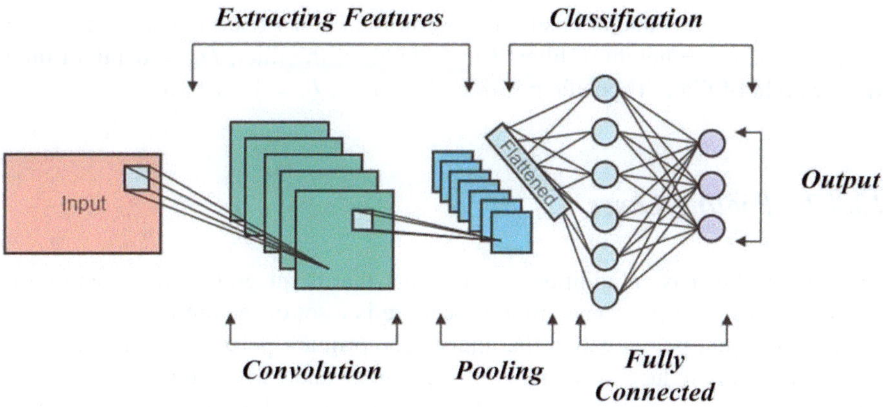

Fig. 13.2 Architectural framework of CNN

size and enhancing computational efficiency. The insertion of this component into CNN serves as a primary purpose for minimizing the occurrence of over fitting. Pooling layers may be implemented using either max pooling or average pooling techniques. The fully connected layer, also known as a normal neural network layer receives input from the preceding layer. The primary aim of this system is to calculate the results for each class resulting in a one-dimensional array with a size equivalent to the number of classes. The overall architecture of the CNN is represented in Fig. 13.2. The architectural framework of CNN is illustrated as follows.

13.4.1 Convolutional Layer

The layer consists of a combination of convolutional filters, whereby each neuron functions as a kernel. However, if the kernel exhibits uniformity, the convolution process will effectively become a correlation operation. The primary image has been partitioned into smaller segments known as receptive fields by use of the convolutional kernel. The previously identified division procedure plays a crucial role in the stage of feature extraction. The kernel performs convolution on images by applying a certain set of weights, which involves multiplying the appropriate elements with the multiplying components of the receptive field. In contrast to fully connected networks, CNNs may extract more information from a given image with fewer parameters by using a sliding kernel with the same set of weights. Different types of convolution operations exist depending on the number of filters used, the padding used, and the direction in which the convolution is performed. The convolution process could be approximated using the following equation (13.7).

$$Cov_{p,q,r} = \sum_{i=0}^{Fh-1} \sum_{j=0}^{Fw-1} \sum_{k=0}^{Dp-1} \left(W_{i,j,k,r} * I_{p.s+i,q.s+j,k}\right) + B_k \tag{13.7}$$

where, $Cov_{p,q,r}$ = value at position $(p.q)$ in the $r - th$ feature map of the output convolution, Fh = height of filter, Fw = width of the filter, Dp = depth of input, W = weight of filter at specific position, I = input, B_k = bias term.

13.4.2 Pooling Layer

The pooling layer is an additional component that is integral to the structure of a CNN. The result of the convolution procedure is a set of feature patterns that may occur at different points within the image. The primary purpose of this function is to systematically reduce the quantity of parameters and calculations. Therefore, it is referred to as down-sampling in a similar manner. The pooling layer can be expressed in equation (13.8).

$$Pool_{p,q,r} = Max_{i=0}^{Ph-1} Max_{j=0}^{Pw-1} \left(I_{p.s+i,q.s+j,r} \right) \tag{13.8}$$

where, Ph = height of pooling window, Pw = width of pooling window, i = input, s = stride.

13.4.3 Fully Connected Layer

This layer constitutes the last layers inside the network, which is employed for the purpose of data classification. The outcomes of the pooling and convolutional layer is transformed into a flat shape and then delivered into the fully connected layer.

13.4.4 Activation Function

This node is situated either at the endpoint position or inside the interconnections of Neural Networks. The selection of a suitable activation function has the potential to enhance the efficiency of the learning process. There are several forms of activation functions, including ReLU, Logistic (Sigmoid), Tanh and softmax. The ReLU function is often used in hidden layers because of its ease in implementation and effectiveness in mitigating the limitations of other activation functions, such as Tanh and Sigmoid. It is essential to recognize that the model exhibits considerable less sensitivity to vanishing gradients, so effectively minimizing potential training issues.

There are many distinct forms of cyber-assaults, such as Denial of Service (DoS) attacks, Nmap attacks, ARP Spoofing and Smurf assaults have been observed in the environment of the MCPS. The two stages included in this study are the data preprocessing phase and the CNN-based assault detection phase. The following sections describe the sequential steps involved in the implementation and evaluation of the

Table 13.2 Parameters of proposed CNN

No. of Conv-2D Layer—2
No. of Pooling Layer—2
No. of Filters in convolution layer—(32,64)
Filter size—5 × 5
Pooling size—2 × 2
Optimizer—Adam
Learning rate—0.01
No. of neurons in hidden layer—128
Dropout rate—0.30

CNN model: (i) The ECU IoHT dataset is employed for the analysis of different cyber-attacks. (ii) Preprocessing techniques such as missing value imputation, elimination of duplicate records, normalization, and managing imbalanced data have been implemented. (iii) The dataset has been labeled with categories including Normal, DoS attack, ARP Spoofing, Smurf attack and Nmap attack in order to prepare for multi-class classification. (iv) The dataset is partitioned into two subsets as the training dataset and the testing dataset with proportions of 80% and 20% respectively. (v) The CNN is evaluated on the training samples by selecting these labels as target features using multiclass classification which produces a good examined model. (vi) The CNN model that has been trained is evaluated using a separate testing dataset in order to provide predictions about normal or other forms of assaults.

The suggested CNN has a deep architecture consisting of four hidden layers. These hidden levels include two convolutional layers and two pooling layers. The network comprises of two convolutional layers, each of which is trained using 64 and 128 convolution kernels respectively. The size of these kernels is 3 x 3. The deep architecture includes a completely linked layer that employs five distinct neurons to facilitate the process of categorization. In order to accomplish average pooling with a factor of 2, two pooling layers are used. The objective of intrusion detection in MCPS may be seen as a classification problem, so the deep architecture incorporates the softmax activation function. Table 13.2 illustrates the configuration of hyper parameters used in the proposed model. Figure 13.3 depicts the proposed model framework.

13.5 Dataset Description and Simulation Setup

This section presents a comprehensive explanation of the dataset used for the proposed study, as well as the experimental setup employed for developing the model.

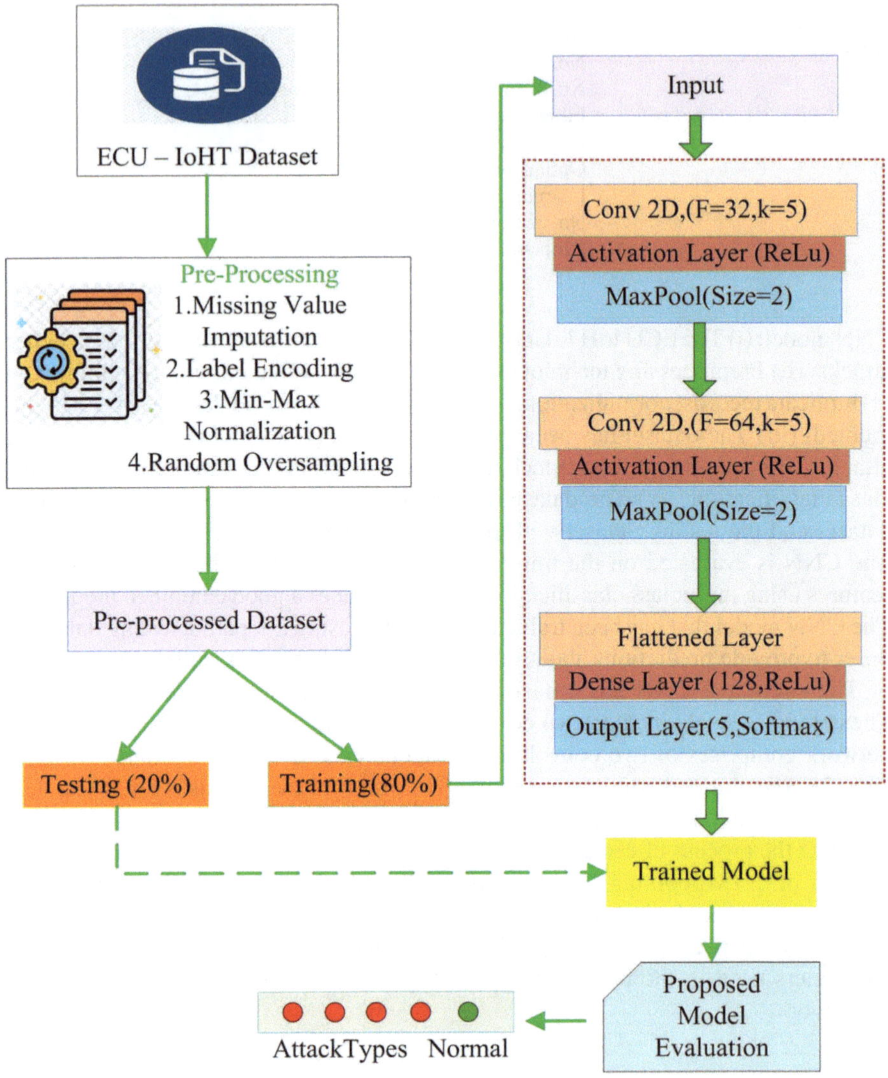

Fig. 13.3 Proposed CNN architecture

13.5.1 Dataset Description

The ECU-IoHT dataset has been created inside an Internet of Healthcare Things (IoHT) setting with the objective of supporting the healthcare security community in the detection and prevention of cyber-attacks against IoHT systems. The dataset was generated using a simulated IoHT network that included several medical sensors. Many different kinds of assaults were executed against the network in consideration.

Table 13.3 Features details of ECU-IoHT dataset

Feature Name	Meaning	Type
Source	Source IP Address of the system	Numerical
Destination	Destination IP address of the system	Numerical
Protocol	Protocol used	Categorical
Length	Packet length found	Numerical
Info	Packet information	Categorical
Type	Generic classification (Attack or Normal)	Categorical
Type of attack	Defining specific attack	Categorical

Table 13.4 Attack type details of ECU-IoHT dataset

Attack type	Counts
Smurf attacks	77,920
Nmap Port Scans	6836
ARP Spoofing	2359
DoS attacks	639
Normal	23,453

The dataset included collected and stored network activities, which were represented as attributes that characterized each network flow. Tables 13.3 and 13.4 present overall summary of features and various attack type of the dataset. The dataset comprises 23,453 instances of normal activities and 87,754 instances of attacks, which are further classified into four distinct categories including Smurf attacks, Nmap Port Scans, ARP Spoofing, and DoS attacks. There are 111,207 observations in the dataset, including both normal and different attacks type.

13.5.2 Dataset Preprocessing

This work extensively used several data preparation approaches, such as missing value imputation, oversampling, label encoding, and normalization. IoT sensors in the network environment may provide missing values or erroneous data for a short span of time due to sensor failure. A strategy for imputing missing values has been used to improve the data's dependability for the model. Moreover, this dataset has a textual property that provides a description of the network activity linked to each input. The label encoding approach has been used due to the presence of string values in each feature vector. The dataset contains many categories of attacks. Nevertheless, several assault categories have a lower frequency in comparison to the prevailing Smurf attack. Therefore, the dataset demonstrates a inequality in class distribution. In order to address this issue, the random oversampling approach has been used for this study. The min-max scalar approach has been applied to standardize the dataset by ensuring that all attribute values are in the same scale.

13.5.3 Experimental Setup

The current research used a Python notebook accessible on the Google Colab platform, which derived from the computational capabilities of GPU-based servers. Furthermore, the experiment used the Keras and Tensor Flow libraries. The experimental setup consist of a system configuration of an Intel Core i7 central processing unit (CPU), 16 gigabytes (GB) of random access memory (RAM), and a 64-bit Windows 10 operating system. The CPU operated at a clock speed of 2.20 gigahertz (GHz). The process of data analysis is performed using the Python packages Pandas, Imblearn, and Numpy. Data visualisation is conducted via the use of Matplotlib and Mlxtend.

13.5.4 Evaluation Measures

The suggested model's detection accuracy has been evaluated using several metrics. Common metrics used for evaluations are accuracy, recall, precision, ROC-AUC and F1-score. The following four variables affect these metrics:

- True Positive (Tp): It shows the percentage of fraudulent network traffic observations in MCPS that were successfully identified by the methodology.
- True Negative (Tn): It indicates the proportion of regular network traffic observations in MCPS that the model accurately categorized as normal.
- False Positive (Fp): It indicates how many apparently normal occurrences in MCPS network traffic were wrongly identified as harmful by the methodology.
- False Negative (Fn): It reveals how many harmful observations the MCPS network traffic model incorrectly categorized as normal.

Using the above parameters, the following metrics are derived for evaluation:

(a) Accuracy: It specifies how many cases were successfully categorized by the model relative to how many observations are in the testing set. It incorporates both Tp and Tn into its model accuracy calculations as given in Eq. (13.9).

$$\text{Accuracy} = (Tp + Tn)/(Tp + Tn + Fp + Fn) \tag{13.9}$$

(b) Precision: It is the ratio of the number of attacks observed to the number of observations that the model has labeled as attacks as given in Eq. (13.10).

$$\text{Precision} = Tp/Tp + Fp \tag{13.10}$$

(c) Recall: The recall rate is defined as the proportion of correct predictions made for an anomalous event as given in Eq. (13.11).

$$\text{Recall} = Tp/Tp + Fn \tag{13.11}$$

(d) F1-Score: Its primary use is in case of imbalanced class distribution, since it is more beneficial than accuracy because it factors in Fp and Fn as written in Eq. (13.12).

$$F1 - Score = 2((Precision * Recall)/(Precision + Recall)) \tag{13.12}$$

(e) ROC-AUC: This measure indicates the probability that a randomly chosen positive test point would be anticipated more positively than a randomly picked negative test point.

13.6 Result Analysis

This section shows an analysis of the performance of the suggested technique, as well as a comparison with existing advanced models like the DT, RF, AdaBoost, GBoost, XGBoost, and the CatBoost. The suggested methodology is implemented on a dataset that comprises both normal instances and anomalous instances in order to identify numerous types of attacks in the MCPS. The suggested DL based CNN technique utilizes the ReLU and softmax activation functions. A batch size of 1000 has been selected and the Adam optimizer has been implemented with the categorical cross entropy loss function. The training process of the model involves adjusting the epoch values within the range of 100. The model that completed 100 epochs has the best validation accuracy and the lowest loss. Figures 13.4 and 13.5 illustrate the accuracy and loss metrics of the suggested approach throughout the process of 100 epochs.

Figure 13.6 depicts a comparison and analysis of the accuracy of the suggested technique compared to other advanced models. Based on the data shown in the figure, the suggested technique exhibits higher performance in terms of accuracy when compared to the DT, RF, AdaBoost, GBoost, XGBoost, and CatBoost methodologies. Table 13.5 and Figure 13.6a–e provide a comparative study of several performance measures, including precision, recall, AUC-ROC, F1-score, and accuracy. These metrics have been employed to examine the effectiveness of the proposed algorithm and other algorithms. Table 13.5 illustrates significance of inherent unpredictability in the optimal and mean performances of several performance measures used for evaluating all of the models. The table reveals that CNN technique exhibits dominance over ML and ensemble learning approaches when it comes to classifying abnormalities. The CNN method have superior performance compared to other models, achieving recall, precision, AUC-ROC, F1-score, and accuracy values of 99.19%, 99.35%, 99.91%, 99.27%, and 99.74% respectively. The RF classifier achieved a

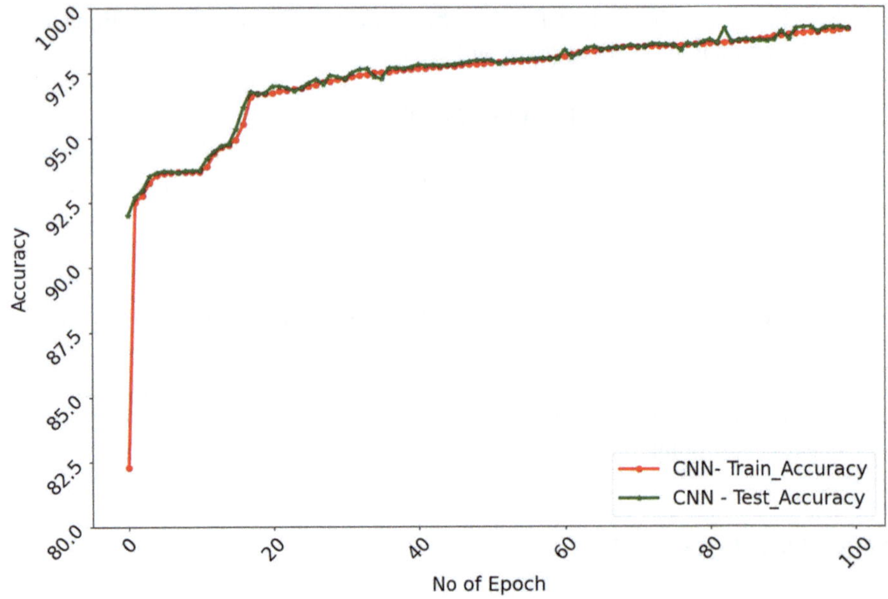

Fig. 13.4 Proposed CNN accuracy curve

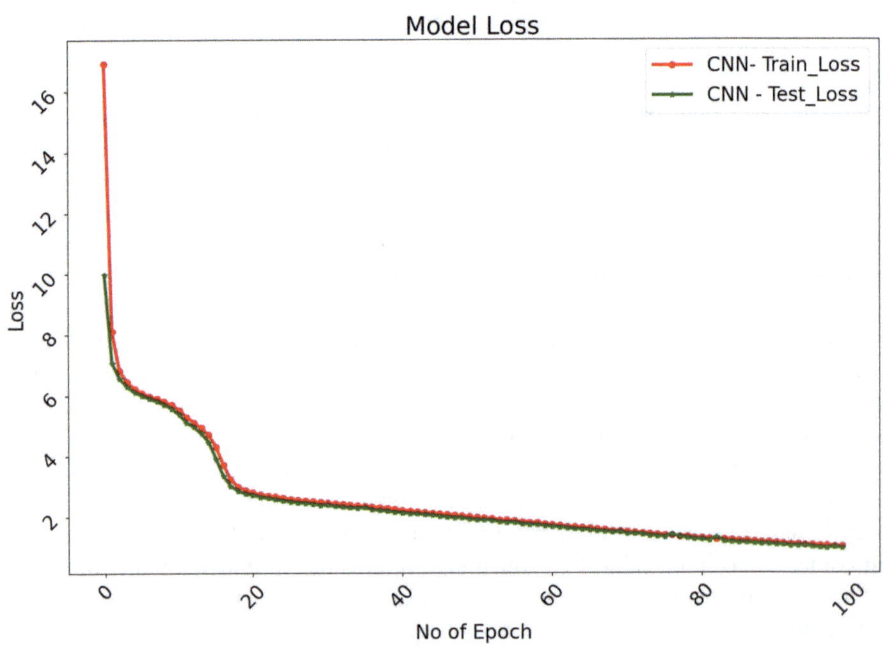

Fig. 13.5 Proposed CNN loss curve

very high accuracy rate of 99.60%, which was the second highest among the many conventional approaches used in the study. The AdaBoost method exhibited the lowest accuracy of 74.89%, while the DT achieved an accuracy of 89.16%.

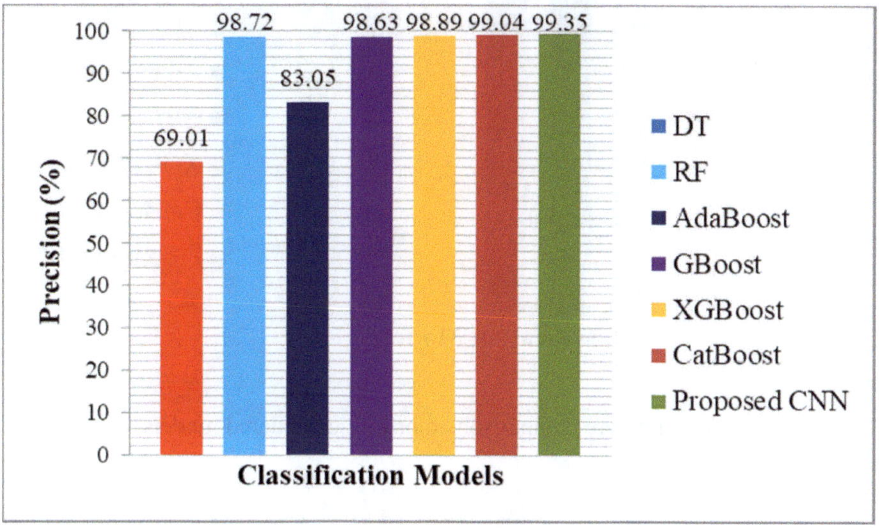

(a) Precision comparison among all considered models

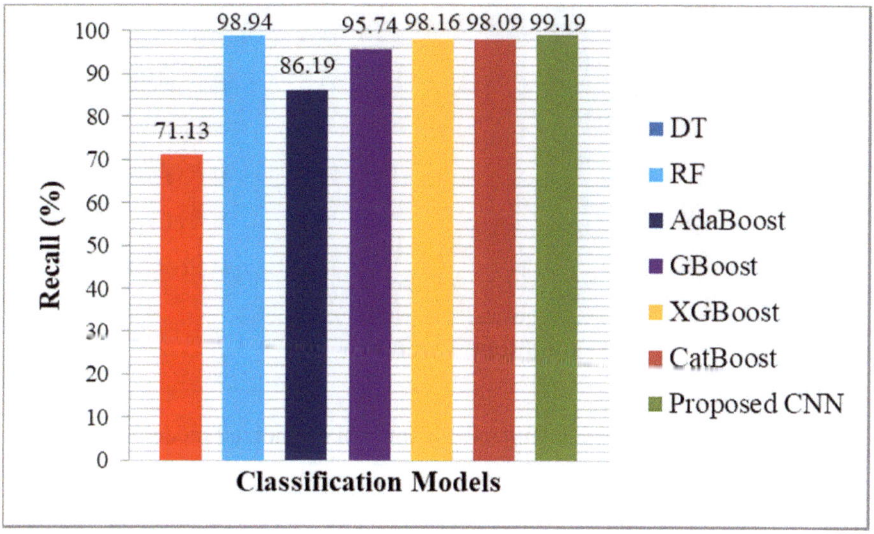

(b) Recall comparison among all considered models

Fig. 13.6 **a–e** Evaluation of the suggested method's metrics in relation to competing models

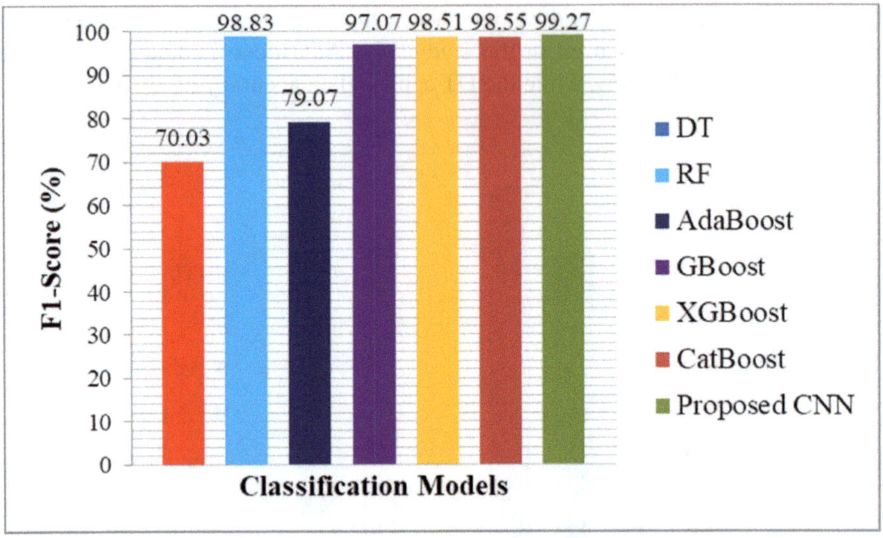

(c) F1-Score comparison among all considered models

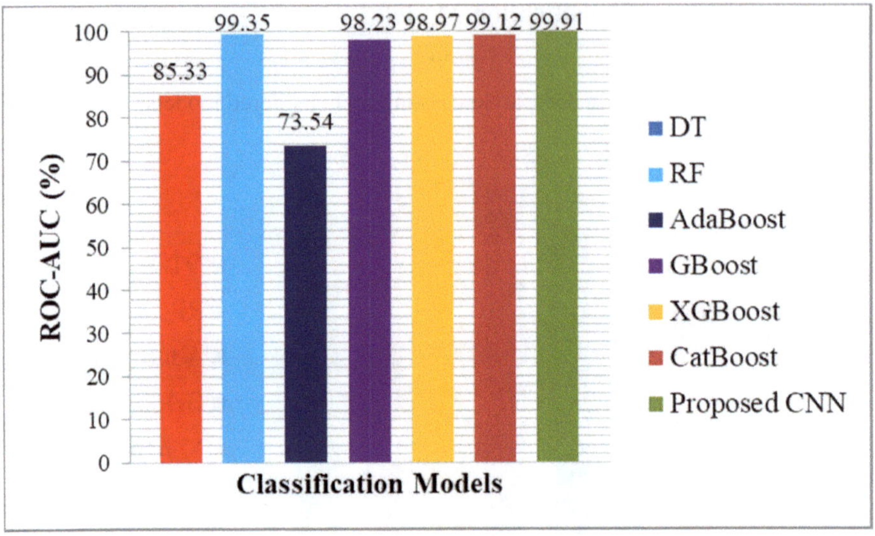

(d) ROC-AUC comparison among all considered models

Fig. 13.6 (continued)

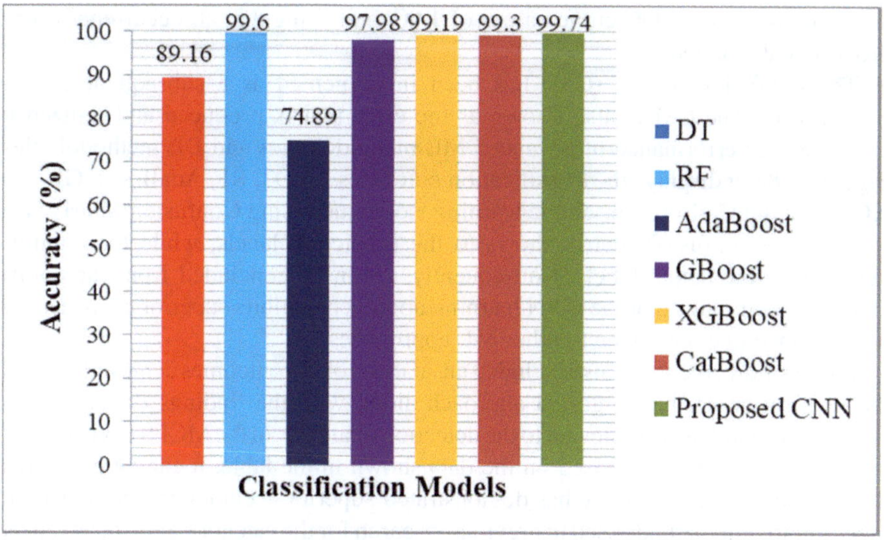

(e) Accuracy comparison among all considered models

Fig. 13.6 (continued)

Table 13.5 Evaluation of the suggested model's metrics in comparison to the results of other competing models

Model	Precision	Recall	F1-score	ROC-AUC	Accuracy
DT	69.01	71.13	70.03	85.33	89.16
RF	98.72	98.94	98.83	99.35	99.60
AdaBoost	83.05	86.19	79.07	73.54	74.89
GBoost	98.63	95.74	97.07	98.23	97.98
XGBoost	98.89	98.16	98.51	98.97	99.19
CatBoost	99.04	98.09	98.55	99.12	99.30
Proposed CNN	99.55	99.59	99.57	99.91	99.74

The findings for the area under the receiver operating characteristic (AUC-ROC) curves of the proposed model and other standard methods are shown in Figure 13.7a–g. Based on the shown figure, it can be noted that the CNN model provided in this study achieved a perfect AUC-ROC score of 1.00 for 0, 1, 2, and 4 class labels and 0.99 for class 3. There are more misclassified data in the RF approach compared to the proposed framework, despite the fact that the RF model has a superior AUC-ROC performance. This performance of proposed CNN surpasses that of other standard techniques, suggesting that the recommended strategy is very successful in accurately classifying all occurrences in the dataset when compared to other methods. The suggested technique achieved perfect classification accuracy, as shown by the micro-

and macro-average ROC curve values of 1.00, indicating that all occurrences were properly categorized.

The confusion matrix (CM) has been implemented as a subsequent evaluation criteria. The CM is also known as the error matrix produced and utilized to examine the performance of advanced ML methodologies and DL methodologies. Figure 13.8a–g displays the classification outcomes for DT, RF, AdaBoost, GBoost, XGBoost and CatBoost ensemble learning techniques using Confusion Matrix. The rows of the confusion matrix shows to the prediction labels, while the columns denoting to the actual labels. Consequently, it can be concluded from the confusion matrix findings that the CNN learning approach exhibits superior classification performance in comparison to other ML approaches.

Furthermore, the study also included an analysis of the outcomes achieved through the application of the suggested approach along with the findings obtained from previous research works relevant to the detection of attacks in the MCPS environment, as shown in Table 13.6. Based on the data shown in the table, it can be concluded that our suggested technique has demonstrated superior accuracy in comparison to other intelligent methods used in previous research for the categorization of advanced attacks in MCPS.

13.7 Conclusions

This study introduces a CNN methodology that improves the identification and minimizing of cyber-attacks in MCPS devices. The proposed strategy aims to enhance the security of healthcare gadgets that use the IoT technology. The proposed system has been developed with a specific emphasis on multi-class classification for the purpose of identifying DoS attacks, ARP Spoofing, Smurf attacks, and Nmap attacks. This is in contrast to the existing system, which is based on binary class classification and is used to detect a range of attack types. Finally, the proposed system has been evaluated using the health care domain dataset (ECU-IoHT), which sets it apart from previous approaches. The experimental findings demonstrate that the suggested CNN approach achieves a significantly greater accurate identification rate and a minimal false detection rate in comparison to the existing method. The suggested model achieved accuracy over 99% after undergoing training with 100 epochs. The recommended method achieved accuracy, recall, and F1 score values of 99.35%, 99.19%, and 99.27%, respectively. This observation highlights the superiority of the suggested system in comparison to existing work. As this study evaluated with less feature data samples, the proposed model might be offer less accuracy with higher dimension dataset as complexity of network leads to over fitting for handling more complex features. In the future, the suggested system has the potential to be implemented for the purpose of evaluating its effectiveness in a real-time MCPS environment with high dimensionality dataset. Additionally, efforts will be made to enhance the scalability of this research in order to identify various forms of attacks on MCPS devices.

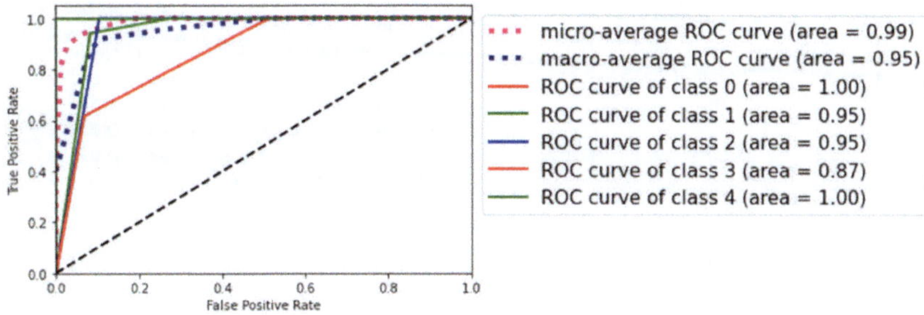

(a) DT

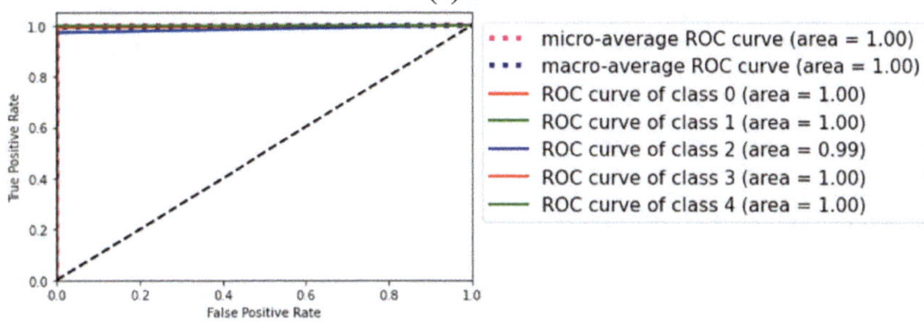

(b) RF

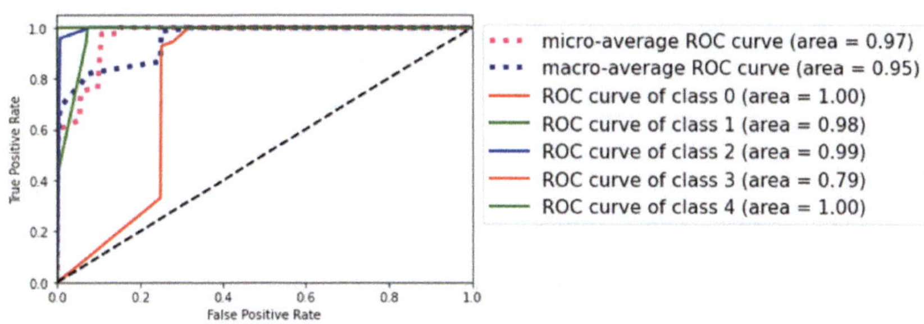

(c) AdaBoost

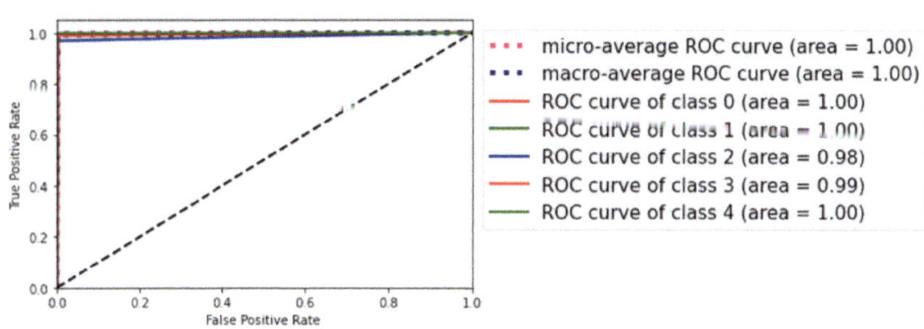

(d) GBoost

Fig. 13.7 ROC-AUC comparisons for **a** DT **b** RF **c** AdaBoost **d** GBoost **e** XGBoost **f** CatBoost and **g** Proposed CNN

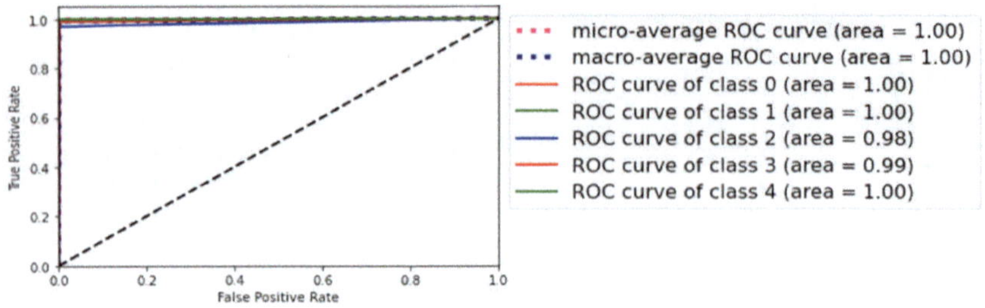

(e) XGBoost

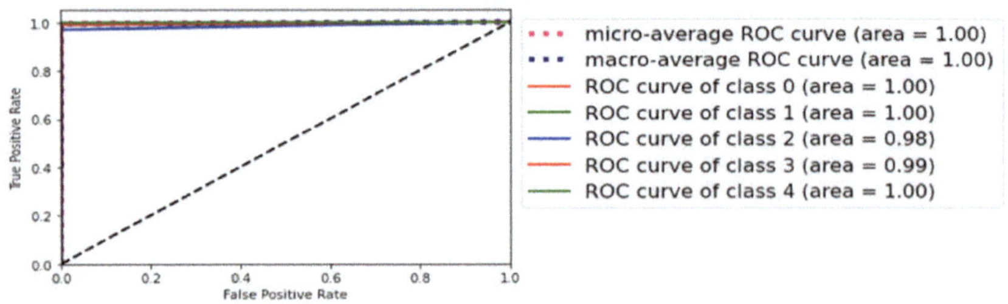

(f) CatBoost

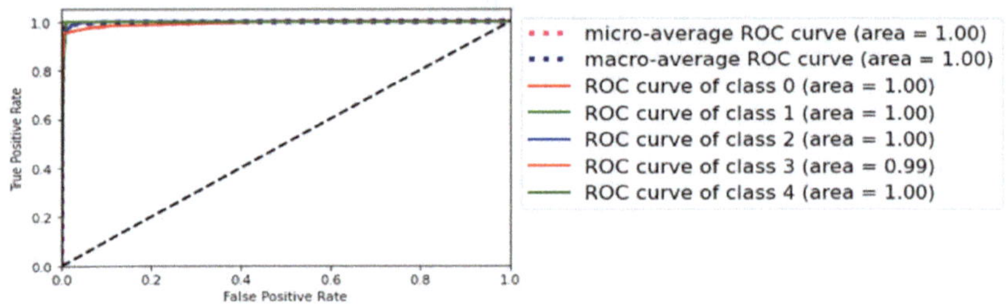

(g) Proposed CNN

Fig. 13.7 (continued)

Fig. 13.8 Confusion Metrics comparison for **a** DT **b** RF **c** AdaBoost **d** GBoost **e** XGBoost **f** CatBoost and **g** Proposed CNN

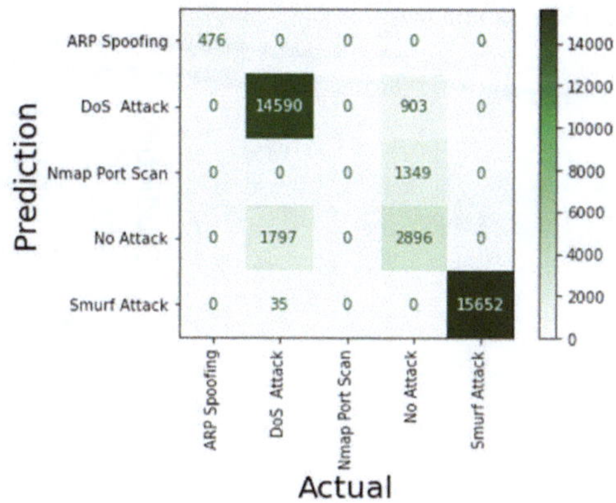

(a) DT

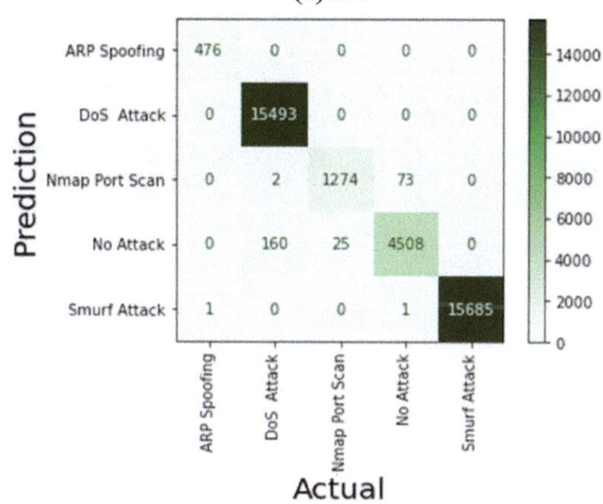

(b) RF

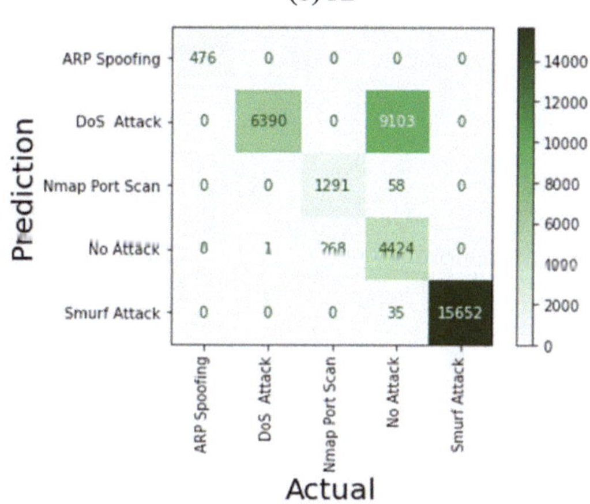

(c) AdaBoost

Fig. 13.8 (continued)

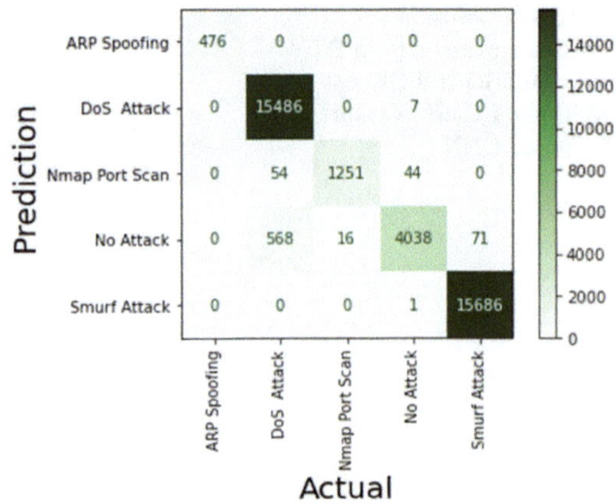

(d) GBoost

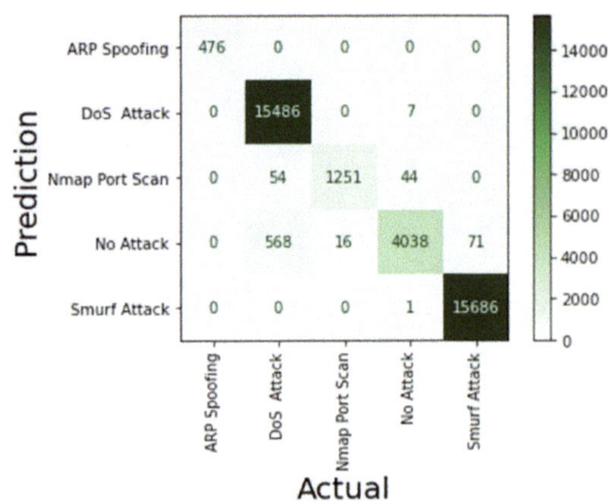

(e) XGBoost

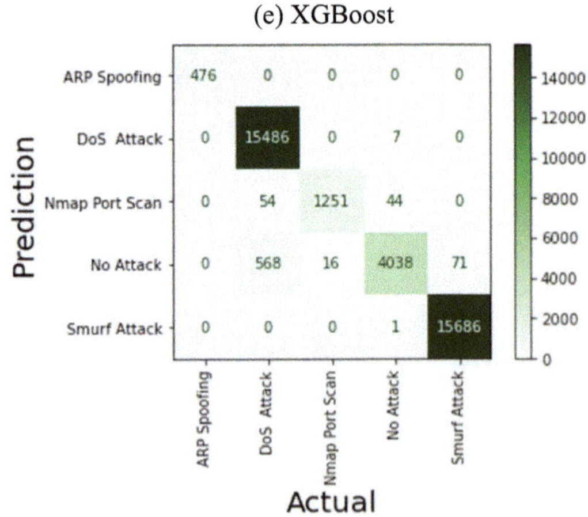

(f) CatBoost

Fig. 13.8 (continued)

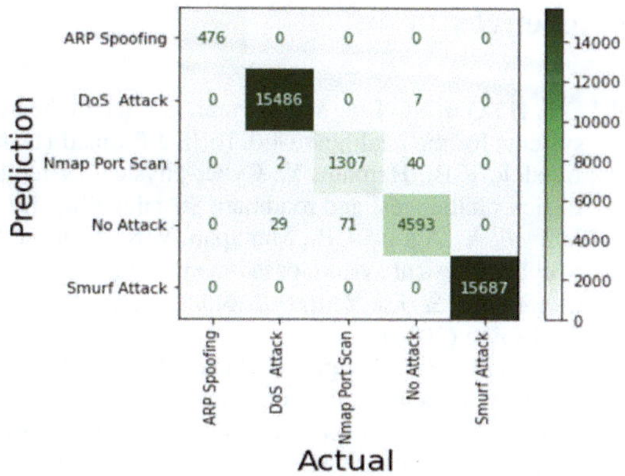

(g) Proposed CNN

Table 13.6 Comparison of proposed CNN with other existing Approaches

Smart Approach used	Dataset used	Performance (F1-Score) (%)	Year	Refs.
DBN	CICIDS 2017 dataset	99.37	2020	[29]
DT	IoT healthcare dataset	99.47	2021	[30]
PSO-RF	NSLKDD datasets	99.46	2021	[31]
DT	ToN-IoT dataset	99	2021	[32]
DNN	ECU-IoHT dataset	99.50	2023	[33]
Proposed CNN	ECU-IoHT dataset	99.57	Our current study	Our current study

References

1. Qiu, H., Qiu, M., Liu, M., Memmi, G.: Secure health data sharing for medical cyber-physical systems for the healthcare 4.0. IEEE J Biomed Health Inform 24(9):2499–2505 (2020)
2. Adedeji, K.B., Hamam, Y.: Cyber-physical systems for water supply network management: Basics, challenges, and roadmap. Sustainability **12**(22), 9555 (2020)
3. Jamwal, A., Agrawal, R., Manupati, V. K., Sharma, M., Varela, L., Machado, J.: Development of cyber physical system based manufacturing system design for process optimization. In *IOP Conference Series: Materials Science and Engineering* (Vol. 997, No. 1, p. 012048). IOP Publishing (2020)
4. Cartwright, R., Cheng, A., Hudak, P., OMalley, M., Taha, W. (2008, November). Cyber-physical challenges in transportation system design. In *National workshop for research on high confidence transportation Cyber-physical systems: automotive, aviation & rail.-2008.*
5. Ahmad, M.O., Ahad, M.A., Alam, M.A., Siddiqui, F., Casalino, G.: Cyber-physical systems and smart cities in India: Opportunities, issues, and challenges. Sensors **21**(22), 7714 (2021)
6. Wang, Eric Ke, et al.: A deep learning based medical image segmentation technique in Internet-of-Medical-Things domain. Future Generation Computer Systems 108 (2020): 135–144
7. Shuwandy, M.L. et al.: mHealth authentication approach based 3D touchscreen and microphone sensors for real-time remote healthcare monitoring system: comprehensive review, open issues and methodological aspects. Comput. Sci. Rev. 38 (2020): 100300
8. Kim, J., Campbell, A.S., de Ávila, B.E.-F., Wang, J.: Wearable biosensors for healthcare monitoring. Nature Biotechnol. **37**(4), 389–406 (2019)
9. Choudhuri, A., Chatterjee, J.M., Garg, S.: Internet of things in healthcare: A brief overview. In: Internet of Things in Biomedical Engineering, Elsevier, pp. 131–160 (2019)
10. Priyadarshini, R., Panda, M.R., Mishra, B.K.: Security in healthcare applications based on fog and cloud computing, Cyber Secur. Parallel Distributed Comput. 231–243 (2019)
11. Yaacoub, J.-P.A., Noura, M., Noura, H.N., Salman, O., Yaacoub, E., Couturier, R., Chehab, A.: Securing internet of medical things systems: Limitations, issues and recommendations, Future Gener. Comput. Syst.. Syst. **105**, 581–606 (2020)
12. Ahmed, M., Byreddy, S., Nutakki, A., Sikos, L., Haskell-Dowland, P.: ECU-IoHT (2020) 10.25958.5f1f97b837aca
13. M. Begli, F. Derakhshan, H. Karimipour, A layered intrusion detection system for critical infrastructure using machine learning, in: 2019 IEEE 7th International Conference on Smart Energy Grid Engineering, SEGE, IEEE, 2019, pp. 120–124.
14. A.I. Newaz, A.K. Sikder, M.A. Rahman, A.S. Uluagac, Healthguard: A machine learning-based security framework for smart healthcare systems, in: 2019 Sixth International Conference on Social Networks Analysis, Management and Security, SNAMS, IEEE, 2019, pp. 389–396.
15. He, D., Qiao, Q., Gao, Y., Zheng, J., Chan, S., Li, J., Guizani, N.: Intrusion detection based on stacked autoencoder for connected healthcare systems. IEEE Netw.Netw. **33**(6), 64–69 (2019)
16. Alrashdi, I., Alqazzaz, A., Alharthi, R., Aloufi, E., Zohdy, M.A., Ming, H., FBAD: Fog-based attack detection for IoT healthcare in smart cities, in,: IEEE 10th Annual Ubiquitous Computing. Electronics & Mobile Communication Conference, UEMCON, IEEE **2019**, 0515–0522 (2019)
17. Hady, A.A., Ghubaish, A., Salman, T., Unal, D., Jain, R.: Intrusion detection system for healthcare systems using medical and network data: a comparison study. IEEE Access. **8**, 106576–106584 (2020). https://doi.org/10.1109/ACCESS.2020.3000421
18. Susilo, B., Sari, R.F.: Intrusion Detection in IoT Networks Using Deep Learning Algorithm. Information **11**, 279 (2020)
19. Ibitoye, O., Shafiq, O.; Matrawy, A. Analyzing Adversarial Attacks against Deep Learning for Intrusion Detection in IoT Networks. In Proceedings of the 2019 IEEE Global Communications Conference (GLOBECOM),Waikoloa, HI, USA, 9–13 December 2019; pp. 1–6
20. Hizal, S., Çavuşoğlu, Ü., Akgün, D.: A new Deep Learning Based Intrusion Detection System for Cloud Security. In: 3rd International Congress on Human-Computer Interaction, Optimization and Robotic Applications (2021)

21. Gopalakrishnan, T. et al.: Deep Learning Enabled Data Offloading With Cyber Attack Detection Model in Mobile Edge Computing Systems. IEEE Access (2020)
22. Xun Y, Qin J, Liu J.: Deep Learning Enhanced Driving Behavior Evaluation Based on Vehicle-Edge-Cloud Architecture. IEEE Transactions on Vehicular Technology (2021)
23. Alkadi, O., Moustafa, N., Turnbull, B., Choo, K.-K.R.: A Deep Blockchain Framework-enabled Collaborative Intrusion Detection for Protecting IoT and Cloud Networks. IEEE Internet Things J. **8**, 1 (2020)
24. Ge, M.; Fu, X.; Syed, N.; Baig, Z.; Teo, G.; Robles-Kelly, A. Deep Learning-Based Intrusion Detection for IoT Networks. In Proceedings of the 2019 IEEE 24th Pacific Rim International Symposium on Dependable Computing (PRDC), Kyoto, Japan, 1–3 December 2019; pp. 256–25609.
25. Samy, A., Yu, H., Zhang, H.: Fog-Based Attack Detection Framework for Internet of Things Using Deep Learning. IEEE Access **8**, 74571–74585 (2020)
26. Parra, G.D.L.T., Rad, P., Choo, K.-K.R., Beebe, N.: Detecting Internet of Things attacks using distributed deep learning. J. Netw. Comput. Appl.Netw. Comput. Appl. **163**, 102662 (2020)
27. Farsi, M.: Application of ensemble RNN deep neural network to the fall detection through IoT environment. Alex. Eng. J. **60**, 199–211 (2021)
28. Shobana, M., Poonkuzhali, S.: A novel approach to detect IoT malware by system calls using Deep learning techniques. In Proceedings of the 2020 International Conference on Innovative Trends in Information Technology (ICITIIT), Kottayam, India, pp. 1–5 (2020)
29. Manimurugan, S., Al-Mutairi, S., Aborokbah, M.M., Chilamkurti, N., Ganesan, S., Patan, R.: Effective attack detection in internet of medical things smart environment using a deep belief neural network. IEEE Access **8**, 77396–77404 (2020)
30. Hussain, Faisal, et al. A framework for malicious traffic detection in IoT healthcare environment. Sensors 21.9 (2021): 3025
31. Saheed, Yakub Kayode, and Micheal Olaolu Arowolo. Efficient cyber-attack detection on the internet of medical things-smart environment based on deep recurrent neural network and machine learning algorithms. IEEE Access 9 (2021): 161546–161554
32. Zachos, G., et al. An Anomaly-Based Intrusion Detection System for Internet of Medical Things Networks. Electronics 2021, 10, 2562." (2021)
33. Vijayakumar, Kedalu Poornachary, et al.: Enhanced Cyber Attack Detection Process for Internet of Health Things (IoHT) Devices Using Deep Neural Network." Processes 11.4 (2023): 1072

Chapter 14
Current Datasets and Their Inherent Challenges for Automatic Vehicle Classification

Sourajit Maity, Pawan Kumar Singh, Dmitrii Kaplun, and Ram Sarkar

Abstract Automatic Vehicle Classification (AVC) systems have become a need of the hour to manage the ever-increasing number of vehicles on roads and thus maintain a well-organized traffic system. Researchers around the world have proposed several techniques in the last two decades to address this challenge. However, these techniques should be implemented on realistic datasets to evaluate their efficiency in practical situations. Hence, it is understood that for the success of this domain, datasets play an important role, mostly publicly accessible by the research community. This article presents a comprehensive survey regarding various datasets available for solving AVC problems such as vehicle make and model recognition (VMMR), automatic license plate recognition, and vehicle category identification during the last decade. The datasets are categorized into two types: still image-based, and video-based. Again, the still image-based datasets are further classified as aerial imagery-based and front image-based datasets. This study has presented a thorough comparison of the different types of datasets with special reference to their characteristics. This study also provides an elaborative analysis of all the datasets and suggests a few fundamental future research scopes toward AVC. This survey can act as a preliminary guideline for researchers to develop a robust AVC system specially designed as per their needs and also to choose suitable datasets for comparing their models.

S. Maity · R. Sarkar
Department of Computer Science and Engineering, Jadavpur University, Kolkata 700032, India
e-mail: sourajit.cse.ju@gmail.com

R. Sarkar
e-mail: ramjucse@gmail.com

P. K. Singh (✉)
Department of Information Technology, Jadavpur University, Kolkata 700106, India
e-mail: pawansingh.ju@gmail.com

D. Kaplun
Department of Automation and Control Processes, Saint Petersburg Electrotechnical University "LETI", Saint Petersburg, Russian Federation 197022
e-mail: dikaplun@etu.ru

J. Nayak et al. (eds.), *Machine Learning for Cyber Physical System: Advances and Challenges*, Intelligent Systems Reference Library 60,
https://doi.org/10.1007/978-3-031-54038-7_14

Keywords Automatic vehicle classification · Survey · Aerial image-based vehicle datasets · Frontal image-based vehicle datasets · Video-based vehicle datasets

14.1 Introduction

As a consequence of the increasing number of vehicles and the evolution of road scenarios, vehicles and their management systems also require upgrades to make our daily lives hassle-free. Plenty of research articles have been published in various domains related to traffic-management systems such as vehicle classification [1], localization, detection, make and model recognition [2], segmentation [3, 4], lane detection, pedestrian detection, etc. [5] Working on such issues based on real-life traffic scenarios is quite challenging in terms of training, testing, and validating the model. In addition to this, very few datasets are available in these domains, and many of them are based on non-realistic scenarios. Moreover, datasets with good images sometimes lack proper annotations, and the well-known datasets are mostly paid, making them difficult to utilize for research work. On the other hand, an ample amount of data is required to make an efficient model, which has good accuracy as well as capable of working in a real-life scenario.

Smart city promotes automatic traffic management, thereby allowing only authorized users to access car functionalities, thus, providing data security. It refers to using communication buses to ensure data security even when connecting from external devices [6]. As a result, security strategies that boost confidentiality and enhance authentication in new cars must be designed [7]. Modern developments for in-car communication technologies allow for more sophisticated connections centred on the dashboard, such as those with a laptop or iPod, roadside devices, smart phones, and sensors. In a nutshell, data is a crucial component of any security network in-vehicle communication system.

In the vehicle detection and recognition domain, the number of datasets available for classification and segmentation purposes is much less compared to the number of datasets available for vehicle localization [8]. In addition, most of them do not adequately capture real-life scenarios. For example, overlapping vehicles within a single image frame is a very common traffic scenario in densely populated countries such as India, Pakistan, Bangladesh, and many other countries in South Asia. This makes the classification, localization, detection, and segmentation processes very difficult [9]. To overcome these challenges, a few datasets have been published recently, such as JUVDsi v1, consisting of nine different vehicle classes by Bhattacharyya et al. [10], IRUVD with 14 vehicle classes by Ali et al. [11]. Factors such as sensing range, size of the target, and similarities present among different vehicle classes [10] must be considered during vehicle classification.

This study presents a comprehensive survey of the datasets available for AVC and vehicle model and make recognition (VMMR) published in the last 10 years highlighting their inherent challenges. Also, we have given a comparative study of the different types of datasets used for classification along with their pros and cons.

14.1.1 Reason of Interest

Although the field of AVC has recently attracted the attention of several researchers, significant improvements in the vicinity of designing systems that are resilient to real-world situations are yet to be made. Taking all these into account, it is necessary to ensure the sound performance of this system in real-life scenarios. Only a few AVC-related research attempts have been made such as Siddiqui et al. [9] Kansas et al. [12], Yuan et al. [13], Sochor et al. [14] and Bharadwaj et al. [15]. However, all these studies hardly provide any specific instruction for selecting a dataset for an individual model. Besides, the foremost state-of-the-art for the specific dataset is not available in these research articles.

14.2 Categorization of AVC-Based Datasets

Datasets used for classification in AVC purposes can be further categorized in terms of the types of captured images and videos into the following: (a) aerial image-based vehicle datasets, (b) frontal image-based vehicle datasets, and (c) video-based vehicle datasets. All the datasets based on aerial images and videos of cars, buses, vans, motorbikes, and many other vehicles that are taken from any front/rear camera on public roads are mentioned here. Figure 14.1 shows the distribution of the datasets, available in the AVC domain.

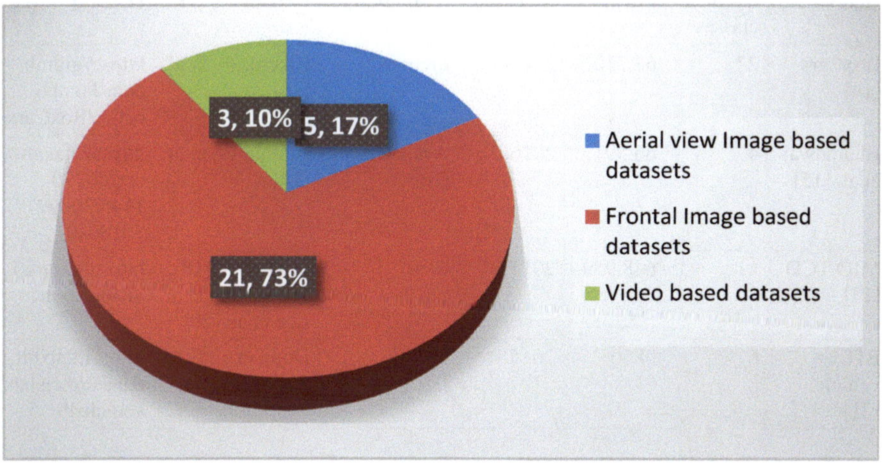

Fig. 14.1 Illustrating the distribution of datasets available in the AVC domain

14.2.1 *Aerial Image-Based Vehicle Datasets*

These types of datasets are composed of images captured by surveillance cameras and CCTV cameras, mounted near the streets and drone cameras. These are intended to be useful for traffic surveillance applications. Table 14.1 shows the list of datasets used for solving the aerial image-based AVC problem.

14.2.1.1 BoxCars Dataset

Sochor et al. [14] developed a dataset, called BoxCars, consisting of 63,750 images (21,250 vehicles of 27 different makes) collected from surveillance cameras. This dataset contained images captured from the front side of the vehicle (similar to images presented in Fig. 14.2) as well as images of passing vehicles, collected from surveillance cameras mounted near streets. They collected three images for each correctly detected vehicle as the vehicle passed the surveillance camera. The vehicles were divided into 3 distinct categories such as (a) 102 make and model classes, (b) 126 make and model + sub-model classes, and (c) 148 make and model + sub-model + model year classes. A few sample images from this dataset are shown in Fig. 14.2. Elkerdawy et al. [16] achieved a classification accuracy of 86.57% using ResNet152 + co-occurrence layer (COOC) model in 2019 on this dataset.

Table 14.1 List of aerial image-based datasets available for developing AVC systems

Dataset	#Vehicle classes	#Images	Released	Availability	Research work	Download link
BoxCars [14]	27	63,750	2019	Free	Elkerdawy et al. [16]	https://github.com/JakubSochor/BoxCars
Bharadwaj et al. [15]	4	66,591	2016	Available on request	Bharadwaj et al. [15]	https://dl.acm.org/doi/10.1145/3009977.3010040
MIO TCD [17]	11	648,959	2017	Free	Jung et al. [18], Kim et al. [19], Lee et al. [20]	https://tcd.miovision.com/
BIT vehicle [21]	6	9850	2015	Free	Dong et al. [21, 22]	http://iitlab.bit.edu.cn/mcislab/vehicledb

Fig. 14.2 Sample images of the novel BoxCars dataset [14]

Fig. 14.3 Sample RGB images under the 4 vehicle classes—Clockwise from top—'Auto Rickshaws', 'Heavy Vehicles', 'Two Wheelers', and 'Light Vehicles' (taken from [15])

14.2.1.2 Dataset by Bharadwaj et al. [15]

Bharadwaj et al. [15] compiled a dataset with surveillance-quality images that were collected from video clips of a traffic junction in an Indian city. They used a widely-accepted classification scheme where the vehicles were classified as 'Auto Rickshaw', 'Light', and 'Two-wheeler'. However, there was not any proper distinction between the vehicles of 'Light' and 'Heavy' classes due to the interchangeability of the vehicle after customizations. Vehicles of the 'Three-wheeler' class with minor modifications were classified as 'freight' vehicles, which should fall under the 'Heavy' category. Figure 14.3 represents some sample images taken from this dataset comprising various vehicle classes. The average F-score measure for this dataset using Caffenet + SVM method was found to be 87.75%.

14.2.1.3 MIO-TCD Dataset

Luo et al. [17] introduced the "MIO vision Traffic Camera Dataset" (MIO-TCD), which is based on classification of motor vehicles as well as localization from a single video frame. It is a compilation of 7,86,702 annotated images with 11 traffic object classes, collected by traffic surveillance cameras deployed across Canada and the United States at different times of the day in different periods of the year. Figure 14.4 shows 11 traffic object classes featured in this dataset. The class-wise distribution of each category in the classification dataset is given in Table 14.2. The dataset had two sections: a "localization dataset", with 1,37,743 full video frames with bounding boxes around traffic objects, and a "classification dataset", with 6,48,959 crops of traffic objects from the 11 object classes. The dataset was divided into an 80% training set (5,19,164 images) and a 20% testing set (1,29,795 images). By applying top-performing methods, the localization dataset achieved a mean-average precision of 77%. Using joint fine-tuning strategy with the DropCNN method, Jung et al. [18] obtained an accuracy of 97.95%.

Fig. 14.4 Few images from each of the 11 classes taken from the MIO-TCD dataset

Table 14.2 Class-wise distribution of vehicles present in the MIO-TCD dataset

Category	No. of training samples	No. of test samples
Articulated Truck	10,346	2587
Bicycle	2284	571
Bus	10,316	2579
Car	260,518	65,131
Motorcycle	1982	495
Non-motorized vehicle	1751	438
Pedestrian	6262	1565
Pickup truck	50,906	12,727
Single unit truck	5120	1280
Work van	9679	2422
Background	160,000	40,000
Total	**5,19,164**	**1,29,795**

Fig. 14.5 Sample images from BIT-vehicle dataset [Images taken from [21]]

14.2.1.4 BIT-Vehicle Dataset [21]

BIT-Vehicle dataset, developed by Dong et al. [21], includes 9,850 vehicle images with approximately 10% of images taken under night conditions. Figure 14.5 shows some sample images of this dataset. The images (with sizes of 1600 × 1200 and 1920 × 1080) in the dataset were captured from two cameras installed at different times and places. All vehicles in the dataset were divided into six categories as: 'Minivan', 'Sedan', 'SUV', 'Microbus', 'Bus', and 'Truck'. For each vehicle class, 200 samples were randomly selected for training the Softmax parameters, and 200 images were used as test samples. Dong et al. [21] obtained an accuracy of 96.1% using the sparse Laplacian filter learning (SLFL) method [23].

14.2.1.5 Summarization

Vehicle classification in aerial images has emerged out to be a fundamental need across the world for both categorization and tracking of vehicles. This is done for both security purposes as well as for maintaining traffic congestion on roads. However, there are very few studies available in this field. In their BoxCars dataset, Sochor et al. [14] studied the classification of components and models with 3D bounding boxes and captured images from roadside surveillance cameras. While dealing with surveillance cameras, the re-identification of vehicles is an essential task apart from the classification of vehicles. Classification accuracy is not satisfactory as required in road scenarios. Bharadwaj et al. [15] proposed a dataset with only four types of classes, including passenger vehicles and heavy vehicles within the same class. This dataset did not include accident-prone situations such as foggy weather conditions. In the MIO-TCD dataset [17], the images were not captured in a real-time scenario.

Therefore, the authors achieved an accuracy close to about 100%, which is difficult to achieve in a real-life scenario.

14.2.2 *Frontal Image-Based AVC Datasets*

Vehicle images can be classified on the basis of type, model, make, or mix of all these characteristics. The datasets related to vehicles that are taken from any front or rear camera on public roads are mentioned here. In Table 14.3, the list of datasets used for developing a front-view image-based AVC system is given and detailed information related to each dataset is discussed below.

14.2.2.1 Stanford Cars Dataset

The Stanford Cars dataset was designed by Krause et al. [24] to recognize the make and model of cars. It is a compilation of 16,185 rear images of cars (of size 360 × 240) divided into 196 classes. The entire data is almost equally divided into a train/test split with 8,144 images for training and 8,041 images for testing purposes. Using the domain adaptive transfer learning technique model on this dataset Ngiam et al. [25] achieved an accuracy of 96.8%. Some images taken from the Stanford Cars dataset are presented in Fig. 14.6.

14.2.2.2 CompCars Dataset

Yang et al. [27] developed the "CompCars" dataset, which covered different car views, showing different internal as well as external parts. The dataset has two types of image sets, a surveillance image set and a web image set. The web image set is a collection of images, taken from car forums, search engines, and public websites, and the surveillance set images were collected by surveillance cameras. The web-image data contained 1,36,727 images of the entire car and 27,618 images featuring car parts of 161 car makes with 1,687 car models. The surveillance-image data had 50,000 car images captured from the front view. The dataset can be used for (a) Fine-grained classification, (b) Attribute prediction, and (c) Car model verification and also for image ranking, multi-task learning, and 3D reconstruction. Yu et al. [30] obtained an accuracy of 99% using K-means with the VR-COCO method.

14.2.2.3 Frontal-103 Dataset

Lu et al. [31] provided an elaborative analysis of the Frontal-103 dataset, which consisted of a total of 1,759 vehicle models in 103 vehicle makes and 65,433 images. Here, the images are assigned to four main viewpoints namely left and right, front

Table 14.3 Frontal image-based datasets available for developing AVC systems

Dataset	#Vehicle classes	#Images	Released	Availability	Research work	Download link
Stanford Cars [24]	196	16,185	2013	Free	Ngiam et al. [25], Ridnik et al. [26]	https://ai.stanford.edu/~jkrause/cars/car_dataset.html
CompCars [27]	163	1,36,726	2015	Free	Hu et al. [28], Tanveer et al. [29], Yu et al. [30]	http://mmlab.ie.cuhk.edu.hk/datasets/comp_cars/index.html
Frontal-103 [31]	103	65,433	2022	Free	Lu et al. [31],	https://github.com/vision-insight/Frontal-103
Liao et al. [32]	8	1482	2015	Paid	Liao et al. [32]	https://en.whu.edu.cn/Research1/Research_Centres.htm
Side Profile dataset [33]	86	10,000	2015	Free	Boyle et al. [33]	http://www.cvg.reading.ac.uk/rvd
Novel car type [34]	14	1904	2011	Free	Stark et al. [34]	https://www.mpi-inf.mpg.de/departments/computer-vision-and-machine-learning/publications
FG3DCar [35]	30	300	2014	Free	Lin et al. [35]	https://www.cmlab.csie.ntu.edu.tw/~yenliang/FG3DCar/

(continued)

Table 14.3 (continued)

Dataset	#Vehicle classes	#Images	Released	Availability	Research work	Download link
VMMR dataset [36]	9170	2,91,752	2017	Free	Tafazzoli et al. [36]	https://github.com/faezetta/VMMRdb
BR Cars [37]	427	3,00,000	2022	Free	Kuhn et al. [37]	https://github.com/danimtk/brcars-dataset
Poribohon BD [38]	15	9058	2021	Free	Tabassum et al. [38]	https://data.mendeley.com/datasets/pwyyg8zmk5/2
Deshi BD [39]	13	10,440	2021	Free	Hasan et al. [39]	https://www.kaggle.com/datasets/nazmultakbir/vehicle-detection-bangladeshi-roads
DriveU Traffic Light Dataset (DTLD) [40]	9	10,000	2021	Free	Deshmukh et al. [40]	https://github.com/deshmukh15/dataset_complete/blob/main/test_1.zip
LSUN + Stanford [41]	196	2,06,7710	2020	Free	Abdal et al. [42]	https://github.com/Tin-Kramberger/LSUN-Stanford-dataset
IRD [43]	13	8520	2022	Free	Gautam et al. [43]	https://sites.google.com/view/ird-dataset/home?pli=1

(continued)

Table 14.3 (continued)

Dataset	#Vehicle classes	#Images	Released	Availability	Research work	Download link
CAR-159 [44]	159	7998	2021	Paid	Sun et al. [44]	https://en.nuist.edu.cn/4251/list.htm
Butt et al. [45]	6	10,000	2021	Paid	Butt et al. [45]	https://www.hindawi.com/journals/complexity/2021/6644861/
IRVD [46]	5	1,10,366	2021	Available on request	Gholamalinejad et al. [46]	https://shahaab-co.com/en/iranian-vehicle-dataset-irvd-demo/
Yu Peng [47]	5	4924	2012	Available on request	Peng et al. [47]	http://dl.dropbox.com/u/52984000/Database1.rar
Fine-Grained Vehicle Detection (FGVD) [48]	6	5502	2022	Free	Khoba et al. [48]	https://zenodo.org/record/7488960#.Y9qzhXZBxdg
Indonesian Vehicle Dashboard Dataset (InaV-Dash) [49]	14	4192	2022	Paid	Avianto et al. [49]	https://www.mdpi.com/2313-433X/8/11/293
Abnormal Traffic Object Classification (ATOC) [50]	12	840	2020	Free	Wang et al. [50]	https://deepgaze.bethgelab.org/

and rear. Therefore, after the annotation, there were eight groups of vehicle images in total from each of the viewpoints. Lu et al. [51] achieved an accuracy of 91.2% using both the pre-trained ResNet 50 and DenseNet121 models. Some sample images collected from the Frontal-103 dataset are presented in Fig. 14.7.

Fig. 14.6 Some sample images taken from the Stanford Cars dataset

Fig. 14.7 Sample image samples found in the Frontal-103 dataset. The dataset includes images of frontal view under variable weather and lighting conditions

14.2.2.4 Dataset by Liao et al. [32]

Liao et al. [32] presented a large-scale dataset compiling vehicle images that were captured from the front view using monitoring cameras fixed on the road. A total of 1482 vehicles were annotated from the images into eight categories and the number of images present in each category are shown in Fig. 14.8. Some sample images of all eight classes of vehicles are shown in Fig. 14.9. Liao et al. [32] achieved an accuracy of 93.3% with a part-based fine-grained vehicle categorization method.

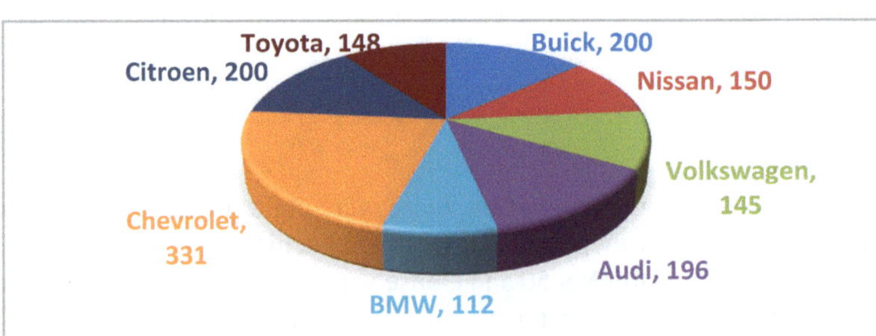

Fig. 14.8 Number of annotated images for different vehicles proposed by Liao et al. [32]

Fig. 14.9 Image samples of all 8 vehicle categories proposed by Liao et al. [32]

14.2.2.5 Side Profile Dataset [33]

Boyle et al. [33] proposed a public vehicle dataset, which has more than 10,000 side profile images of cars divided into 86 make/model and 9 subtype classes. The vehicle subtypes and the total number of labeled images for each vehicle class are represented as a pie chart in Fig. 14.10. They achieved high classification rates of 98.7% for subtypes and 99.7–99.9% for VMMR.

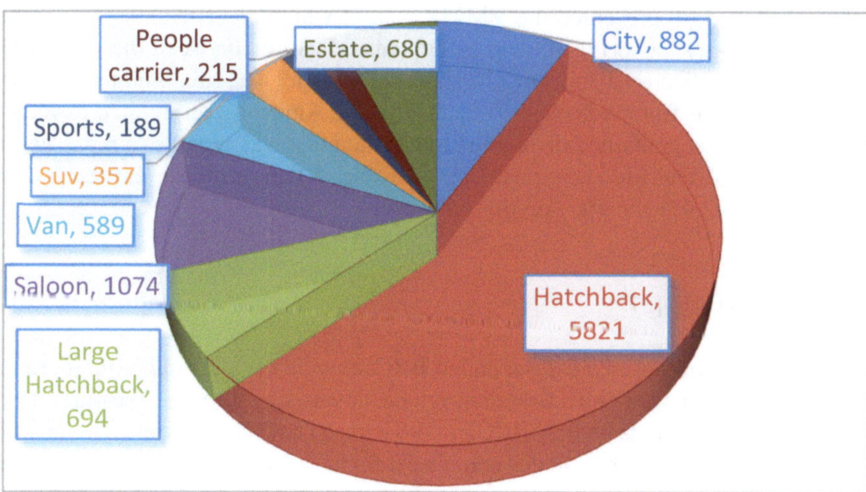

Fig. 14.10 Number of total vehicle subtypes for the Side Profile dataset [33]

14.2.2.6 Cartypes Dataset [34]

Stark et al. [34] introduced a novel dataset of fine-grained car types, compiling 1904 images of cars of 14 different vehicle classes, with class label and annotations, 2D bounding boxes, and a viewpoint estimate. They used the Ford campus vision and Lidar dataset [52, 53] for testing. Stark et al. [34] obtained an accuracy of 90.3% with an ensemble of Histogram of oriented gradients (HOG), Locality-constrained linear coding [54] (LLC), and struct DPM method.

14.2.2.7 FG3DCar Dataset [35]

Lin et al. [35] developed a fresh fine-grained 3D car dataset (FG3DCar), which includes 300 images of 30 various automobile models under various viewing angles, including those of a 'pickup truck', 'hatchback', 'SUV' and 'crossover'. They manually included 64 landmark places in each car image. They manually annotated the correspondences between the 2D projections of the visible 3D landmarks on the image and changed the shape as well as posture parameters iteratively to reduce the distance errors between the correspondences. The authors achieved an accuracy of 95.3% with the ensemble of GT alignment and (HOG/FV) feature vector method.

14.2.2.8 VMMR Dataset [23]

Tafazzoli et al. [36] presented the VMMR dataset which contains 2,91,752 images with 9,170 classes, covering vehicle models manufactured between the years 1950 to 2016. They collected data from 712 areas covering all 412 subdomains of United States metro areas from web pages (like Wikipedia, Amazon, etc.) related to vehicle sales. This dataset contained diversified image data that was capable of representing a wide range of real-life scenarios. Using ResNet-50 architecture, Tafazzoli et al. [36] obtained 92.9% accuracy on this dataset.

14.2.2.9 BRCars [37]

Kuhn et al. [37] proposed a dataset called BRCars which is a compilation of around 300 K images gathered from a Brazilian vehicle advertising website. The dataset was segregated into two parts namely BRCars-196 with 2,12,609 images and BRCars-427 with 3,00,325 images. The images contain 52 K car instances including views from both the exterior as well as interior parts of the cars and have a skewed distribution among 427 different models. Using the InceptionV3 architecture, Kuhn et al. [37] obtained accuracies of 82% on the BRCars-427 dataset and 92% on the BRCars-196 dataset.

Fig. 14.11 Sample images of the Poribohon-BD dataset

14.2.2.10 Poribohon-BD [38]

Poribohon-BD dataset used for vehicle classification in Bangladesh, developed by Tabassum et al. [38], is a compilation of 9,058 labeled and annotated images of 16 native Bangladeshi vehicles. The vehicle images were collected using smartphone cameras and from social media. To maintain a balance between the number of images for each vehicle type, data augmentation techniques were applied. The dataset is compatible with various CNN architectures such as YOLO [55], VGG-16 [38], R-CNN [56], and DPM [38]. Sample images taken from this dataset are shown in Fig. 14.11. Tabassum et al. [38] achieved an accuracy of 98.7% with ResNet-152 and DenseNet-201 models.

14.2.2.11 Deshi-BD Dataset [39]

For the classification of Bangladeshi native vehicle types, Hasan et al. [39] developed a dataset consisting of 10,440 images of 13 common vehicle classes and also designed a model based on transfer learning, incorporating data augmentation. Despite the changing physical properties of the vehicles, the proposed model achieved progressive accuracy. The highest accuracy on this dataset is 98%. Sample images of this dataset are shown in Fig. 14.12, whereas a bar-chart illustrating the data description is shown in Fig. 14.13.

Fig. 14.12 Sample images from the Deshi-BD dataset representing each class [39]

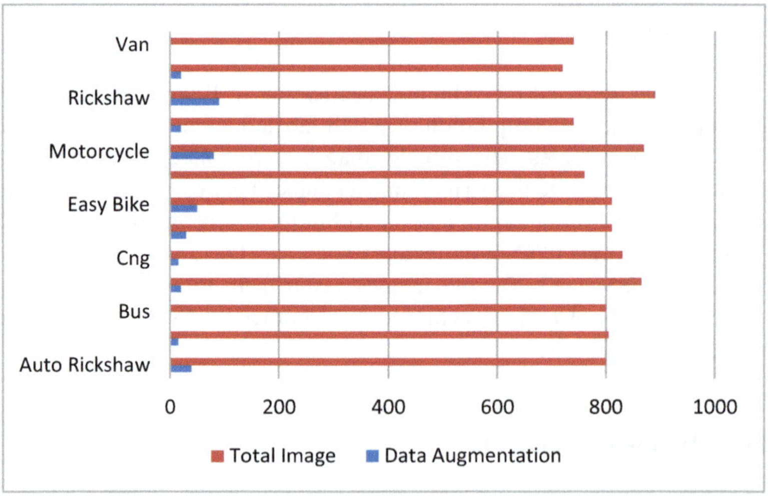

Fig. 14.13 Data description of Deshi-BD vehicle dataset

Fig. 14.14 Sample images of the combined LSUN + Stanford cars dataset [41]

14.2.2.12 DTLD Dataset [40]

Deshmukh et al. [40] proposed a DTLD that contained around 10,000 unordered images including 9 types of Indian vehicles of traffic scenarios taken from different camera angles. The images were captured under rainy and noisy weather. It has 10,000 images out of which 20% are used for model testing. This dataset yielded an accuracy of 96.3% using the STVD [57] method with ST backbone.

14.2.2.13 LSUN Car Dataset [41]

To overcome the shortcomings of the LSUN car dataset containing 55,20,753 car images, Kramberger et al. [41] created a dataset by combining the LSUN and the Stanford car datasets. After the pruning, the new dataset had about 20,67,710 images of cars with enhanced quality. The StyleGAN training on the combined LSUN-Stanford car dataset was about 3.7% more advanced than training with just the LSUN dataset. Therefore, it can be inferred that the LSUN-Stanford car dataset is more consistent and better suited for training GAN neural networks than other currently available large car datasets. Abdal et al. [42] achieved an accuracy of 99% using the Detectron2 model on this dataset. Figure 14.14 shows the sample images of the LSUN + Stanford cars dataset [41].

14.2.2.14 Car-159 Dataset [44]

The car-159 dataset, developed by Sun et al. [44], comprised images of different vehicle types captured either by camera or taken from online sources. The images were captured from five viewpoints such as on the right ahead, on the rear side, on the side, on the front side, right behind, and right side. The dataset had 8 vehicle brands, 159 types of vehicle types, and 7998 images. The training set contained 6042 images, and the validation set had 1956 images. The authors obtained an accuracy

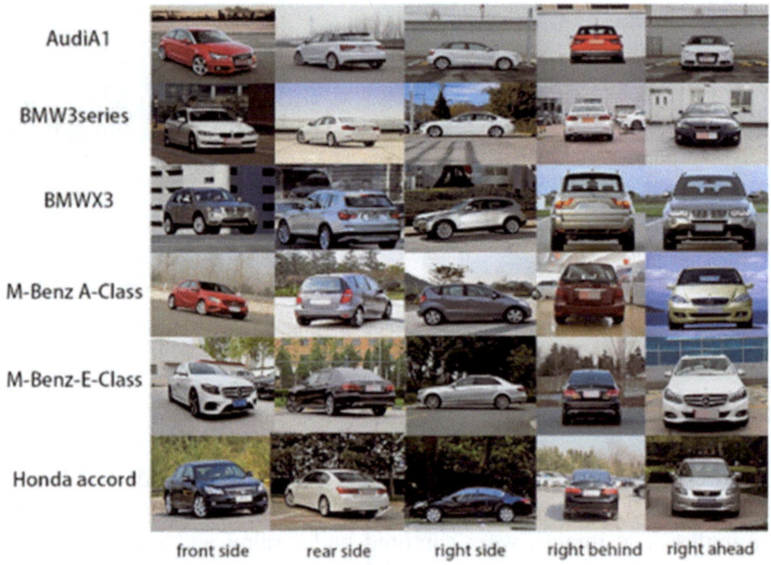

Fig. 14.15 Sample images from the Car-159 dataset [44]

of 85.8% using the fine-grained VTC [44] method. Some sample images from the Car159 dataset are shown in Fig. 14.15.

14.2.2.15 Dataset by Butt et al. [45]

Butt et al. [45] proposed a dataset different from the existing CompCars and Stanford car datasets, which were mainly region-specific and were difficult to employ in a real-time AVC system. To overcome these issues, vehicle images were extracted from road surveillance and driving videos collected from various regions, and a dataset, comprising 10,000 images with six common road vehicle classes was compiled through manual labeling using the Windows editing tool, as presented in Fig. 14.16. Sample images from this dataset are shown in Fig. 14.16. On this dataset, Butt et al. [45] achieved an accuracy of 99.6% with a modified CNN model.

14.2.2.16 IRVD Dataset [46]

Gholamalinejad et al. [46] developed IRVD, an image-based dataset of Iranian vehicles, appropriate for the classification of vehicles as well as for the recognition of license plates. IRVD was categorized into training and testing parts. In the second proposed structure, they used some popular pre-trained CNN models among which the ResNet18 model performed best and this achieved an accuracy of 99.50%. Their

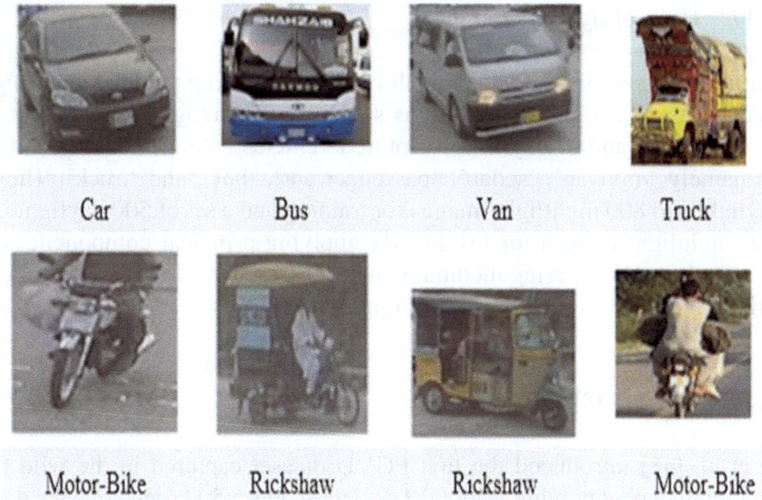

Fig. 14.16 Sample images of the dataset proposed by Butt et al. [45]

dataset, IRVD6, compiled a range of lighting conditions, weather conditions as well and variations in road conditions. Some images can be found in Fig. 14.17.

Fig. 14.17 Sample images of the IRVD dataset [46]

14.2.2.17 Dataset by Pen et al. [47]

Peng et al. [47] compiled a dataset with images of passing vehicles on a highway captured in different lighting conditions such as both daylights with sunny, partly cloudy conditions and at night. All captured vehicles belonged to any of the five classes namely 'minivan', 'sedan', 'passenger car', 'bus', and 'truck'. They used 800 daylight and 800 nightlight images for training and a set of 500 daylight images and 500 nightlight images for testing. By applying principal component analysis (PCA) with the self-clustering method, Peng et al. [47] achieved an accuracy of 90% in daylight and 87.6% for the cases of nightlight.

14.2.2.18 FGVD [48]

Khoba et al. [48] introduced the first FGVD dataset captured in the wild from a camera mounted on a moving vehicle. The dataset has 5,502 scene images with 210 unique fine-grained labels of multiple vehicle types organized in a three-level hierarchy. FGVD dataset introduced the new class labels for categorizing 'two-wheelers', 'autorickshaws', and 'trucks'. The dataset also presented difficulties since it included cars in complicated traffic situations with *intra-class* and *inter-class* changes in type, scale, position, occlusion, and lighting. The images of each of the vehicle classes of the FGVD dataset are shown in Fig. 14.18. This dataset has three levels of hierarchies for classification. Using the fine-tuned Hierarchical Residual Network (HRN) model, Khoba et al. [48] obtained an accuracy of 96.6% in the level 1.

Fig. 14.18 Sample images of FGVD dataset

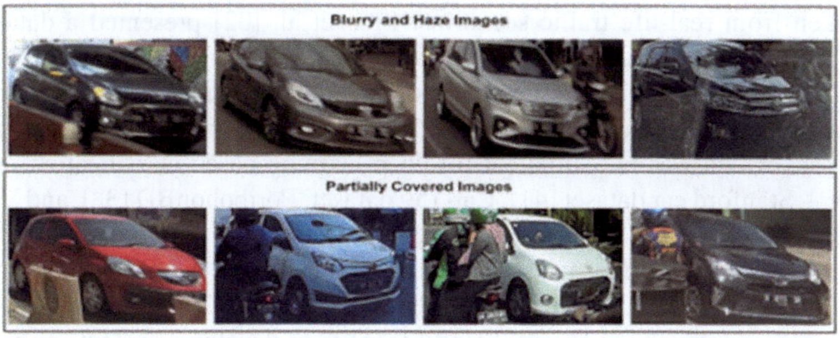

Fig. 14.19 Sample images from the Inav-Dash dataset [3]

14.2.2.19 InaV-Dash Dataset [49]

Avianto et al. [49] developed a dataset called the InaV-Dash dataset, consisting of a total of 4,192 images with four vehicle makes and 10 vehicle models. The dashboard camera was set to run at 60 frames per second with a full HD resolution of 1920 by 1080 pixels. The dataset was partitioned into a training set with 2934 images and a testing set with 1258 images. The authors obtained an accuracy of 95.3% using the ResNet50 CNN architecture. The blurry and hazed, as well as partially covered images of the InaV-Dash dataset, are illustrated in Fig. 14.19.

14.2.2.20 ATOC [50]

Wang et al. [50] developed the ATOC dataset, which consisted of 840 images with 12 vehicle classes and 70 images for all classes. This dataset contained objects under both normal and damaged status. The authors attained an accuracy of 87.08% using a pre-trained deep convolutional network for feature extraction and SVM for classification along with Saliency Masking (SM) + Random Sample Pairing (RSP) methods.

14.2.2.21 Summarization

Vehicle classification using front images has become a fundamental need across the world for the categorization and tracking of vehicles for security purposes and for maintaining traffic congestion on roads. Many studies are available in this area, among which several methods have also reached an accuracy of 100%. Though the Stanford Cars dataset [24] presented a large number of classes, the training images in the dataset were collected from online sources, and the number of images was also very few. Hence, it may not be useful for deep learning models as, in general, such models require a huge amount of data for better training. Additionally, the images in the Frontal-103 [31] dataset that studied the VMMR problem, were not taken from real-life traffic scenarios. Liao et al. [32] presented a dataset for car type categorization. However, they have considered very few types of vehicles, which might not be useful for categorization in practical scenarios. FG3D car dataset [35] considered 3D models for vehicles (cars), but there were only 300 images for 30 classes which was not sufficient for deep learning-based classification models. LSUN + Stanford car dataset [41], Car-159 dataset, PoribohonBD [38], and VMMR dataset had taken many images, but all of them were collected from online sources. Additionally, there is no further scope for research if we use datasets with near about 100% accuracy such as the IRVD [46] dataset. The FGVD [48] dataset has a few vehicle classes, which is not very helpful in AVC to develop a system for practical purposes.

14.2.3 Video-Based AVC Datasets

It has been already mentioned that vehicles can be classified based on type, model, make, or a mix of all these characteristics. There are some important video datasets commonly used to classify vehicles in terms of their type, make, and model. Some research attempts based on videos of 'cars', 'buses', 'vans', 'motorbikes', and many other vehicles that are taken from any rear camera on public roads are discussed in this section. Table 14.4 lists the datasets used for video-based AVC.

14.2.3.1 I-Lids Dataset [58]

Branch et al. [58] provided a public dataset on a 500 GB USB2/Firewire external hard drive, in either NTFS or Mac format, as required. The video was rendered in "Quicktime MJPEG" format, whereas Apple's free "Quicktime" viewer was required to view the video. Apart from AVC work, this dataset can also be used for baggage detection, doorway surveillance, and parked vehicle detection. This dataset is used in video analysis as well as event detection to provide effective assistance to policing and anti-terrorist operations.

14.2.3.2 CDnet 2014 [59]

CDnet2014 dataset, developed by Wang et al. [60], is a compilation of 11 video categories with 4–6 video sequences in each category to assess the performance of moving object detection. A range of scenarios such as baseline camera jitter, intermittent motion object, bad weather, dynamic background, and shadow, were used to evaluate different methods. The duration of the videos ranged from 900 to 7,000 frames. Wang et al. [60] achieved an accuracy of 99% using a 3-scale spatio-temporal color/luminance Gaussian pyramid background model.

Table 14.4 List of video-based classification datasets available for developing AVC system

Dataset	#Vehicle classes	Total videos	Released	Availablity	Research work	Download link
iLids [58]	4	24-h video	2006	Paid	Branch et al. [58]	http://scienceandresearch.homeoffice.gov.uk/hosdb
CDnet 2014 [59]	6	11 video category	2014	Free	Wang et al. [60]	http://www.changedetection.net/
Carvideos [61]	10	129	2019	Paid	Alsahafi et al. [61]	https://link.springer.com/chapter, https://doi.org/10.1007/978-3-030-14070-0_63

14.2.3.3 Carvideos [61]

Alsahafi et al. [61] proposed a dataset, containing over a million frames and 10 different vehicle classes belonging to specific make and model recognition of cars, including 'sedans', 'SUVs', 'convertibles', 'hatchbacks' and 'station wagons'. To make the video dataset suitable for fine-grained car classification, they selected the specific models based on the availability of review videos (both professional and amateur) of recent car models. Each bounding box was labeled with one of vehicle classes and those that did not fit any of the classes were labeled as 'other'. Alsahafi et al. [61] obtained an accuracy of 76.1% for RGB—25 frames using the Single Shot Multibox Detector (SSD) + CNN.

14.2.3.4 Summarization

Vehicle classification with video datasets is a challenging task for researchers. However, the number of such datasets [73], available for vehicle classification, is very small. One such dataset is i-lids [58], which came into the picture in the year 2006. Therefore, the videos are very old and the dataset is also not freely available for the research community. The CDnet [59] dataset is much better than the i-lids dataset, but the accuracy of this dataset is about 100%. Therefore, it hardly gives any scope for further improvement.

14.3 Research Gaps and Challenges Limitations Related to AVC Datasets

In recent times, deep learning-based models have mostly been used for image and video classification purposes. This is also valid for the AVC task. These methods have been generating state-of-the-art results. This section summarizes a few issues of deep learning-based approaches along with some limitations of the existing datasets are mentioned below:

- In general, a huge number of samples is required for training purposes in deep learning-based models, and it needs a specialized GPU (graphics processing unit) to train the model. Additionally, the processing of the data is also a difficult task due to the unavailability of the required resources.
- Deep learning-based models take longer training time due to larger data density.
- Even if datasets are available, in many cases it is not perfectly processed and annotated as it requires extreme human labor to appropriately annotate the data.
- In countries such as India, Bangladesh, or Pakistan, all roads are not as good as those seen in developed countries of Europe or America. There is a lot of traffic congestion and unnecessary traffic rules. Due to this, we see the overlap of one

vehicle with another, and this makes the task of vehicle type classification from the still images of such vehicles very difficult.

- Developed and developing countries have different traffic scenario conditions. However, some datasets are available for developing countries that are not quite suitable for all conditions.
- Multi-view or multi-modal datasets are not available for the classification of vehicles. Much research and such data are required to develop a practical solution for AVC.
- There are datasets where the images are collected from different websites or Google. Sometimes, these images are significantly different from the real ones. This issue makes such datasets irrelevant to research work.
- Some datasets have an accuracy of almost 100%, so there is no further scope for research work with these datasets.
- Some datasets have very few classes, which are not appropriate for research works.
- The number of video datasets available for vehicle classification is very small. Therefore, video datasets are needed for further research in vehicle classification with video data.

With the above analysis, Table 14.5 lists the advantages and limitations of the datasets used for solving the AVC problem.

Table 14.5 Advantages and limitations of the datasets used for AVC

Dataset type	Advantage	Limitation
Aerial image-based datasets	Aerial image-based datasets provide a bird's eye view. Aerial images record the general flow and patterns of traffic, including information on congestion, how the roads are used, and even how many and what kinds of vehicles are present	Limited details can be found with this type of image. Aerial images depend on weather conditions
Frontal image-based datasets	The frontal view offers an unobscured and unambiguous perspective of the vehicles, which facilitates precise identification and classification of various vehicle types	The side and rear-view of the vehicles are not shown in frontal images, which also gives a limited viewpoint. Frontal views can be impeded by other vehicles, objects, and the surroundings, which makes some areas of vehicles harder to visualize
Video-based AVC datasets	Static images do not offer the same dynamic context that videos do. Videos record how vehicles move over time, giving important temporal information. However, knowledge of traffic patterns, dynamics of congestion, as well as variations in vehicle density throughout the day need to be specified	Compared to static images, video data often requires more bandwidth and storage space. This might provide difficulties, particularly when utilizing either streaming apps or extensive surveillance systems

14.3.1 Future Scope

It has already been mentioned that an ample amount of data is required for training, testing, and validation purposes to obtain high accuracy by deep learning models. Not only the availability of data but also correctly annotated and precisely processed data as per model requisition are essential criteria for data collection. However, in the case of AVC, very few datasets are available, and among them, the well-known datasets are not freely available to the research community. Also, video-based classification datasets are rarely available in the vehicle classification domain. Based on the above discussion, some future research directions regarding AVC are highlighted in this section.

(1) Lightweight models can be thought of considering the demand for IoT-based technologies, as such models can be easily deployed to edge devices.
(2) Semi-supervised and/or Few-shot learning approaches can be used when we have fewer annotated data for AVC.
(3) Data should be captured in all weather conditions and at various times of the day.
(4) Images/videos should be taken from different angles to deal with overlapped vehicles in heavy traffic regions.
(5) The diversity of data in terms of vehicle types, road conditions, and traffic congestion is very important.
(6) The availability of video data is very much required to develop realistic AVC systems.
(7) Availability of multi-modal data would help in designing practically applicable systems.
(8) Data of the same vehicles at multiple locations are required for vehicle re-identification for surveillance purposes.
(9) Frontal view-based datasets are specifically useful for license plate recognition; driver behavior analysis and classification of vehicle make and model.
(10) Aerial image-based datasets are specifically useful for vehicle counting and classification, parking lot management, and route planning of vehicles.

14.4 Conclusion

This study is an attempt to weigh the benefits and drawbacks of various datasets available for vehicle classification. Although this survey is not exhaustive, researchers may find it useful, as a guide to implement new methods or to update their past methods to meet the needs of realistic AVC systems. Following our discussion of AVC methods and available datasets, we have discussed some open challenges as well as some intriguing research directions. In this survey paper, we have discussed the datasets that are useful for vehicle classification. We have classified the datasets into two parts based on still image data and video data. Still image datasets are further classified into two parts aerial image-based datasets and front image-based datasets. We have estimated the accuracy of each dataset. We have also summarized the advantages and drawbacks of using these datasets. We are trying to work on the detection and segmentation of datasets in our future studies. Our findings suggest that AVC research is still an unexplored domain that deserves more attention. We believe that reviewing previous research efforts focusing on datasets will be beneficial in providing a comprehensive and timely analysis of the existing inherent challenges and potential solutions to this problem.

References

1. Kumar, C.R., Anuradha, R.: Feature selection and classification methods for vehicle tracking and detection, J. Ambient Intell. Humaniz Comput, pp. 1–11 (2020)
2. Lee, H.J., Ullah, I., Wan, W., Gao, Y., Fang, Z.: Real-time vehicle make and model recognition with the residual SqueezeNet architecture. Sensors **19**(5), 982 (2019)
3. Maity, S., Bhattacharyya, A., Singh, P.K., Kumar, M., Sarkar, R.: Last Decade in Vehicle Detection and Classification: A Comprehensive Survey. Archives of Computational Methods in Engineering, pp. 1–38 (2022)
4. Zhang, J., Yang, K. and Stiefelhagen, R.: ISSAFE: Improving semantic segmentation in accidents by fusing event-based data. In: 2021 IEEE/RSJ International Conference on Intelligent Robots and Systems (IROS), IEEE, pp. 1132–1139 (2021)
5. Buch, N., Cracknell, M., Orwell, J., Velastin, S.A.: Vehicle localisation and classification in urban CCTV streams. Proceedings of 16th ITS WC, pp. 1–8 (2009)
6. Martínez-Cruz, A., Ramírez-Gutiérrez, K.A., Feregrino-Uribe, C., Morales-Reyes, A.: Security on in-vehicle communication protocols: Issues, challenges, and future research directions. Comput. Commun.. Commun. **180**, 1–20 (2021)
7. Rathore, R.S., Hewage, C., Kaiwartya, O., Lloret, J.: In-vehicle communication cyber security: challenges and solutions. Sensors **22**(17), 6679 (2022)
8. El-Sayed, R.S., El-Sayed, M.N.: Classification of vehicles' types using histogram oriented gradients: comparative study and modification. IAES International Journal of Artificial Intelligence **9**(4), 700 (2020)
9. Siddiqui, A.J., Mammeri, A., Boukerche, A.: Towards efficient vehicle classification in intelligent transportation systems. In: Proceedings of the 5th ACM Symposium on Development and Analysis of Intelligent Vehicular Networks and Applications, pp. 19–25 (2015)
10. Bhattacharyya, A., Bhattacharya, A., Maity, S., Singh, P.K., Sarkar, R.: JUVDsi v1: developing and benchmarking a new still image database in Indian scenario for automatic vehicle detection. Multimed. Tools Appl. pp. 1–33 (2023)
11. Ali, A., Sarkar, R., Das, D.K.: IRUVD: a new still-image based dataset for automatic vehicle detection. Multimed Tools Appl, pp. 1–27 (2023)
12. Kanistras, K., Martins, G., Rutherford, M.J., Valavanis, K.: PA survey of unmanned aerial vehicles (UAVs) for traffic monitoring. In: 2013 International Conference on Unmanned Aircraft Systems (ICUAS), IEEE, pp. 221–234 (2013)
13. Yuan, C., Zhang, Y., Liu, Z.: A survey on technologies for automatic forest fire monitoring, detection, and fighting using unmanned aerial vehicles and remote sensing techniques. Can. J. For. Res. **45**(7), 783–792 (2015)
14. Sochor, J., Herout, A., Havel, J.: Boxcars: 3d boxes as cnn input for improved fine-grained vehicle recognition. In: Proceedings of the IEEE conference on computer vision and pattern recognition, pp. 3006–3015 (2016)

15. Bharadwaj, H.S., Biswas, S., Ramakrishnan, K.R.A.: large scale dataset for classification of vehicles in urban traffic scenes. In: Proceedings of the Tenth Indian Conference on Computer Vision, Graphics and Image Processing, pp. 1–8 (2016)
16. Elkerdawy, S., Ray, N., Zhang, H.: Fine-grained vehicle classification with unsupervised parts co-occurrence learning. In: Proceedings of the European Conference on Computer Vision (ECCV) Workshops, p. 0 (2018)
17. Luo, Z., et al.: MIO-TCD: A new benchmark dataset for vehicle classification and localization. IEEE Trans. Image Process. **27**(10), 5129–5141 (2018)
18. Jung, H., Choi, M.K., Jung, J., Lee, J.H., Kwon, S., Young Jung, W.: ResNet-based vehicle classification and localization in traffic surveillance systems. In: Proceedings of the IEEE conference on computer vision and pattern recognition workshops, pp. 61–67 (2017)
19. Kim, P.K., Lim, K.T.: Vehicle type classification using bagging and convolutional neural network on multi view surveillance image. In: Proceedings of the IEEE conference on computer vision and pattern recognition workshops, pp. 41–46 (2017)
20. Taek Lee, J., Chung, Y.: Deep learning-based vehicle classification using an ensemble of local expert and global networks. In: Proceedings of the IEEE conference on computer vision and pattern recognition workshops, pp. 47–52 (2017)
21. Dong, Z., Wu, Y., Pei, M., Jia, Y.: Vehicle type classification using a semisupervised convolutional neural network. IEEE Trans. Intell. Transp. Syst.Intell. Transp. Syst. **16**(4), 2247–2256 (2015)
22. Dong, H., Wang, X., Zhang, C., He, R., Jia, L., Qin, Y.: Improved robust vehicle detection and identification based on single magnetic sensor. Ieee Access **6**, 5247–5255 (2018)
23. Sunderlin Shibu, D., Suja Priyadharsini, S.: Multimodal medical image fusion using L0 gradient smoothing with sparse representation. Int J Imaging Syst Technol, vol. 31, no. 4, pp. 2249–2266 (2021)
24. Krause, J., Stark, M., Deng, J., Fei-Fei, L.: 3d object representations for fine-grained categorization. In: Proceedings of the IEEE international conference on computer vision workshops, pp. 554–561 (2013)
25. Ngiam, J., Peng, D., Vasudevan, V., Kornblith, S., Le, Q.V., Pang, R.: Domain adaptive transfer learning with specialist models. *arXiv preprint* arXiv:1811.07056 (2018)
26. Ridnik, T., Ben-Baruch, E., Noy, A., Zelnik-Manor, L.: Imagenet-21k pretraining for the masses. arXiv:2104.10972 (2021)
27. Yang, L., Luo, P., Change Loy, C., Tang, X.: A large-scale car dataset for fine-grained categorization and verification. In: Proceedings of the IEEE conference on computer vision and pattern recognition, pp. 3973–3981 (2015)
28. Hu, Q., Wang, H., Li, T., Shen, C.: Deep CNNs with spatially weighted pooling for fine-grained car recognition. IEEE Trans. Intell. Transp. Syst.Intell. Transp. Syst. **18**(11), 3147–3156 (2017)
29. Suhaib Tanveer, M., Khan, M.U.K., Kyung, C.-M.: Fine-Tuning DARTS for Image Classification. p. arXiv-2006 (2020)
30. Yu, Y., Liu, H., Fu, Y., Jia, W., Yu, J., Yan, Z.: Embedding pose information for multiview vehicle model recognition. IEEE Trans. Circuits Syst. Video Technol. **32**(8), 5467–5480 (2022)
31. Lu, L., Wang, P., Huang, H.: A large-scale frontal vehicle image dataset for fine-grained vehicle categorization. IEEE Transactions on Intelligent Transportation Systems (2020)
32. Liao, L., Hu, R., Xiao, J., Wang, Q., Xiao, J., Chen, J., "Exploiting effects of parts in fine-grained categorization of vehicles. In: IEEE international conference on image processing (ICIP). IEEE 2015, 745–749 (2015)
33. Boyle, J., Ferryman, J.: Vehicle subtype, make and model classification from side profile video. In: 2015 12th IEEE International Conference on Advanced Video and Signal Based Surveillance (AVSS), IEEE, pp. 1–6 (2015)
34. Stark, M., et al.: Fine-grained categorization for 3d scene understanding. Int. J. Robot. Res. **30**(13), 1543–1552 (2011)
35. Lin, Y.-L., Morariu, V.I., Hsu, W., Davis, L.S.: Jointly optimizing 3d model fitting and fine-grained classification. In: European conference on computer vision, Springer, pp. 466–480 (2014)

36. Tafazzoli, F., Frigui, H., Nishiyama, K.: A large and diverse dataset for improved vehicle make and model recognition. In: Proceedings of the IEEE conference on computer vision and pattern recognition workshops, pp. 1–8 (2017)
37. Kuhn, D.M., Moreira, V.P.: BRCars: a Dataset for Fine-Grained Classification of Car Images. In: 2021 34th SIBGRAPI Conference on Graphics, Patterns and Images (SIBGRAPI), IEEE, pp. 231–238 (2021)
38. Tabassum, S., Ullah, S., Al-nur, N.H., Shatabda, S.: Poribohon-BD: Bangladeshi local vehicle image dataset with annotation for classification. Data Brief **33**, 106465 (2020). https://doi.org/10.1016/j.dib.2020.106465
39. Hasan, M.M., Wang, Z., Hussain, M.A.I., Fatima, K.: Bangladeshi native vehicle classification based on transfer learning with deep convolutional neural network. Sensors **21**(22), 7545 (2021)
40. Deshmukh, P., Satyanarayana, G.S.R., Majhi, S., Sahoo, U.K., Das, S.K.: Swin transformer based vehicle detection in undisciplined traffic environment. Expert Syst. Appl. **213**, 118992 (2023)
41. Kramberger, T., Potočnik, B.: LSUN-Stanford car dataset: enhancing large-scale car image datasets using deep learning for usage in GAN training. Appl. Sci. **10**(14), 4913 (2020)
42. Abdal, R., Zhu, P., Mitra, N.J., Wonka, P.: Labels4free: Unsupervised segmentation using stylegan. In: Proceedings of the IEEE/CVF International Conference on Computer Vision, pp. 13970–13979 (2021)
43. Gautam, S., Kumar, A.: An Indian Roads Dataset for Supported and Suspended Traffic Lights Detection. arXiv:2209.04203 (2022)
44. Sun, W., Zhang, G., Zhang, X., Zhang, X., Ge, N.: Fine-grained vehicle type classification using lightweight convolutional neural network with feature optimization and joint learning strategy. Multimed Tools Appl **80**(20), 30803–30816 (2021)
45. Butt, M.A. et al.: Convolutional neural network based vehicle classification in adverse illuminous conditions for intelligent transportation systems. Complexity, **2021** (2021)
46. Gholamalinejad, H., Khosravi, H.: Irvd: A large-scale dataset for classification of iranian vehicles in urban streets. Journal of AI and Data Mining **9**(1), 1–9 (2021)
47. Peng, Y., Jin, J.S., Luo, S., Xu, M., Cui, Y.: Vehicle type classification using PCA with self-clustering. In: 2012 IEEE International Conference on Multimedia and Expo Workshops, IEEE, pp. 384–389 (2012)
48. Khoba, P.K., Parikh, C., Jawahar, C.V., Sarvadevabhatla, R.K. Saluja, R.: A Fine-Grained Vehicle Detection (FGVD) Dataset for Unconstrained Roads. arXiv:2212.14569 (2022)
49. Avianto, D., Harjoko, A.: CNN-Based Classification for Highly Similar Vehicle Model Using Multi-Task Learning. J Imaging **8**(11), 293 (2022)
50. Wang, C., Zhu, S., Lyu, D., Sun, X.: What is damaged: a benchmark dataset for abnormal traffic object classification. Multimed Tools Appl **79**, 18481–18494 (2020)
51. Huang, G., Liu, Z., Van Der Maaten, L., Weinberger, K.Q.: Densely connected convolutional networks. In: Proceedings of the IEEE Conference on Computer Vision and Pattern Recognition, pp. 4700–4708 (2017)
52. Bao, S.Y., Savarese, S.: "Semantic structure from motion", in *CVPR* IEEE **2011**, 2025–2032 (2011)
53. Pandey, G., McBride, J.R., Eustice, R.M.: Ford campus vision and lidar data set. Int J Rob Res **30**(13), 1543–1552 (2011)
54. Wang, J., Yang, J., Yu, K., Lv, F., Huang, T., Gong, Y., "Locality-constrained linear coding for image classification", in,: IEEE computer society conference on computer vision and pattern recognition. IEEE **2010**, 3360–3367 (2010)
55. Shafiee, M.J., Chywl, B., Li, F., Wong, A.: Fast YOLO: A fast you only look once system for real-time embedded object detection in video. arXiv:1709.05943 (2017)
56. Girshick, R.: Fast r-cnn, In: Proceedings of the IEEE International Conference on Computer Vision, pp. 1440–1448 (2015)
57. Atieh, A.M., Epstein, M.: The method of spatio-temporal variable diffusivity (STVD) for coupled diffusive processes. Mech. Res. Commun.Commun. **111**, 103649 (2021)

58. Branch, H.O.S.D.: Imagery library for intelligent detection systems (i-lids). In: 2006 IET Conference on Crime and Security, IET, pp. 445–448 (2006)
59. Wang, Y., Jodoin, P.M., Porikli, F., Konrad, J., Benezeth, Y., Ishwar, P.: CDnet 2014: An expanded change detection benchmark dataset. In: Proceedings of the IEEE Conference on Computer Vision and Pattern Recognition Workshops, pp. 387–394 (2014)
60. Wang, Y., et al.: Detection and classification of moving vehicle from video using multiple spatio-temporal features. IEEE Access **7**, 80287–80299 (2019)
61. Alsahafi, Y., Lemmond, D., Ventura, J., Boult, T.: Carvideos: a novel dataset for fine-grained car classification in videos. In: 16th International Conference on Information Technology-New Generations (ITNG 2019), Springer, pp. 457–464 (2019)

MIX
Papier aus verantwortungsvollen Quellen
Paper from responsible sources
FSC® C105338

If you have any concerns about our products,
you can contact us on
ProductSafety@springernature.com

In case Publisher is established outside the EU,
the EU authorized representative is:
Springer Nature Customer Service Center GmbH
Europaplatz 3, 69115 Heidelberg, Germany

Printed by Libri Plureos GmbH
in Hamburg, Germany